ENZYMES AND ISOENZYMES

STRUCTURE, PROPERTIES AND FUNCTION

Fifth FEBS Meeting

Volume 15

GAMMA GLOBULINS Structure and Biosynthesis

Volume 16

BIOCHEMICAL ASPECTS OF ANTIMETABOLITES AND OF DRUG HYDROXYLATION

Volume 17

MITOCHONDRIA Structure and Function

Volume 18

ENZYMES AND ISOENZYMES Structure, Properties and Function

FEDERATION OF EUROPEAN BIOCHEMICAL SOCIETIES
FIFTH MEETING, PRAGUE, JULY 1968

ENZYMES AND ISOENZYMES

STRUCTURE, PROPERTIES AND FUNCTION

Volume 18

Edited by

D. SHUGAR

Department of Biophysics, University of Warsaw;
and Institute of Biochemistry and Biophysics
Academy of Sciences, Warsaw, Poland

1970

ACADEMIC PRESS · London and New York

ACADEMIC PRESS INC. (LONDON) LTD.
Berkeley Square House
Berkeley Square
London, W1X 6BA

U.S. Edition published by
ACADEMIC PRESS INC.
111 Fifth Avenue,
New York, New York 10003

SBN: 12-640860-2
Library of Congress Catalog Card Number: 75-107930

Printed in Great Britain by
Spottiswoode, Ballantyne and Co. Ltd
London and Colchester

This volume contains the Proceedings of two Symposia held at the Fifth Annual Meeting of the Federation of European Biochemical Societies, Prague, July 1968. Papers 1-10 comprise the Symposium entitled "Relation of Enzyme Structure and Activity" (organised by J. I. Harris and B. Keil); Papers 11-35 comprise the Symposium entitled "Isoenzymes, their Properties, Structure and Function" (organised by G. Pfleiderer and B. Večerek).

Editorial note. A major portion of the editorial work of this volume was carried out while the editor was Visiting Scientist at the Department of Biochemistry, Faculty of Medicine, University of Laval, Quebec, Canada, with the support of the Medical Research Council of Canada.

Contents

FEBS Symposium, Volume 18, 1970, pp. 1–15

The Primary Structure and Activity of Glyceraldehyde 3-Phosphate Dehydrogenase

J. I. HARRIS, B. E. DAVIDSON, M. SAJGÓ and H. F. NOLLER

Medical Research Council Laboratory of Molecular Biology,
Cambridge, and

R. N. PERHAM

Medical Research Council Laboratory of Molecular Biology
and Department of Biochemistry, University of Cambridge,
Cambridge, England

Recent work on a number of monomeric enzymes such as lysozyme [1], ribonuclease [2] and chymotrypsin [3] has shown how a knowledge of the molecular structure of an enzyme aids in the understanding of its mechanism of action. At present the detailed structure of an enzyme in three dimensions is best obtained by interpretation of the results of X-ray crystallographic analysis in conjunction with a knowledge of the linear order of the amino acid residues in its polypeptide chain(s).

Some five to six years ago, encouraged by the progress that was then being made in solving the detailed structure of haemoglobin [4], we began to consider how the scope of these methods could be extended to study the detailed structures of larger and metabolically important enzymes such as the NAD-linked dehydrogenases that are involved in the carbohydrate metabolism of most living organisms (for review see [5, 6]).

Several of these enzymes, e.g. glyceraldehyde 3-phosphate dehydrogenase, lactic dehydrogenase, and yeast alcohol dehydrogenase, had been reported to possess molecular weights in the range of 120,000-150,000, and there was evidence to suggest that the active molecules were composed of sub-units of lower molecular weight. Thus, the molecular weight of active rabbit muscle glyceraldehyde 3-phosphate dehydrogenase (GPDH) was reported to be between 118,000 and 140,000 and this apparent variability in molecular weight was also reflected in the variable values reported for other parameters of the active enzyme, such as the number of catalytically-active sites and bound coenzyme molecules [5]. These values were based on the assumption that the enzyme

1

preparations under study were composed wholly of fully active molecules, and in order to eliminate this possible source of error we have investigated the chemical anatomy of the enzyme by methods which do not necessarily depend upon the availability of fully active enzyme.

Among the glycolytic cycle enzymes, GPDH appeared to be particularly suitable for a study of this kind. The reaction which it catalyses, namely the reversible oxidative phosphorylation of glyceraldehyde 3-phosphate, represents an important step in the intermediary metabolism of virtually all living organisms. Consequently GPDH occurs widely and abundantly throughout nature, and is readily obtained in gramme quantities and in pure crystalline form from a variety of different species ranging from yeasts and bacteria to mammals. The enzyme is inhibited by stoichiometric amounts of iodoacetic acid and this reaction provided a convenient means of introducing a radioactive probe into the active site of the enzyme, which led to the identification and chemical estimation of the cysteine residues which are involved in the catalytic reaction [7]. Moreover, uniquely among the NAD-linked dehydrogenases, crystals of muscle GPDH contain firmly bound NAD, a property which should eventually facilitate the identification by X-ray crystallographic methods of the amino acid side chains which are involved in binding the coenzyme in the three-dimensional structure of the molecule. These studies are being undertaken by Dr. H. C. Watson and his colleagues and are described elsewhere in this volume (see p. 51).

THE PRIMARY STRUCTURE OF GPDH

GPDH was prepared in pure crystalline form from pig and lobster muscle as described previously [8, 9]. The inactive derivative formed by reaction of native muscle enzyme with $[2\text{-}^{14}C]$ iodoacetic acid was found to contain from 3.0 to 3.5 moles of $S\text{-}[2\text{-}^{14}C]$ carboxymethylcysteine per mole (140,000 g) of protein. Digestion with trypsin revealed that these reactive cysteines occur in a unique sequence in the primary structure, suggesting that the enzyme molecule could be composed of either three or four similar, and possibly identical, polypeptide chains [7, 10]. In a more detailed study of the pig muscle enzyme (involving end-group and amino acid analysis, and the identification of four unique cysteines in peptides isolated from a trypsin digest of $S\text{-}[2\text{-}^{14}C]$ carboxymethy-lated enzyme), Harris and Perham [11] showed decisively that native GPDH with a molecular weight of 146,000 [12] was composed of *four* similar, and probably identical, polypeptide chains with a molecular weight of about 36,000. Similar results [13, 14] have been obtained with enzyme from lobster muscle, and additional support for the tetrameric structure of this enzyme was provided by the X-ray crystallographic studies of Watson and Banaszak [15].

Table 1. Amino acid compositions of the protein chains in glyceraldehyde 3-phosphate dehydrogenase from pig and lobster muscle.

Amino acid	Pig	Lobster
Lysine	26	27
Histidine	11	5
Arginine	10	9
Cysteine	4*	5*
Aspartic acid	39	31
Threonine	22	20
Serine	17	26
Glutamic acid	19	25
Proline	13	13
Glycine	34	30
Alanine	33	31
Valine	30	37
Methionine	9*	10*
Isoleucine	18	18
Leucine	18	19
Tyrosine	9	9
Phenylalanine	14	15
Tryptophan	3†	3†

*Estimated as cysteic acid, and methionine sulphone, respectively.
†Estimated by sequence analysis of peptides containing tryptophan.

The amino acid compositions of the S-carboxymethylated chains of pig and lobster GPDHs are given in Table 1. Further evidence for the identity of the polypeptide chains in each of the two enzymes has now been obtained by amino acid sequence analysis.

AMINO ACID SEQUENCE OF PIG MUSCLE GPDH

S-carboxymethylation with [2-^{14}C]iodoacetic acid was carried out in 8M urea as previously described [11]. The S-[2-^{14}C]carboxymethylated derivative was digested with trypsin and the resulting peptides were fractionated by gel-filtration on Sephadex G-25 and G-50, followed by suitable combinations of high voltage electrophoresis and chromatography on paper. In this way a total of 31 unique peptides were obtained representing 259 amino acid residues out of an expected total of about 330 residues. These peptides accounted for the 10 residues of arginine and 22 of a probable 26 lysine residues in the protein. The remaining tryptic peptides were not amenable to purification by paper methods and these experiments were finally abandoned in favour of a more indirect approach to the problem.

This involved the separation of peptide fragments produced from S-[2-^{14}C]-carboxymethylated protein in which the lysine residues had been substituted by reaction in 8M urea at pH 9.5 with either S-ethyltrifluorothioacetate [16] or maleic anhydride [17]. Digestion of these lysine-blocked derivatives with trypsin was thus limited to the ten arginyl bonds in the protein. The tryptic digest of N-trifluoroacetyl-S-[2-^{14}C]carboxymethyl-protein was completely soluble in 0.1% ammonium bicarbonate and was separated into five fractions (S1-S5, Fig. 1) by gel-filtration on Sephadex G-50. Fractions S2-S5 gave rise to eight pure

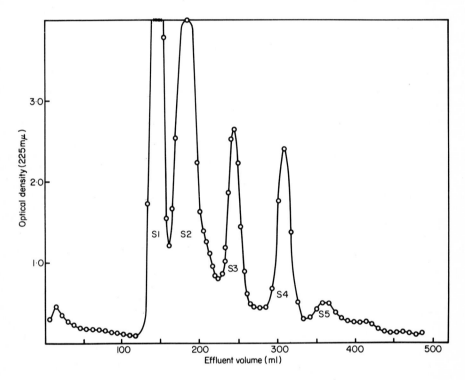

Figure 1. Gel-filtration of a tryptic digest (1% trypsin, 4 h, 37 °C) of N-trifluoroacetyl-S-[2-^{14}C]carboxymethyl-pig GPDH (100 mg) on Sephadex G-50 (140 cm x 2 cm) in 0.1M NH$_4$HCO$_3$.

peptides by further gel-filtration and paper electrophoresis. Fraction S1, which was totally excluded on G-50 as well as on G-75, proved to be a highly aggregated mixture of several large peptides which could not be satisfactorily separated even in the presence of 8M urea. The corresponding S1 fraction from the N-maleyl-S-[2-^{14}C]carboxymethyl-protein was not totally excluded on G-50; this fraction was redigested with trypsin in order to ensure complete hydrolysis of arginyl bonds, and the remaining lysine-blocked fragments were then obtained in pure form as their N-maleyl derivatives by a combination of gel-filtration and chromatography on DEAE-cellulose.

In this way twelve unique peptides, which together represent all the amino acid residues (332) in the protein chain of pig GPDH, were obtained from tryptic digests of the lysine-blocked protein (cf. Table 2). As expected, ten of these peptides contained C-terminal arginine and another, the C-terminal peptide

Table 2. Peptides isolated from tryptic digests of lysine-blocked derivatives of S-[2-^{14}C] carboxymethyl-pig GPDH.

Sephadex (G-50) fraction*	Peptide residue no.†	Number of residues	N- and C-terminal groups
TFA:S5	(i) 11–13	3	Ile ---Arg
	(ii) 14–17	4	Leu--- Arg
	(iii) 195–197	3	Asp---Arg
S4	(i) 1–10	10	Val---Arg
	(ii) 321–332	12	Val--- Glu
	(iii) 232–245	14	Val---Arg
S3	198–231	34	Gly---Arg
S2	246–320	75	Leu ---Arg
MA:S1‡	(i) 116–194	79	Val---Arg
	(ii) 18–42	25	Ala ---Tyr
	(iii) 43–77	35	Met ---Arg
	(iv) 78–115	38	Asp---Arg
Total no.	12	332	

*See Fig. 1.
†See Fig. 2.
‡MA:S1 was redigested with trypsin and refractionated by gel-filtration and by chromatography on DEAE-cellulose.

in the protein chain, contained C-terminal glutamic acid. The presence of an additional peptide containing C-terminal tyrosine was unexpected but the subsequent isolation of the N-terminal fragment (residues 1-40, Fig. 2), produced by cyanogen bromide cleavage of the protein, served to establish its position in the molecule and to show that it had been formed as the result of a chymotrypsin-like cleavage between Tyr-42 and Met-43 during redigestion of fraction MA-S1 with trypsin.

Removal of the trifluoroacetyl and maleyl blocking groups [16, 17], followed by redigestion of the unblocked peptides with trypsin, led to the

Val-Lys-Val-Gly-Val-Asp-Gly-Phe-Gly-ARG-Ile-Gly-ARG-Leu-Val-Thr-ARG-Ala-
10

Ala-Phe-Asn-Ser-Gly-Lys-Val-Asp-Ile-Val-Ala-Ile-Asn-Asp-Pro-Phe-Ile-Asp-
20   30

Leu-His-Tyr-Met-Val-Tyr-Met-Phe-Gln-Tyr-Asp-Ser-Thr-His-Gly-Lys-Phe-His-
40   50

Gly-Thr-Val-Lys-Ala-Glu-Asp-Gly-Lys-Leu-Val-Ile-Asp-Gly-Lys-Ala-Ile-Thr-
60   70

Ile-Phe-Gln-Glu-ARG-Asp-Pro-Ala-Asn-Ile-Lys-Trp-Gly-Asp-Ala-Gly-Thr-Ala-
80   90

Tyr-Val-Val-Glu-Ser-Thr-Gly-Val-Phe-Thr-Thr-Met-Glu-Lys-Ala-Gly-Ala-His-
100

Leu-Lys-Gly-Gly-Ala-Lys-ARG-Val-Ile-Ile-Ser-Ala-Pro-Ser-Ala-Asp-Ala-Pro-
110   120

Met-Phe-Val-Met-Gly-Val-Asn-His-Glu-Lys-Tyr-Asp-Asn-Ser-Leu-Lys-Ile-Val-
130   140

Ser-Asn-Ala-Ser-CÝS-Thr-Thr-Asn-Cys-Leu-Ala-Pro-Leu-Ala-Lys-Val-Ile-His-
150   160

Asp-His-Phe-Gly-Ile-Val-Glu-Gly-Leu-Met-Thr-Thr-Val-His-Ala-Ile-Thr-Ala-
170   180

Thr-Gln-LÝS-Thr-Val-Asp-Gly-Pro-Ser-Gly-Lys-Leu-Trp-ARG-Asp-Gly-ARG-Gly-
190

Ala-Ala-Gln-Asn-Ile-Ile-Pro-Ala-Ser-Thr-Gly-Ala-Ala-Lys-Ala-Val-Gly-Lys-
200   210

Val-Ile-Pro-Glu-Leu-Asp-Gly-Lys-Leu-Thr-Gly-Met-Ala-Phe-ARG-Val-Pro-Thr-
220   230

Pro-Asn-Val-Ser-Val-Val-Asp-Leu-Thr-Cys-ARG-Leu-Glu-Lys-Pro-Ala-Lys-Tyr-
240   250

Asp-Asp-Ile-Lys-Lys-Val-Val-Lys-Gln-Ala-Ser-Glu-Gly-Pro-Leu-Lys-Gly-Ile-
260   270

Leu-Gly-Tyr-Thr-Glu-Asp-Gln-Val-Val-Ser-Cys-Asp-Phe-Asn-Asp-Ser-Thr-His-
280

Ser-Ser-Thr-Phe-Asp-Ala-Gly-Ala-Gly-Ile-Ala-Leu-Asn-Asp-His-Phe-Val-Lys-
290   300

Leu-Ile-Ser-Trp-Tyr-Asp-Asn-Glu-Phe-Gly-Tyr-Ser-Asn-ARG-Val-Val-Asp-Leu-
310   320

Met-Val-His-Met-Ala-Ser-Lys-Glu
330

Figure 2. Amino acid sequence of glyceraldehyde 3-phosphate dehydrogenase from pig muscle [20]. Note that residue 45 is Gln and not Glu as given in [20].

identification of those peptides containing C-terminal lysine that occur between any two consecutive arginine residues in the protein chain. Moreover, in this way four tryptic peptides containing C-terminal lysine, which it had not been possible to purify from total trypsin digests of the S-[2-^{14}C] carboxymethylated-protein, were also obtained in pure form for the first time. Overlaps of lysine and of arginine residues were established from peptides obtained by pepsin, chymotrypsin and cyanogen bromide cleavage of appropriate lysine-blocked fragments, and of intact S-[2-^{14}C] carboxymethylated-protein. The resulting peptide mixtures were purified by suitable combinations of gel-filtration, paper electrophoresis at different pHs and paper chromatography; their sequences were established by standard methods, much use being made of the dansyl-Edman procedure [18]. Amide groups in peptides containing aspartic and glutamic acids have been assigned from considerations of peptide mobility during electrophoresis [19], and wherever possible by the direct identification of asparagine and glutamine in these peptides by digestion with aminopeptidase and/or carboxypeptidases A and B.

On the basis of these results Harris and Perham [20] were able to derive a unique amino acid sequence for the polypeptide chain of pig muscle GPDH. This provisional sequence given in Fig. 2 shows that it contains 332 amino acids corresponding to a molecular weight of 35,740. The active enzyme is thus shown to consist of four chains of *identical* sequence.

AMINO ACID SEQUENCE OF LOBSTER MUSCLE GPDH

The total trypsin digest of the S-[2-^{14}C] carboxymethylated derivative of lobster GPDH contains about forty peptide fragments. Several attempts were made to isolate all of these tryptic peptides in pure form but with only partial success [13]. More recently, however, Allison [14] has succeeded in isolating all but two of these peptides, which together account for 297 of the anticipated total of about 330 residues (cf. Table 1) in the protein chain of the enzyme. In the meantime we sought to limit the action of trypsin to bonds involving the nine arginine residues in the chain by reacting the lysine residues with either S-ethylthiotrifluoroacetate [16] or maleic anhydride [17]. This procedure proved to be highly successful; trypsin digests of the resulting lysine-blocked proteins, like their counterparts from the pig muscle enzyme [20], proved to be completely soluble in 0.1% ammonium bicarbonate and were readily amenable to fractionation by gel-filtration, chromatography on DEAE-cellulose, and paper electrophoresis [21]. For example, trypsin digests of both N-trifluoroacetyl and N-maleyl derivatives of S-[2-^{14}C] carboxymethylated-protein were each separated into six fractions by gel-filtration on Sephadex G-50. These six fractions from the N-maleyl-protein were further purified by a combination of gel-filtration on Sephadex G-75 and G-100, chromatography on DEAE-cellulose and, in the case of the smaller peptides present in fractions S4 and S5, by paper

Table 3. Peptides isolated from a tryptic digest of N-maleyl-S-[2-^{14}C] carboxy-methyl-lobster GPDH.

Sephadex (G-50) fraction	Peptide residue no.*	Number of residues	N- and C-terminal groups
MA:S6	(i) 11–13	3	Ile --- Arg
	(ii) 14–17	4	Leu --- Arg
	(iii) 194–196	3	Gly --- Arg
S5	1–10	10	Ac. Ser ---Arg
S4	(i) 231–244	14	Val --- Arg
	(ii) 320–333	14	Val --- Ala
S3	(i) 197–230	34	Gly --- Arg
	(ii) 288–319	32	Ser ---Arg
S2	245–287	43	Leu --- Arg
S1ai	(i)† 18–193	176	Ala --- Arg
Total no.	10	333	

*See Fig. 3.
†Isolated by gel-filtration of MA:S1 (G-50) on Sephadex G-100.

electrophoresis of the unblocked peptides. In this way it proved possible to isolate ten unique peptides, nine with C-terminal arginine and one, the C-terminal peptide in the protein chain, with C-terminal alanine, which together account for the 333 amino acid residues in the protein chain of lobster GPDH. These peptides are listed in Table 3 and it should be noted that peptide MAS1ai (containing 176 amino acids including sixteen lysines), which was difficult to obtain as its N-trifluoroacetyl derivative [21], was obtained pure and in good yield as its N-maleyl derivative simply by gel-filtration of fraction MA-S1a on G-100 (W. Kenney, unpublished results).

Trypsin digests of the unblocked peptides (with the exception of MAS1ai which contains sixteen lysines, cf. [21]) were relatively easy to fractionate, since in each case they contained only those peptides that occur between any two consecutive arginines in the protein chain. In this manner, therefore, the isolation of *all* the tryptic peptides from lobster GPDH was accomplished more simply in a series of separate fractionation steps from individual lysine-blocked fragments than by direct fractionation of the complex mixture of more than forty peptides present in the trypsin digest of the whole protein chain. The amino acid sequence of each lysine-blocked fragment could then be determined

Ac.Ser-Lys-Ile-Gly-Ile-Asp-Gly-Phe-Gly-ARG-Ile-Gly-ARG-Leu-Val-**Leu**-ARG-Ala-
10

Ala-Leu-Ser-Cys-Gly-Ala-Gln-Val-Val-Ala-Val-Asn-Asp-Pro-Phe-Ile-Ala-Leu-
20   30

Glu-Tyr-Met-Val-Tyr-Met-Phe-Lys-Tyr-Asp-Ser-Thr-His-Gly-Val-Phe-Lys-Gly-
40   50

Glu-Val-Lys-Met-Glu-Asp-Gly-Ala-Leu-Val-Val-Asp-Gly-Lys-Lys-Ile-Thr-Val-
60   70

Phe-Asn-Glu-Met-Lys-Pro-Glu-Asn-Ile-Pro-Trp-Ser-Lys-Ala-Gly-Ala-Glu-Tyr-
80   90

Ile-Val-Glu-Ser-Thr-Gly-Val-Phe-Thr-Thr-Ile-Glu-Lys-Ala-Ser-Ala-His-Phe-
100

Lys-Gly-Gly-Ala-Lys-Lys-Val-Val-Ile-Ser-Ala-Pro-Ser-Ala-Asp-Ala-Pro-Met-
110   120

Phe-Val-Cys-Gly-Val-Asn-Leu-Glu-Lys-Tyr-Ser-Lys-Asp-Met-Thr-Val-Val-Ser-
130   140

Asn-Ala-Ser-C*ỸS-Thr-Thr-Asn-Cys-Leu-Ala-Pro-Val-Ala-Lys-Val-Leu-His-Glu-
150   160

Asn-Phe-Glu-Ile-Val-Glu-Gly-Leu-Met-Thr-Thr-Val-His-Ala-Val-Thr-Ala-Thr-
170   180

Gln-L*ỸS-Thr-Val-Asp-Gly-Pro-Ser-Ala-Lys-Asp-Trp-ARG-Gly-Gly-ARG-Gly-Ala-
190

Ala-Gln-Asn-Ile-Ile-Pro-Ser-Ser-Thr-Gly-Ala-Ala-Lys-Ala-Val-Gly-Lys-Val-
200   210

Ile-Pro-Glu-Leu-Asp-Gly-Lys-Leu-Thr-Gly-Met-Ala-Phe-ARG-Val-Pro-Thr-Pro-
220   230

Asp-Val-Ser-Val-Val-Asp-Leu-Thr-Val-ARG-Leu-Gly-Lys-Glu-Cys-Ser-Tyr-Asp-
240   250

Asp-Ile-Lys-Ala-Ala-Met-Lys-Thr-Ala-Ser-Glu-Gly-Pro-Leu-Gln-Gly-Phe-Leu-
260   270

Gly-Tyr-Thr-Glu-Asp-Asp-Val-Val-Ser-Ser-Asp-Phe-Ile-Gly-Asp-Asn-ARG-Ser-
280

Ser-Ile-Phe-Asp-Ala-Lys-Ala-Gly-Ile-Gln-Leu-Ser-Lys-Thr-Phe-Val-Lys-Val-
290   300

Val-Ser-Trp-Tyr-Asp-Asn-Glu-Phe-Gly-Tyr-Ser-Gln-ARG-Val-Ile-Asp-Leu-Leu-
310   320

Lys-His-Met-Gln-Lys-Val-Asp-Ser-Ala
330

Figure 3. Amino acid sequence of glyceraldehyde 3-phosphate dehydrogenase from lobster muscle [21].

in a series of separate operations, and the sequence of the entire protein chain established by determining sequences around each of the nine arginine residues in peptides obtained from other suitable digests of the S-[2-^{14}C] carboxymethylated-protein (see refs. [20, 21]).

These results enabled Davidson *et al.* [21] to derive a unique sequence for the protein chain of lobster muscle GPDH. This provisional sequence is given in Fig. 3 and shows that the protein sub-unit in the enzyme consists of a single polypeptide chain of 333 amino acid residues corresponding to a molecular weight of 35,780. It follows that the active tetramer is composed of four polypeptide chains of identical amino acid sequence.

RELATIONSHIPS BETWEEN STRUCTURE AND ACTIVITY

The amino acid sequences of pig and lobster muscle GPDH, given in Figs. 2 and 3, provide strong evidence for the chemical identity of the four protein chains in each of the two enzymes. A comparison of the two sequences (Fig. 4) shows that 241 (72%) of the residues occur in indentical sequence in the two chains. Of the ninety differences, fifty-four can be ascribed to single base changes, thirty-five to two base changes, and one (Met → Cys at position 130) requires three base changes in the codons from the *E. coli* code [22]. The most common interchange, that of isoleucine for valine, occurs twelve times and in the great majority of cases polar residues are substituted by other polar residues, and non-polar residues by other non-polar residues, in the sequence. It has been found necessary to postulate lysine-24 as an insertion in the pig sequence, this being done to maximize homology elsewhere in the two molecules. Apart from this, and the two additional residues at the C-terminus of the lobster chain, the two sequences are strictly homologous and no other additions or deletions are observed in the sequence comparison. Moreover, the information that is available on the sequences of the rabbit and ox muscle enzymes shows that the sequences of the three mammalian enzymes are almost identical. This degree of amino acid sequence homology in proteins derived from sources so phylogenetically distant as lobster and mammal is considerably greater than has been found to occur in other proteins such as haemoglobin [23] and cytochrome-*c* [24]. In the haemoglobins extensive differences are found to occur between any two given haemoglobin α or β chains, and in the case of the cytochromes more than 60% of the residues have been shown to vary among the species so far studied. Moreover in *B. subtilis* two subtilisins have been found to differ in as many as 83 out of 274 residues, indicating a sequence homology of only 70% in enzymes from related strains of the same organism [25]. It should be pointed out that the high level of sequence identity between mammalian and lobster GPDHs is partially hidden in peptide maps of tryptic digests of the enzyme proteins owing to the frequently different disposition of lysine and arginine residues. Thus, whereas

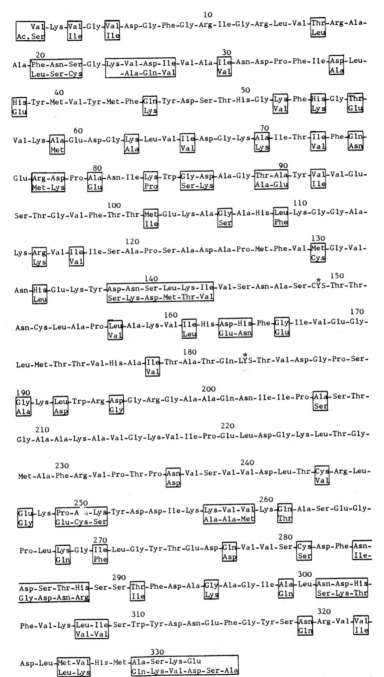

Figure 4. Comparative sequence of glyceraldehyde 3-phosphate dehydrogenase from pig and lobster muscle [20]. The sequence of the lobster enzyme is identical with that of the pig enzyme except where indicated in the boxed areas below the continuous line of the pig sequence. Residue 24 is represented as a deletion in the lobster sequence in order to maximize the sequence homology between the two polypeptide chains; this involves a change in the residue numbers given in ref. 21.

the pig and lobster enzymes contain 26 and 28 lysines respectively, only 17 of these occupy homologous positions in the two proteins. Similarly, although eight of the arginines are homologous, the presence of additional residues of arginine at positions 77 and 115 in pig, and at position 288 in lobster, has a marked effect on the size and properties of the peptide fragments that are produced by trypsin digestion of the respective lysine-blocked proteins. These observations merit particular emphasis in view of the widespread use of peptide mapping techniques in comparisons of proteins.

Despite the large variations in the amino acid sequences of haemoglobins, most of these variations are accommodated within the framework of a basic tetrameric structure for the molecule. It would thus seem reasonable to suppose that GPDH from lobster and pig (and by analogy from other sources too) will possess a common three-dimensional structure. While the chemical evidence indicates that the active enzyme molecule comprises four identical protein chains, X-ray diffraction studies [15] of the lobster enzyme-NAD complex suggest that these four chains could be related in pairs to form a tetrameric molecule analogous in structure to haemoglobin [4]. The resolution of this particular problem will be of great interest in understanding the nature and effect of the interactions of *apo*enzyme with NAD. In this connection there is evidence to show that the binding sites for NAD are not equivalent [26a, b] and that marked conformational changes occur when NAD interacts with the *apo*enzyme in solution [27-29]. A comparison of the crystal structure of NAD-enzyme with that of the *apo*enzyme has not as yet been possible because the *apo*enzyme, unlike that of lactic dehydrogenase [30], does not crystallize in the absence of the coenzyme.

In previous work [7, 10, 13] Cys-149, which participates in the catalytic reaction by forming a thioester bond with the substrate, has been shown to be the sole site of reaction of iodoacetic acid with native GPDH—in the presence or absence of NAD. The rate of this reaction is, moreover, significantly increased in the presence of NAD, an effect that is in sharp contrast to the shielding effect of NAD on the reaction of iodoacetic acid with essential cysteine residues in the alcohol dehydrogenases [31, 32]. Cys-153, although present in all GPDHs so far examined [13, 14], does not react with iodoacetic acid under these conditions despite its close proximity to Cys-149 in the primary structure. In this connection it is of interest that the reaction of the enzyme-NAD complex with sodium iodosobenzoate leads to the formation of an intra-chain disulphide bond between Cys-149 and Cys-153 [33]. The other cysteines are variable both in presence and in position (Cys-244 and 281, present in pig, are absent in lobster, while Cys-22, 130, and 250, present in lobster, are absent in pig) and are therefore unlikely to play any fundamental role in the catalytic reaction.

A glimpse of the three-dimensional structure of the active enzyme is provided by the observation that the acetyl group, bound initially in thioester linkage to

Cys-149 during the enzyme-catalysed hydrolysis of p-nitrophenylacetate [7], and acetyl phosphate [34], respectively, are transferred to a specific lysine residue in the protein at higher pH [34, 35]. This particular lysine has been shown to be Lys-183 in the pig [36], rabbit [34], and lobster [37] muscle enzymes. It is of interest to note that, although NAD does not prevent the formation of S-acetyl enzyme from acetyl phosphate, its formation from p-nitrophenylacetate (as well as the $S \rightarrow N$ transacetylation reaction itself) is effectively inhibited in the presence of the coenzyme [38]. These observations suggest that there is steric competition at the active site between NAD and p-nitrophenylacetate; and that the coenzyme catalyses the deacylation of the S-acetylenzyme [38]. Nevertheless, the nature of the transacetylation reaction does show that Cys-149 and Lys-183 must be spatially close in the three-dimensional structure of the tetrameric *apo*enzyme. However, it remains to be established whether this acetyl transfer occurs within or between monomers; that is, whether the active site of the enzyme is composed of portions of only one polypeptide chain, or comprises elements of more than one chain within the tetramer.

In this connection it is interesting to note that two long sequences in the middle portion of the protein chain (residues 93-137, and 144-243 which contain Cys-149 and Lys-183) have been conserved to an appreciably greater extent than other similarly large sequences towards the N- and C-terminal parts of the molecule. It is also of interest that the nine tyrosine and three tryptophan residues, as well as eleven of the twelve proline residues, occur in corresponding positions in the pig and lobster chains. Of the three tryptophans, only that in position 310 is in a highly conserved sequence, which might argue that it is this tryptophan which participates in the suggested (see [6]) charge-transfer complex with NAD. In contrast, only five histidines and two cysteines are common to both chains, indicating that *at least* six of the eleven histidines and two of the four cysteines in the pig sequence are not essential for maintaining the tertiary and quaternary structure of the enzyme molecule.

The amino acid sequence of the protein monomer in GPDHs from two different species provides a framework for the interpretation of the results of further chemical, kinetic and X-ray crystallographic studies on the active tetrameric molecule in solution and in the crystalline state. The present results, reinforced by as yet incomplete sequence information on the rabbit, ox [39] and yeast [40] ezymes, indicate that the amino acid sequence of glyceraldehyde 3-phosphate dehydrogenase has been strongly conserved during the evolution of the species. This, in turn, implies a conservation of three-dimensional structure and enzymic mechanism of action. It may be hoped that studies such as these on GPDHs from different sources, and the detailed study of the three-dimensional structure of the lobster muscle enzyme, will help to delineate those areas of the molecule essential for the enzymic reaction and contribute decisively to the analysis of reaction mechanism. It is also to be hoped that the results of work in progress on other related glycolytic cycle enzymes such as lactic dehydrogenase, alcohol dehydrogenase, aldolase and triosephosphate isomerase will lead to the discovery of structural and evolutionary interrelationships between enzymes that catalyse successive steps in the same cycle of biochemical reactions.

ACKNOWLEDGEMENT

We thank Miss Judith Thompson, Mrs. Beryl Preston and Mrs. Louise Oliver for their careful and highly skilled experimental assistance.

REFERENCES

1. Phillips, D. C., *Proc. natn Acad. Sci. U.S.A.* **57** (1967) 484.
2. Wyckoff, H. W., Hardman, K. D., Allewell, N. M., Inagami, T., Johnson, L. N. and Richards, F. M., *J. biol. Chem.* 242 (1967) 3984.
3. Matthews, B. W., Sigler, P. B., Henderson, R. and Blow, D. M., *Nature, Lond.* **214** (1967) 652.
4. Cullis, A. F., Muirhead, H., Perutz, M. F., Rossmann, M. G. and North, A.C.T., *Proc. R. Soc.* **A265** (1961) 15.
5. Velick, S. F. and Furfine, C., *in* "The Enzymes", Vol. 7 (edited by P. D. Boyer, H. Lardy and K. Myrbäck), Academic Press, New York, 1963, p. 243.
6. Colowick, S. P., Van Eys, J. and Park, J. H., *in* "Comprehensive Biochemistry", Vol. 14 (edited by M. Florkin and E. H. Stotz), Elsevier, Amsterdam, 1966, p. 1.
7. Harris, J. I., Meriwether, B. P. and Park, J. H., *Nature, Lond.* **197** (1963) 154.
8. Elödi, P. and Szörényi, E. T., *Acta physiol. hung.* 9 (1956) 339.
9. Allison, W. S. and Kaplan, N. O., *J. biol. Chem.* 239 (1964) 2140.
10. Perham, R. N. and Harris, J. I., *J. molec. Biol.* 7 (1963) 316.
11. Harris, J. I. and Perham, R. N., *J. molec. Biol.* 13 (1965) 876.
12. Harrington, W. F. and Karr, G. M., *J. molec. Biol.* 13 (1965) 885.
13. Allison, W. S. and Harris, J. I., *Second FEBS Meeting, Vienna,* 1965, **A205**, 140.
14. Allison, W. S., *Ann. N. Y. Acad. Sci.* **148** (1968) 180.
15. Watson, H. C. and Banaszak, L. J., *Nature, Lond.* **204** (1964) 918.
16. Goldberger, R. F. and Anfinsen, C. B., *Biochemistry* 1 (1962) 401.
17. Butler, P. J. G., Harris, J. I., Hartley, B. S. and Leberman, R., *Biochem. J.* **103** (1967) 78P.
18. Gray, W. R., *in* "Methods in Enzymology", Vol. 11 (edited by C. H. W. Hirs), Academic Press, New York, 1967, p. 469.
19. Offord, R. E., *Nature, Lond.* **211** (1966) 591.
20. Harris, J. I. and Perham, R. N., *Nature, Lond.* **219** (1968) 1025.
21. Davidson, B. E., Sajgó, M., Noller, H. F. and Harris, J. I., *Nature, Lond.* **216** (1967) 1181.
22. Crick, F. H. C., *Cold Spring Harb. Symp. quant. Biol.* 31 (1966) 1.
23. Braunitzer, G., *J. cell. comp. Physiol.* 67 (Suppl. 1) (1966) 1.
24. Margoliash, E. and Smith, E. L., *in* "Evolving Genes and Proteins". (edited by V. Bryson and H. J. Vogel), Academic Press, New York, 1965, p. 221.

25. Smith, E. L., Markland, F. S., Kasper, C. B., De Lange, R. J., Landon, M. and Evans, W. H., *J. biol. Chem.* **241** (1966) 5974.
26a. de Vijlder, J. J. M. and Slater, E. C., *Biochim. biophys. Acta* **132** (1967) 207.
26b. Conway, A. and Koshland, D. E., *Biochemistry* **7** (1968) 4011.
27. Listowsky, I., Furfine, C. S., Betheil, J. J. and England, S., *J. biol. Chem.* **240** (1965) 4253.
28. Havsteen, B. H., *Acta chem. scand.* **19** (1965) 1643.
29. Kirschner, K., Eigen, M., Bittman, R. and Voigt, B., *Proc. natn Acad. Sci. U.S.A.* **56** (1966) 1661.
30. Rossman, M. G., Jeffery, B. S., Main, P. and Warren, S., *Proc. natn Acad. Sci. U.S.A.* **57** (1967) 515.
31. Whitehead, E. P. and Rabin, B. R., *Biochem. J.* **90** (1964) 532.
32. Li, T-K. and Vallee, B. L., *Biochem. biophys. Res. Commun.* **12** (1963) 44.
33. Perham, R. N., Ph.D. Thesis, 1964, University of Cambridge.
34. Park, J. H., Agnello, C. F. and Matthew, E., *J. biol. Chem.* **240** (1965) P. C. 3232; **241** (1966) 769.
35. Polgàr, L., *Acta physiol. hung.* **25** (1964) 1; *Biochim. biophys. Acta* **118** (1966) 276.
36. Harris, J. I. and Polgàr, L., *J. molec. Biol.* **14** (1965) 630.
37. Davidson, B. E. (1967) Unpublished results.
38. Matthew, E., Meriwether, B. P. and Park, J. H., *J. biol. Chem.* **242** (1967) 5024.
39. Perham, R. N., *Biochem. J.* **99** (1966) 14C.
40. Jones, G. M. T. and Harris, J. I., *Fifth FEBS Meeting, Prague*, 1968, **A740**, 185.

FEBS Symposium, Volume 18, 1970, pp. 17-26

Localization of Functional Groups in Dehydrogenases

P. ELÖDI, S. LIBOR and S. MÓRA

*Institute of Biochemistry, Hungarian Academy of
Sciences, Budapest, Hungary*

The structural basis of the biological activity of proteins, i.e. the chemical basis of unusual reactivity of particular side chains, is still one of the most exciting problems in protein chemistry. It is known that some side chains of the same chemical composition can have both enhanced or diminished affinity towards particular chemical reagents. For example, in an enzyme molecule only one or a few side chains located in the active center can bind substrate or coenzyme while other groups of the same chemical composition do not react with these substances. This is illustrated by the presence of one sulfhydryl group per subunit in glyceraldehyde 3-phosphate dehydrogenase [1], of a lysine residue in aldolase [2], and by the active serine in proteolytic enzymes, etc.

The selective reactivity of amino acid residues in proteins is not limited to a few side chains of the active center since it is observable as a general phenomenon when the specific chemical modification of different amino acid residues is studied. The difference in the reactivity of the side chains cannot be simply explained in terms of organic chemistry, because it is restricted to polypeptide chains of organized (native) steric structure, and disappears when the structure is destroyed.

In the highly organized and complex structure of a globular protein, the side chains occupy different positions either on the surface or in the interior of the molecule. It follows that, in some cases, the selective reactivity of the side chains can be simply due to *steric effects*. Reactive side chains are commonly located at or near the surface of the molecule, whereas the non-reactive ones are either situated in the interior, or are shielded by interaction with other side chains.

Moreover, in the native protein molecule a wide spectrum of *chemical interactions* may also develop between neighbouring side chains in the range of van der Waals forces to covalent bond formation, depending on the chemical composition of the interacting side chains. This may create, for a given side

Enzymes. Glyceraldehyde 3-phosphate dehydrogenase or D-glyceraldehyde-3-phosphate: NAD oxidoreductase (phosphorylating) (EC. 1.2.1.12): Lactic dehydrogenase or L-lactate: NAD-oxidoreductase (EC. 1.1.1.27)

chain, a specific local environment favourable for one reaction and unfavourable for another.

Accordingly, the difference in reactivity of certain groups may reflect both their *localization* within the polypeptide chain and their *status* with reference to a particular conformation of the polypeptide chain. Our basic premise was to make use of the selective reactivity of certain side chains in dehydrogenases for the investigation of their localization within the protein molecule.

Table 1. Accessible side chains in dehydrogenases determined by different methods (number of groups per 140,000 molecular weight).

Side chain	Method	Glyceraldehyde 3-phosphate dehydrogenase		Lactic dehydrogenase	
		Total	Accessible	Total	Accessible
Tryptophan	perturbation	16	8	20	8
	perturbation		16		20
Tyrosine		36		32	
	iodination		24		30
Histidine	carbethoxy-				
	lation[5]	40	20	44	44

Table 1 summarizes the data from which the distribution of tyrosine, tryptophan and histidine residues in swine sketetal muscle glyceraldehyde 3-phosphate and lactic dehydrogenases can be deduced [3-6]. In these studies we applied a solvent perturbation technique for tryptophan and tyrosine [7], iodination for tyrosine [8], and a spectrophotometric method [5] for the determination of carbethoxylated histidine.

As may be seen in Table 1, pronounced differences are observed between the two dehydrogenases under study. Nearly all the histidine and tyrosine residues in native lactic dehydrogenase are in accessible positions at neutral pH, while a significant proportion of these side chains in glyceraldehyde 3-phosphate dehydrogenase are found to be inaccessible under these conditions.

The difference observed between the total number of tyrosines and those reacting with iodine in glyceraldehyde 3-phosphate dehydrogenase appeared to be a good starting point for the study of their distribution within the molecule. It seemed likely that those tyrosine residues which do not react with iodine without previous urea denaturation (three per subunit in glyceraldehyde 3-phosphate dehydrogenase, and less than one in lactic dehydrogenase) might be rather

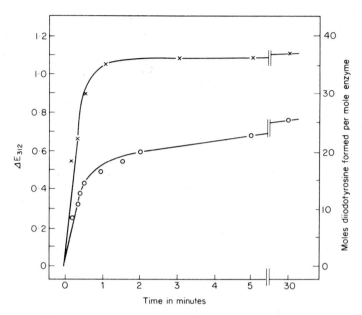

Figure 1. Time dependence of diiodotyrosine formation in glyceraldehyde 3-phosphate dehydrogenase. Iodine was added to a protein solution of 5×10^{-6}M concentration in a 130-fold molar ratio, and the excess iodine was removed by addition of 0.1M arsenite at different time intervals indicated in the figure. o – in 0.1M borate buffer, pH 9.2; x – in 7M urea, pH 9.2.

deeply embedded in the hydrophobic interior of the molecule [9]. Our attention was then focused on glyceraldehyde 3-phosphate dehydrogenase through the assumption that a study of the iodination of tyrosines could lead to a reasonable picture concerning their localization in the steric structure of the protein and that they might also be identified in the known primary structure [10] of the molecule. Lactic dehydrogenase was certainly of less interest in this respect since it exhibited a very low content of non-reacting tyrosine residues.

The differential reactivity of tyrosines in glyceraldehyde 3-phosphate dehydrogenase towards iodine can be demonstrated when the time-course of the reaction is studied. When the enzyme is incubated with iodine in 0.1M borate buffer, pH 9.2 ["native enzyme"], and in 7M urea, pH 9.2 ["denatured enzyme"], three groups of tyrosines can be distinguished on the basis of their differential reactivity with iodine (Fig. 1). Generally speaking, there are three rapidly-reacting and three slowly-reacting tyrosines in each subunit, and a further three residues which do not react at all without previous denaturation of the molecule in urea. An exact kinetic analysis could not be undertaken because of the very complex nature of the reaction and we have therefore focused our attention on a study of the mechanism and products of the reaction.

Spectrophotometric titration of iodine-treated protein provides a means for the identification of the end-product of the iodination reaction, because tyrosine, mono-, and diiodotyrosine have different absorption maxima as well as different pK values for ionization (Table 2). Under the experimental conditions used, iodination of glyceraldehyde 3-phosphate dehydrogenase in urea results in the conversion of all its tyrosines into diiodotyrosine (Fig. 2). However, in the

Table 2. Characteristics of tyrosine and its derivatives.

Compound	pK	λmax (mμ) ionized
Tyrosine	10.0	293
Monoiodotyrosine	8.2	305
Diiodotyrosine	6.6	311

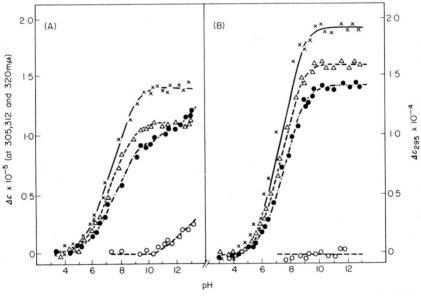

Figure 2. Spectrophotometric titration of iodine-treated glyceraldehyde 3-phosphate dehydrogenase in 7M urea. To a 5 x 10^{-6}M enzyme solution, a 130-fold molar excess of iodine was added in 0.1M borate buffer, pH 9.2 (A), and in 7M urea, pH 9.2 (B). The pH of the samples were adjusted to the desired values, and the molar absorption differences at 295 (o), 305 (●), 312 (x) and 325 (△) mμ were calculated. Reference solution–iodinated samples at pH 4.0.

Table 3. Formation of diiodotyrosine in glyceraldehyde 3-phosphate dehydrogenase upon addition of 130 moles I_2 per mole protein.

Experimental conditions	$\Delta\epsilon_{325}$ x 10^{-5}	Moles diiodotyrosine formed	$\Delta\epsilon_{295}$ x 10^{-4}	Moles tyrosine unchanged
In 0.1M borate buffer, pH 9.2	1.1	24.5	2.5	10.9
In 7M urea, pH 9.2	1.6	35.5	0	0

absence of urea, about 24-25 tyrosines are converted to diiodotyrosine, while 11 to 12 remain unchanged (Fig. 2A, Table 3). Monoiodotyrosine was not detected in either case.

The modified protein was found to be homogeneous by ultra-centrifugation and by gel-electrophoresis. Some physico-chemical parameters, such as optical rotatory dispersion and viscosity, showed a shift towards the parameters of the denatured state (Table 4). We might say that glyceraldehyde 3-phosphate dehydrogenase iodinated in borate buffer pH 9.2 represents a metastable state of this enzyme.

Thus, the conversion of six tyrosyl side chains per subunit to diiodotyrosine in borate buffer results in additional conformational changes. It appears likely that iodination of some of the tyrosines leads to a structural change in the protein, which enables additional tyrosyl residues to react with iodine. Moreover, it might be assumed that in the original, unaltered protein even less than six tyrosyls per subunit are accessible to iodine without alteration of the conformation of the polypeptide chain.

Table 4. Physico-chemical constants of native and iodinated glyceraldehyde 3-phosphate dehydrogenase.

Experimental conditions	Optical rotatory dispersion		Viscosity $[\eta]$ dl/g
	$-a_0$	$-b_0$	
Control, in 0.1M phosphate buffer, pH 8.2	150	210	0.036
Iodine treated*	305	120	0.062
7M urea	710	0	0.325

*Contains about 6 moles diiodotyrosine per subunit.

In order to facilitate the separation of peptides containing tyrosines of different reactivity, glyceraldehyde 3-phosphate dehydrogenase was labelled with radioactive iodine, both in the presence and absence of urea.

When the iodination was carried out in the absence of urea ("native protein"), and exposure to iodine was brief, three strongly labelled spots could be detected in the fingerprint of tryptic peptides (Fig. 3B). We have assumed that these peptides contain the most reactive tyrosyl side chains, the reactivity of which is not dependent upon an induced structural alteration.

This assumption is supported by the finding that four accessible tyrosines per subunit (i.e. sixteen per mole) were detected by solvent perturbation (see Table 1), a method which does not alter the structure of the enzyme. Eight radioactive spots were found in the fingerprint of the protein iodinated in urea for about 30 min (Fig. 3A). Since glyceraldehyde 3-phosphate dehydrogenase would be expected to give six tryptic peptides containing tyrosine [10, 12], it is very probable that non-specific binding of iodine to side chains other than tyrosine may also occur in 7M urea.

The radioactive peptides were separated in several steps of electrophoresis at pH 6.5 and 1.9 as well as by chromatography in different solvent systems. The yield of peptides was low, but sufficient to determine their amino acid composition.

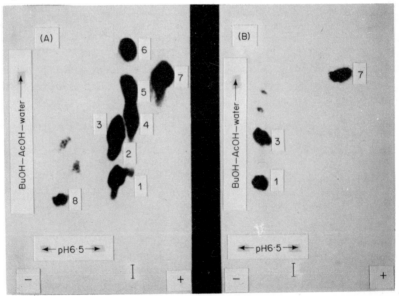

Figure 3. Radioautograms of the fingerprints of ^{131}I-labelled tryptic peptides of glyceraldehyde 3-phosphate dehydrogenase. 5×10^{-5}M solution was exposed to 130 moles ^{131}I$_2$ in 7M urea, pH 9.2, for 30 min (A); short exposure (2 min) with iodine in 0.1M borate buffer, pH 9.2 (B).

Table 5. Amino acid composition of tyrosyl peptides labelled with iodine upon a 2-min exposure in 0.1M borate buffer, pH 9.2.

| | No. of peptide* | | |
	3	1	7
Lys	1	2	1
Asp	2	2	2
Ser	1		
Gly	(1)		
Leu	1		
Ile		1	1
Tyr	1	1	1

*The numbers refer to the radioautograms shown in Fig. 3.

It turned out from the amino acid analyses, performed by Dr. T. Dévényi with a Beckman Unichrom analyzer, that three small, rather hydrophilic peptides contain the most reactive tyrosines (Table 5). Two of these peptides, Nos. 1 and 7, are analogous, differing in one lysine residue only. These peptides contain tyrosines 137 and 252, respectively, according to the sequence presented by Dr. Harris in the preceding paper [10].

Three tyrosines do not react with iodine in the "native protein". As mentioned before, these residues react only when the protein is denatured with urea. These tyrosines belong to three hydrophobic peptides of which we have hitherto completed the analysis of two. One of these was found to be identical with a large cysteine-containing peptide isolated by Harris and Perham in 1965 [11]. This peptide contains tyrosine 273 [10, 12]. The other peptide has an N-terminal arginine and contains tyrosine 318. The investigation of the tyrosine peptides not mentioned here is still in progress in our laboratory.

From these data we conclude that the differential reactivity of tyrosines is related to their location in the protein fabric. Those side chains which react rapidly during a short exposure to iodine are situated on the surface while the non-reactive residues are embedded in the hydrophobic interior. Whether the iodination of embedded residues is only hindered by steric factors, or whether some form of secondary bond between tyrosine and other groups also contributes to the inhibition, cannot be decided from these experiments.

The study of differential reactivity of tyrosines with iodine proved to be a useful tool in attempting to localize these residues in the tertiary structure of glyceraldehyde 3-phosphate dehydrogenase, and it should be noted that iodine can also be applied for a study of other side chains.

Iodination at slightly alkaline pH is considered to be specific for tyrosyl residues, but other groups such as cysteine, histidine and tryptophan may also react under these conditions. A reaction of limited specificity is often considered to be a disadvantage in protein chemistry. We have, however, been able to make use of this situation by exploiting the limited specificity of the reaction for the study of sulfhydryl groups.

In fact, at slightly alkaline pH sulfhydryl groups react faster than tyrosine. Iodine in low concentration does not react at all with tyrosine at neutral pH although it does oxidize sulfhydryls specifically to disulfide. When all four sulfhydryls per subunit of glyceraldehyde 3-phosphate dehydrogenase are oxidized with iodine, the protein readily dissociates into subunits in 6M guanidine hydrochloride, indicating that only *intra*-chain disulfide formation has occurred upon iodine treatment [13]. This means that in this enzyme the pairs of sulfhydryl groups are sufficiently close to form disulfide bonds [13-15]. In addition to the oxidation of sulfhydryl groups, another "non-specific" reaction of iodine was

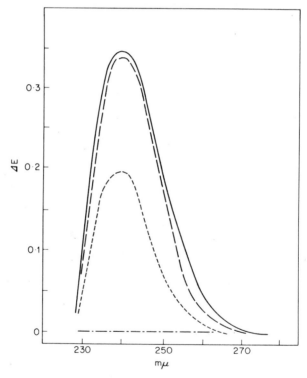

Figure 4. Spectrophotometric determination of histidine in glyceraldehyde 3-phosphate dehydrogenase following carbethoxylation with diethylpyrocarbonate [5]. Difference spectra were recorded with samples containing 2.5×10^{-6}M protein against a reference in 0.1M phosphate buffer, pH 6.0: ——— control (41 histidines); - - - reacted with 130 moles I_2 per mole protein in 0.1M borate buffer, pH 9.2 (40 histidines); · · · · · · and ------ reacted with 130 and 260 moles I_2, respectively, in 7M urea, pH 9.2 (29 and 0 histidines).

also observed. A substitution reaction of iodine in groups other than tyrosine can be detected when radioactive iodine is added to the protein in the presence of 7M urea, although such a reaction cannot be detected in the absence of urea. As mentioned earlier, the fingerprint of glyceraldehyde 3-phosphate dehydrogenase (see Fig. 3A) also suggests that non-specific iodine binding can occur.

In the presence of urea about 48-50 moles of iodine were bound per mole protein, which reflects a non-specific binding of about 12–14 moles iodine. This "non-specifically bound" iodine was found to have reacted with histidine. In the enzyme iodinated in 7M urea only about 28–29 histidines were detected by amino acid analysis, as compared to 40 in the untreated protein.

Table 6. Reaction of histidine with iodine in glyceraldehyde 3-phosphate dehydrogenase (calculated for 140,000 molecular weight).

Moles iodine added per mole GAPD	In borate, pH 9.2		In 7M urea, pH 9.2	
	$\Delta\epsilon_{240}$ x 10^{-5}	Moles histidine found	$\Delta\epsilon_{240}$ x 10^{-5}	Moles histidine found
0	1.35	42.0	1.34	41.6
130	1.34	41.6	0.94	29.0
260	—	—	0.00	0.0

Finally, we should like to mention that, for the investigation of histidyl residues, a direct spectrophotometric determination has been developed in our laboratory [5] which is much simpler than amino acid analysis. The procedure is based on the carbethoxylation of histidine with diethylpyrocarbonate at pH 6. The carbethoxylated histidine exhibits a difference spectrum with a maximum at about 240 mμ and the absorption difference at this wavelength is proportional to the amount of histidine in the sample.

The disappearance of histidyl residues upon iodine treatment in urea can be followed by measuring the decrease in absorption at 240 mμ (Fig. 4).

Practically no histidine disappeared upon iodination of the native protein (Table 6), whereas in urea about one-third of the histidines disappeared after the addition of 130 moles iodine per mole protein. No histidine could be detected after addition of a further excess of iodine. Finally, when a large excess of iodine was added, the oxidation of tryptophyl residues could also be detected.

The complete and detailed description of the conformation of polypeptide chains, as well as the determination of the spatial distribution of amino acid side chains in the crystalline proteins, can only be achieved by X-ray crystallography. Nevertheless, when interpreted in conjunction with a knowledge of the amino acid sequence, we believe that the investigations described above are useful in understanding what we call the structure of an active enzyme. The differential reactivity of side chains provides a possibility not only for visualizing their localization within the enzyme molecule, but also to explain the importance of their interactions in the maintenance of the biologically active form.

REFERENCES

1. Harris, J. I., Meriwether, B. P. and Park, J. H., *Nature, Lond.* **198** (1963) 154.
2. Lai, C. Y., Tchola, O., Cheng, T. and Horecker, B. L., *J. biol. Chem.* **240** (1965) 1347.
3. Libor, S., Elödi, P. and Nagy, Z., *Biochim. biophys. Acta* **110** (1965) 484.
4. Elödi, P., *in* "Biochemical Evolution and Homologous Enzymes" (edited by N. V. Thoai and J. Roche), Gordon and Breach, New York, 1968, p. 110.
5. Ovádi, J., Libor, S. and Elödi, P., *Acta biochim. Biophys. Acad. Sci. Hung.* **2** (1967) 455.
6. Móra, S. and Elödi, P., *Eur. J. Biochem.* **5** (1968) 574.
7. Herskovits, T. T. and Laskowski, M. Jr., *J. biol. Chem.* **235** (1960) PC 56; **237** (1962) 248.
8. Hughes, W. L. and Streassle, R., *J. Am. chem. Soc.* **72** (1950) 452.
9. Libor, S. and Elödi, P., *Eur. J. Biochem.* in press.
10. Harris, J. I., Davidson, B. E., Sajgó, M., Noller, H. F. and Perham, R. N., this volume, p. 1.
11. Harris, J. I. and Perham, R. N., *J. molec. Biol.* **13** (1965) 876.
12. Harris, J. I. and Perham, R. N., *Nature, Lond.* **219** (1968) 1025.
13. Móra, S., Hüvös, P. and Elödi, P., "Abstracts of the Fifth FEBS Meeting, Prague, 1968", Academic Press, London and New York, 1968, p. 197.
14. Harris, J. I., *in* "Structure and Activity of Enzymes" (edited by T. W. Goodwin, J. I. Harris and B. S. Hartley), Academic Press, London and New York, 1964, p. 97.
15. Móra, S., Hüvös, P., Libor, S. and Elodi, P., *Acta biochim. Biophys. Acad. Sci. Hung.* **4** (1969) 151.

FEBS Symposium, Volume 18, 1970, pp. 27-38

The Active Site Sulphydryl Groups of Alcohol Dehydrogenases: A Mechanism for the Enzymic Catalysis*

B. R. RABIN, N. EVANS and N. RASHED

Department of Biochemistry, University College London, London, England.

Enzymes catalysing the oxidation of alcohols by pyridine nucleotide coenzymes fall into two broad categories depending on whether or not they have zinc at their active sites. Yeast and liver alcohol dehydrogenases, which are the principal subjects of discussion in this paper, are both examples of the zinc-containing group [1–4]. Each molecule of yeast alcohol dehydrogenase is composed of four subunits [5, 6], contains four atoms of zinc [2], and binds four molecules of coenzyme [7]; it probably contains four active sites. Of the large number of sulphydryl groups in the molecule, four react much faster then the rest with alkylating agents and these four groups are protected against alkylation by the coenzymes [8, 9]. Liver alcohol dehydrogenase has a molecular weight approximately half that of the yeast enzyme, binds two molecules of coenzyme [10, 11] and contains two reactive sulphydryl groups per mole [12]. The number of subunits is controversial [5, 13–15] and the enzyme contains four zinc atoms [13, 14], two of which appear to be associated with catalytic activity [14]. It is reasonable to suppose that the liver enzyme contains two active sites and it is probable that for both enzymes each active site contains a zinc atom and a sulphydryl group. Other functional groups must, of course, also be present.

For liver alcohol dehydrogenase a considerable body of information is available on the nature of the enzyme-coenzyme interactions. Recent spectrophotometric evidence [15] indicates that the adenine ring is bound to a proton-donating group of the protein, whose chemical nature is as yet unknown. The evidence suggests that the adenine ring of the coenzyme is located on the surface of the enzyme rather than buried in the interior which is probably hydrophobic. Interaction between the enzyme and coenzyme is weakened, but not abolished, by alkylation of the active centre thiol group [12]. This group is probably required for catalysis rather than binding and it probably operates in the hydride transfer process in the ternary enzyme-coenzyme substrate

*This work was supported by grants from the Medical Research Council and the Science Research Council.

complexes. Orthophenanthroline complexes with the zinc at the active centre [16, 17] and prevents the formation of the binary enzyme-coenzyme complexes [18, 19]. By contrast, ADP-ribose, which inhibits by competition with the coenzymes, forms a ternary enzyme-orthophenanthroline-ADP-ribose complex [20]. Thus, the indications are very strong that the dihydronicotinamide ring of the coenzyme reacts with, or is bound close to, the zinc. If the dihydronicotinamide ring is replaced by 3-acetyl dihydropyridine, the binding constant for the reduced coenzyme is reduced by a factor of twenty-four [21], but is still greater than the binding content for ADP-ribose. All these facts can be explained by the simple postulate that the amide or acetyl oxygen of the 3-substituent of the dihydropyridine ring of the reduced coenzyme is coordinated to zinc in the binary complexes. Although it might be suspected that an analogous interaction occurs with the oxidized coenzyme, NAD^+, evidence has been presented which suggests that this is not so. Indeed, there are very strong indications that repulsive forces operate between the positively-charged pyridine ring and the zinc, and at pH 7.0 ligand binding constants decrease in the following order [20]:NADH > ADP-ribose > NAD^+. In sharp contrast to the binding constants for NADH and ADP-ribose, which are virtually independent of pH between 6 and 9 [20, 23], the binding constant for NAD^+ increases with pH and approaches that of NADH at pH 10 [22, 23]. This effect must be due to the ionization of a group on the protein which, it has been suggested [24], is a water molecule coordinated to the zinc at the active site. Removal of a proton from this water molecule presumably neutralizes the charge on the zinc (which would need to be bound to the protein in such a manner as to be effectively singly positively charged) and abolishes the electrostatic repulsion term.

Powerful additional evidence has been obtained in support of this idea [23]. Imidazole, which coordinates to the active site zinc by displacing bound water, slightly lowers the binding content for NAD^+ at low pHs. Of greater significance, it renders the binding constant for NAD^+ independent of pH, as might be anticipated, since imidazole has eliminated the ionizing group. In the presence of imidazole the binding constant for NADH is lowered but remains independent of pH [23]. As we shall see later, imidazole has no effect on the binding constant for ADP-ribose.

It was shown some years ago that the active site sulphydryl groups of yeast alcohol dehydrogenase possess some extraordinary kinetic properties [8, 9, 25]. Between pH 6 and 10 the rate of reaction of these groups with iodoacetamide is completely independent of pH, as shown in Fig. 1, in sharp contrast to the behaviour of simple mercaptans. Apparently no ionization, of the type $RSH = RS^- + H^+$, occurs for these particular protein thiols and it was suggested that they exist in the native protein hydrogen-bonded to a base group as illustrated in Fig. 2. An arrangement of this sort would be expected to possess a nucleophilic reactivity between that of a mercaptan and a mercaptide ion, as

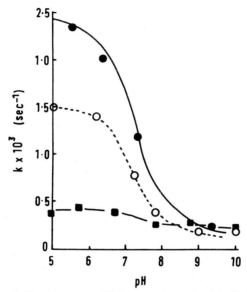

Figure 1. Effect of pH on the apparent first-order rate constant for the inactivation of yeast alcohol dehydrogenase by alkylating agents at 25°C. ●●●, iodoacetate (0.5 mM); ■■■, iodoacetamide (0.5 mM), calculated from [8]; ○○○, bromopyruvate (0.3 mM).

experimentally observed. If the hydrogen bond is sufficiently strong, the pK of the thiol group would be perturbed to high enough values to be outside the range of experimental observations.

More recently we have investigated the chemical reactivity of these thiol groups towards negatively-charged alkylating agents [26, 27], and some of the data for the yeast enzyme are also shown in Fig. 1. The rate constants for alkylation by bromopyruvate and iodoacetate decrease with increasing pH, inflecting about a pK of 7.2. At low pH iodoacetate reacts faster than iodoacetamide but at high pHs the two reagents react at comparable rates. The simplest explanation is that there is near the active site thiol a positively-charged group with a pK of 7.2 which, in the acid form, facilitates the reaction by binding the negatively-charged carboxyl of the reagent. The nature of this group is clearly of great interest and will be discussed later.

Although the sulphydryl groups at the active site of liver alcohol dehydrogenase are less reactive towards alkylating agents than those of the yeast enzyme, the qualitative features of the reactivities of these groups are similar for

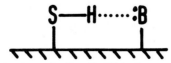

Figure 2. Suggested structure of active site thiol of yeast alcohol dehydrogenase.

the two enzymes. As shown in Fig. 3 the rate of reaction of the thiols of the liver enzyme with the neutral alkylating agent, iodoacetamide, is independent of pH between pH 6 and 10, whereas for iodoacetate the rate constant decreases with pH, following a somewhat skewed titration curve corresponding to a group of pK approximately 7.9. We arrive at the interesting conclusion that for both enzymes an ancillary ionizing group, positively charged in the acid form, is located near the active site thiols and this group has a pK of 7.2 for the enzyme from yeast and 7.9 for the enzyme from liver.

Coenzymes protect both yeast [8] and liver [12] alcohol dehydrogenases against alkylation. It is possible to calculate the binding constants for the

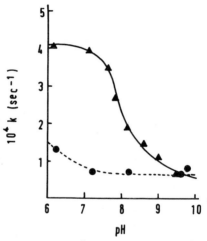

Figure 3. Effect of pH on the apparent first-order rate constant for the inactivation of liver alcohol dehydrogenase by alkylating agents (3.3 mM) at 25°C. ▲▲▲, iodoacetate; ●●●, iodoacetamide.

coenzymes with yeast alcohol dehydrogenase from protection data and the results agree well with other determinations. It is important to know if the nicotinamide or dihydronicotinamide ring is an essential requirement for protection. The data illustrated in Figs. 4 and 5 show that it is not, since both ADP and ADP-ribose protect against alkylation by iodoacetate. The ligand binding constants calculated from the data in Figs. 4 and 5 are collected in Table 1 and it can be seen that they are in good agreement with data obtained by other methods.

For the liver enzyme, data have been obtained which shed light on the chemical nature of the ancillary ionizing group. Since we suspected this group to be the active site zinc, the effects of orthophenanthroline on the rate of alkylation were investigated. The results are clear-cut and are shown in Fig. 6: orthophenanthroline protects against alkylation, suggesting that the zinc and thiol at

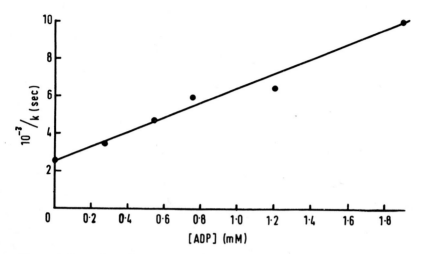

Figure 4. Protection of liver alcohol dehydrogenase by ADP against inactivation by iodoacetate (3.3 mM) at pH 7.2. The reciprocal of the apparent first-order rate constant for inactivation is plotted against the concentration of ADP.

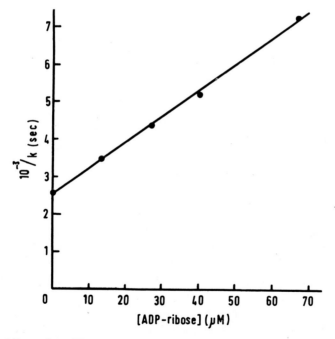

Figure 5. Protection of liver alcohol dehydrogenase by ADP-ribose against inactivation by iodoacetate (3.3 mM) at pH 7.2. The reciprocal of the apparent first-order rate constant for inactivation is plotted against the concentration of ADP-ribose.

Table 1. Dissociation constants of liver alcohol dehydrogenase-ligand complexes at pH 7.2. $I = 0.1$ and 25°C.

	Dissociation Constants (μM)		
Ligand	From promotion of alkylation	From protection against alkylation	Previous values
o-Phenanthroline	—	13	9 [28]
ADP-ribose	—	37	35 [29]
ADP	—	650	390 [28]
Imidazole	0.7	—	0.6 [24]

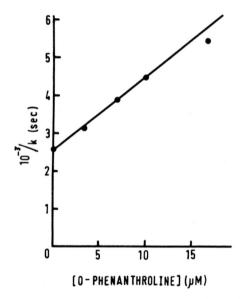

Figure 6. Protection of liver alcohol dehydrogenase by *o*-phenanthroline against inactivation by iodoacetate (3.3 mM) at pH 7.2. The reciprocal of the apparent first-order rate constant for inactivation is plotted against the concentration of *o*-phenanthroline.

the active site are in close proximity. The data in Fig. 6 enable the binding constant to be calculated and this is given in Table 1. To summarize, the evidence produced so far is consistent with the postulate that the variation with pH of both the rate constant for alkylation of the enzyme by iodoacetate, and the binding constant of the enzyme with NAD$^+$, derives from the ionization of a proton from a water molecule coordinated to zinc. This idea can be tested experimentally in a very simple way in view of the effects, already described, of

Table 2. The effect of imidazole on the inactivation of liver alcohol dehydrogenase. $I = 0.1$ and $25°C$.

Inhibitor		Apparent first-order rate constants $(\text{sec}^{-1} \times 10^4)$ at	
		pH 7.2	pH 9.6
Iodoacetate	Imidazole absent	4.1	0.7
(3.3 mM)	Imidazole saturating	18.4	13.2
Iodoacetamide	Imidazole absent	0.7	0.7
(3.3 mM)	Imidazole saturating	3.1	2.4

imidazole on coenzyme binding. If zinc is responsible for the effect of pH on the rate of alkylation by iodoacetate, the rate should become independent of pH in the presence of saturating imidazole. This is indeed observed experimentally, as shown in Table 2. Imidazole promotes the alkylation by both iodoacetamide and iodoacetate: furthermore, it renders the reaction with iodoacetate independent of pH. The promoting effect of imidazole can be precisely interpreted in terms of the following scheme (I = imidazole):

$$E + I \quad \overset{K_I}{\rightleftharpoons} \quad EI$$

$$\downarrow k' \text{ inhibitor} \qquad \downarrow k'' \text{ inhibitor}$$

$$\text{inactive } E \qquad \text{inactive } E$$

The determined values of the kinetic parameters are given in Tables 1 and 2. We can thus draw the important conclusion that the thiol is located near the zinc, which is itself coordinated by the amide oxygen of the dihydropyridine ring of NADH in the enzyme-NADH binary complex. It is entirely reasonable to suppose, as also suggested by the experiments of van Eys *et al.* [30], that the thiol is near the ring, which it activates for hydride transfer.

It is at this point appropriate to consider the possible origin of the promotion effect of imidazole on the alkylation reaction. Two explanations suggest themselves: (1) imidazole binding generates a small conformation change in the region of the active centre, or (2) imidazole becomes directly involved as a catalytic entity. If the first explanation is correct, it casts some doubt on the rationale based on the experimentally observed effects of imidazole. Fortunately it can be eliminated with reasonable certainty. Small conformation changes would be impossible to detect by the usual, and rather crude, physical techniques, apart from X-ray crystallography, and an indirect method must be designed. Since these changes, if they exist at all, must necessarily be in the immediate region of the active centre, they should, almost by definition, affect the binding of ligands in this area. As there is reason to suppose that imidazole interferes directly with the enzyme-dihydronicotinamide interaction, there is not much purpose in using the coenzymes to test this proposition. The best test, therefore, is the binding of ADP-ribose: it would be very surprising if conformation changes at the active site had no effect on the binding of this ligand. The binding constant for ADP-ribose to the free enzyme is easily measured directly from protection experiments and the determined value (actually the dissociation constant) is given in Table 1. The binding constant of ADP-ribose to the enzyme-imidazole complex can be measured from the kinetics of alkylation in the presence of both ligands since one speeds up, and the other slows down, the reaction. The reaction scheme is:

$$
\begin{array}{ccc}
EA & \xrightarrow{\;K_{A,I}\;} & EAI \\[2pt]
\Big\updownarrow K_A & & \Big\updownarrow K_{I,A} \\[2pt]
E & \xrightarrow{\;K_I\;} & EI \\[2pt]
\searrow k'\,[\text{inhibitor}] & & \swarrow k''\,[\text{inhibitor}] \\[2pt]
& \text{Inactivated } E &
\end{array}
$$

where the Ks are dissociation constants and A and I are ADP-ribose and imidazole, respectively.

If k_0 is the apparent first-order rate constant for alkylation, then:

$$
1 + \frac{[I]}{K_I} + \frac{[A]}{K_A} + \frac{[A]}{K_{I,A}}\frac{[I]}{K_I} = \frac{k' + k''\,[I]/K_I}{k_0}.
$$

For constant $[I]$, plots of $1/k_0$ against A should be linear. Sets of such plots, at different values of I should intersect at a common point on the abscissa where

$$
[A] = \frac{k'' - k'}{k'/K_{I,A} - k''/K_A}.
$$

Figure 7 shows that these expectations are entirely in accord with the experimental observations. The value of $K_{I,A}$ obtained was 37 μM. This is the constant for the dissociation of ADP-ribose from the ternary enzyme-imidazole-ADP-ribose complex; it is identical to the dissociation constant of the binary enzyme-ADP-ribose complex. Thus imidazole, like o-phenanthroline [20], is totally without effect on the interaction of the enzyme with ADP-ribose and it is most improbable that the effects of imidazole derive from conformational changes in the protein. As an interesting corollary, it is easy to show that ADP-ribose has no

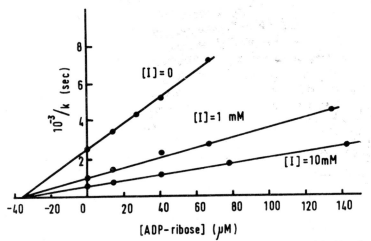

Figure. 7. Effect of the simultaneous presence of ADP-ribose and imidazole on the inactivation of liver alcohol dehydrogenase by iodoacetate (3.3 mM) at pH 7.2. The reciprocal of the apparent first-order rate constant for inactivation is plotted against the concentration of ADP-ribose at three different concentrations of imidazole, [I] .

effect on the interaction of the enzyme with imidazole. Thus, ADP-ribose blocks access to the sulphydryl group at the active centre but not to the zinc. It is thus possible to build up a tentative and somewhat crude picture of the active site region in the total absence of detailed structural information.

A possible explanation for the effect of imidazole in promoting the alkylation of the enzyme is illustrated in Fig. 8. Imidazole is pictured as acting as an acid catalyst for the departure of the leaving group. This hypothesis obviously requires further experimental testing.

It is now possible to formulate a reasonable chemical mechanism for these enzymes, which is in complete accord with all experimental data. It is essentially

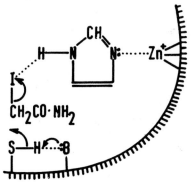

Figure 8. Possible mechanism for the promotion by imidazole of the rate of alkylation of liver alcohol dehydrogenase.

a modification and extension of an earlier proposal [25]. We suggest that the zinc at the active site lies in the bottom of a cleft or depression in the protein. The thiol group lies within a few angstroms at the side, and possibly also slightly above, the zinc. The binding of the coenzyme is of such a nature as to place the amide oxygen of the coenzyme near to the zinc and to bury the dihydronicotinamide ring in a substantially hydrophobic environment, which would account for the changes in the absorption spectrum of the ring on interaction of the coenzyme with the protein. We suggest that the adenine ring lies on the outside of the enzyme with free access to the solvent and it interacts with an acid group of the protein. Its position is such that access of the sulphydryl group to the solvent is blocked; this could be achieved by the adenine lying partially across

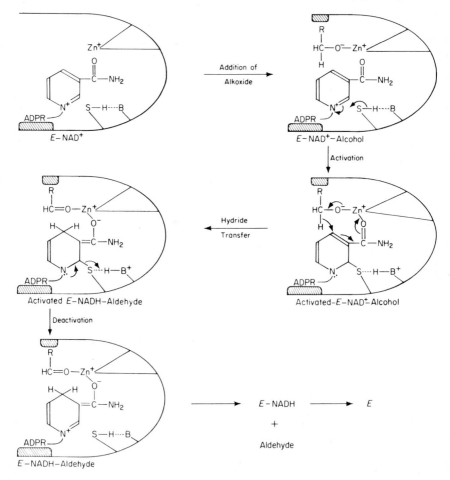

Figure 9. Proposed mechanisms of action of alcohol dehydrogenase.

the cleft. The considerable increase in binding constant in passing from ADP to ADP-ribose could be due to the unfavourable extra negative charge on ADP, since the terminal phosphate would point towards the protein core, or to direct protein ribose interaction. Further experiments are clearly required to decide between these alternatives.

The mechanism we suggest is illustrated schematically in Fig. 9. For the oxidation of ethanol a binary E-NAD$^+$ is pictured as forming initially, though a compulsory binding order is certainly not an *absolute* requirement of our mechanism. Addition of the substrate, to give the ternary E-NAD$^+$-alcohol complex, involves, in addition to hydrophobic interaction of the R group with a site on the protein, the production of an alkoxide ion coordinated to zinc, as previously suggested by Wallenfels and Sund [31] and Theorell and McKee [24]. This involves an analogous ionization to that found for water and is consistent with the demonstration that substrates compete with 2,2'-bipyridine for binding to zinc at the active site [32]. Charge neutralization in a hydrophobic environment would provide a powerful thermodynamic driving force for alkoxide production. We suggest the next step is ring activation by adduct formation between the thiol group and the 2 position of the nicotinamide ring. This creates a substituted 1,2-dihydronicotinamide and removes the charge on the ring, permitting formation of an activated E-NAD$^+$-alcohol complex as shown. The hydride transfer process is pictured as occurring by a modified Meerwein-Ponndorf reaction, with zinc participating in the cyclic flow of electron pairs to produce an activated E-NADH-aldehyde complex. 1,2-NADH is known to be readily reduced [33] and the reaction should occur readily. The reaction is completed by removal of the sulphydryl group followed by dissociation of the products. The reduction of acetaldehyde can be pictured as a complete reversal of the pathway illustrated in Fig. 9. In this reverse direction the activation of NADH is achieved by abolishing the resonance system between the pyridine N and the amide oxygen, thus fixing a double bond in the required position between the amide and pyridine 3 carbon atoms for the hybrid transfer reaction to take place. It should be noted that the activated complexes need only exist at very low concentrations and their detection, though feasible, could require very refined techniques. An extension of this mechanism to the non-zinc-containing alcohol dehydrogenase is easily achieved on paper, but at the moment there is insufficient experimental data to justify this extrapolation. We are at present accumulating data relevant to the problem of the mechanistic relationships between the two groups of enzymes.

REFERENCES

1. Vallee, B. L. and Hoch, F. L., *J. biol. Chem.* **225** (1957) 185.
2. Vallee, B. L. and Hoch, F. L., *J. Am. chem. Soc.* **77** (1955) 821.

3. Wallenfels, K., Sund, H., Faessler A. and Buchard, W., *Biochem. Z.* **329** (1957) 31.
4. Theorell, H., Nyguard, A.P. and Bonnichsen, R., *Acta chem. scand.* **9** (1955) 1148.
5. Harris, J. I., *in* "Structure and Activity of Enzymes" (edited by T. W. Goodwin, J. I. Harris and B. S. Hartley), Academic Press, London and New York, 1964, p. 97.
6. Kagi, J. H. R. and Vallee, B. L., *J. biol. Chem.* **235** (1960) 3188.
7. Hayes, J. E. and Velick, S. F., *J. biol. Chem.* **207** (1954) 225.
8. Whitehead, E. P. and Rabin, B. R., *Biochem. J.* **90** (1964) 532.
9. Rabin, B. R., Ruiz Cruz, J., Watts, D. C. and Whitehead, E. P., *Biochem. J.* **90** (1964) 539.
10. Theorell, H. and Bonnichsen, R., *Acta chem. scand.* **5** (1951) 1105.
11. Theorell, H. and Winer, A. D., *Archs Biochem. Biophys.* **83** (1959) 291.
12. Li, T-K. and Vallee, B. L., *Biochemistry* **4** (1965) 1195.
13. Oppenheimer, H. L., Green R. W. and McKay, R. H., *Archs Biochem. Biophys.* **119** (1967) 552.
14. Drum, D. E., Harrison, J. H., Li, T-K., Bethune, J. L. and Vallee, B. L., *Proc. natn. Acad. Sci. U.S.A.* **57** (1967) 1434.
15. Fisher, R. F., Haine, A. C., Mathias, A. P. and Rabin, B. R., *Biochim. biophys. Acta* **139** (1967) 169.
16. Vallee, B. L., Coombs, T. L. and Williams, R. J. P., *J. Am. chem. Soc.* **80** (1958) 397.
17. Vallee, B. L. and Coombs, T. L., *J. biol. Chem.* **234** (1959) 2615.
18. Vallee, B. L., Williams, R. J. P. and Hoch, F. L., *J. biol. Chem.* **234** (1959) 2621.
19. Plane, R. A. and Theorell, H., *Acta chem. scand.* **15** (1961) 1866.
20. Yonetani, T., *Acta chem. scand.* **17** (Suppl. 1) (1963) 596.
21. Shore, J. D. and Theorell, H., *Eur. J. Biochem.* **2** (1967) 32.
22. Theorell, H. and McKee, J. S., *Acta chem. scand.* **15** (1961) 1797.
23. McKee, J. S., *Prog. Biophys. Mol. Biol.* **14** (1963) 223.
24. Theorell, H. and McKee, J. S., *Acta chem. scand.* **15** (1961) 1834.
25. Rabin, B. R. and Whitehead, E. P., *Nature, Lond.* **196** (1962) 658.
26. Evans, N. and Rabin, B. R., *Eur. J. Biochem.* **4** (1968) 548.
27. Rashed, N. and Rabin, B. R., *Eur. J. Biochem.* **5** (1968) 147.
28. Yonetani, T. and Theorell, H., *Archs Biochem. Biophys.* **106** (1964) 243.
29. Theorell, H. and Yonetani, T., *Archs Biochem. Biophys.* **106** (1964) 252.
30. van Eys, J., Kretszchmar, R., Tseng, N. S. and Cunningham, L. W., *Biochem. biophys. Res. Commun.* **8** (1962) 243.
31. Wallenfels, K. and Sund, H., *Biochem. Z.* **329** (1957) 59.
32. Sigman, D. S., *J. biol. Chem.* **242** (1967) 3815.
33. Chaykin, S., King, L. and Watson, J. G., *Biochim. biophys. Acta* **124** (1966) 13.

FEBS Symposium, Volume 18, 1970, pp. 39-49

Characterization of the Sulfhydryl Groups of Rabbit Muscle Aldolase

G. SZABOLCSI, M. SAJGÓ, E. BISZKU, B. SZAJÁNI,
P. ZÁVODSZKY and L. B. ABATUROV*

*Magyar Tudományos Akadémia Biokémiai Intézete,
Budapest, Hungary*

The most common method used for mapping the active site of an enzyme is the chemical modification of its amino acid side chains and some time ago we pointed out that the data obtained in this way requires careful interpretation. While studying the motility of protein structure, we found that chemical modification of a single amino acid residue might induce profound alteration in the conformation of an enzyme [1-3].

In this paper we shall present results obtained in the course of a study of the active center of aldolase, and discuss the problems encountered in the interpretation of the data.

According to recent data, aldolase appears to consist of four subunits [4, 5, 17]. Two of the subunits are believed to be identical, while the other two have a slightly different primary structure [6, 7]. Aldolase contains 28 cysteinyl residues, 4 of which can be blocked instantaneously with PMB; approximately another 6 are blocked within 20 min at room temperature and another 8 react slowly on prolonged incubation with the mercurial. The remainder of the cysteinyl residues can only be titrated in the presence of urea [8].

We have shown earlier [2] that blocking of the most reactive SH groups with PMB is in itself sufficient to alter the conformation of the enzyme, and that the unmasking of further SH groups is a consequence of this structural modification.

Aldolase, which contains ten mercaptide bonds, is fully active although the conformation of the enzyme is markedly altered. It can be shown by a number of methods that its structure is less compact than that of native aldolase. A

Present address: Insitute of Molecular Biology, Academy of Sciences of the USSR, Moscow, USSR.

Non-standard abbreviations. PMB, *p*-mercuri benzoate; CM, carboxymethyl; 10-CM aldolase, aldolase containing 10 carboxymethyl groups per mole of enzyme; 4-CM aldolase, aldolase containing 4 carboxymethyl groups per mole of enzyme; CMC, carboxymethyl-cysteine.

Enzyme. Aldolase: Fructose-1,6-diphosphate D-glyceraldehyde-3-phosphate lyase (EC 4.1.2.13).

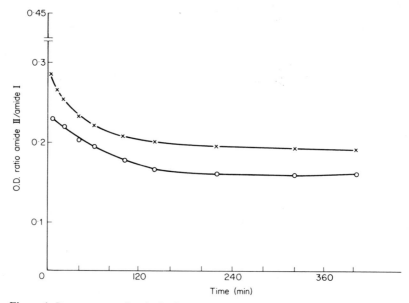

Figure 1. Progress curve for the hydrogen-deuterium exchange of peptide hydrogens in native aldolase and aldolase containing ten mercaptide linkages at pH 7.5 in Tris buffer I, 0.04, as measured by infrared spectrophotometry at 1,550 cm^{-1} (Amide II) using as intrinsic reference the absorption at 1,650 cm^{-1} (Amide I). x--x, native aldolase; o--o, aldolase containing ten mercaptide linkages.

direct method which gives information on the structure of a protein is the measurement of the rate of hydrogen-deuterium exchange of peptide hydrogens. As shown in Fig. 1, the number of slowly exchangeable hydrogens of aldolase which contain ten mercaptide bonds is decreased by about 20%, indicating that the modified enzyme is less compact than the native one [9].

The mercaptidated enzyme is less resistant to heat treatment than native aldolase (Fig. 2), [9]. At 58°C the rate constant for heat inactivation is about twice that of the native enzyme. In the presence of substrate or inorganic phosphate, both the native and the modified enzymes are protected against heat inactivation.

It is generally assumed that fructose-1,6-diphosphate is bound to aldolase through its two phosphate groups. Ginsburg and Mehler [10] have shown that aldolase contains two types (one tight and one loose) of specific phosphate binding sites. This fact is also reflected by the intermediate plateau in the protection of the enzyme by phosphate against heat inactivation (Fig. 2). A comparison of the protective effects of fructose diphosphate (Fig. 2) and inorganic phosphate shows that FDP is effective at much lower concentrations than inorganic phosphate, and that the substrate dependence of protection is a simple rectangular hyperbolic function. This phenomenon may be due to the fact that, in

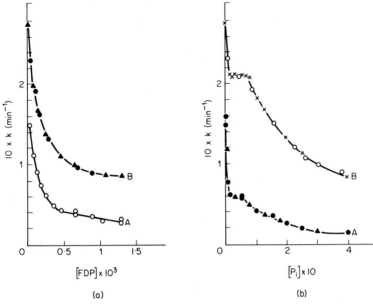

Figure 2. Substrate (a) and inorganic phosphate (b) dependence on heat inactivation at 58°C of native aldolase (A) and aldolase containing ten mercaptide linkages (B) in Tris buffer, pH 7.5; I, 0.04. k(min^{-1}), first-order rate constant of inactivation. When the effect of FDP was tested, the enzyme was preincubated with substrate at a concentration given by the abscissa at 25°C until equilibrium was reached.

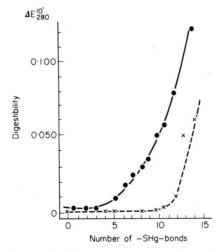

Figure 3. Tryptic digestibility of aldolase containing an increasing number of mercaptide bonds. Digestion at 25°C in 0.1M phosphate buffer, pH 7.5; ●--●, in absence, and x--x, in presence of 3.6 x 10^{-3} M FDP. Trypsin: 2.6 x 10^{-4} [TU]$^{Hb}_{ml}$.

the presence of FDP, the probability of both phosphate binding sites being simultaneously occupied is greatly increased. Therefore, fructose diphosphate may form a junction between two segments of the polypeptide chain and, by restricting some fluctuational possibilities of protein, lead to a more compact structure.

The tryptic digestibility of aldolase derivatives containing an increasing number of mercaptide bonds [2] is shown in Fig. 3. Digestibility increases substantially with the number of mercaptide bonds formed. In the presence of FDP, aldolase containing up to ten mercaptide linkages is resistant to proteolysis, within certain limits of trypsin concentrations.

The above data indicate that blocking of those cysteinyl residues which are readily accessible, or become easily accessible to PMB, does not affect the conformation of the substrate binding site. Moreover, binding of substrate or inorganic phosphate to the modified enzyme restores a structure which is similar to that of the native enzyme. This fact might account for the full activity of aldolase which contains ten mercaptide bonds. However, if more than ten cysteinyl residues are blocked with PMB, the activity of aldolase gradually decreases (Fig. 4).

The observation that modification of cysteinyl residues, which differ from the most reactive ones, leads to the inactivation of the enzyme, is a source of difficulty for biochemists concerned with the study of aldolase. Kowal *et al.* [11] showed that about twelve SH groups of aldolase may be blocked with chlorodinitrobenzene at pH 9.6 with a concomitant loss of activity. The authors assumed that the cysteinyl residues protected by substrate are located at, or near, the substrate binding site of the enzyme.

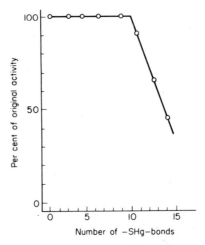

Figure 4. Enzyme activity of aldolase containing an increasing number of mercaptide bonds.

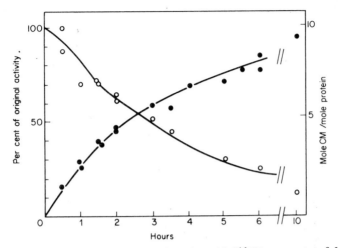

Figure 5. Progress curve for the reaction of aldolase with [^{14}C] bromoacetate followed by enzyme activity and radioactivity measurements. Tris buffer, pH 8.6; I, 0.1, 37°C. Protein concentration, 6.4 x 10^{-5}M; bromoacetate concentration, 4.6 x 10^{-3}M. o–o, enzyme activity; ●–●, radioactivity.

In our laboratory we approached this problem with the aim of identifying the particular cysteinyl residues involved in catalysis. [^{14}C] bromoacetate was used as the blocking agent, and conditions were devised for the selective carboxymethylation of about ten cysteinyl residues in the molecule [12, 13] .

At pH 8.6 in Tris buffer (I, 0.1) and 37°C, carboxymethylation of ten sulfhydryl groups leads to the inactivation of aldolase (Fig. 5). The 10-CM derivative of aldolase retains a residual activity of about 10%, although it has been shown in separate experiments that this activity is not an inherent property of the modified enzyme. Kinetic analysis of the process could have revealed whether the reactive SH groups are blocked at different rates. Unfortunately a proper kinetic analysis could not be performed because of the instability of bromoacetate under the conditions of the reaction.

However, if the loss of enzyme activity is plotted against the number of carboxymethyl groups bound to aldolase, an S-shaped curve is obtained (Fig. 6) indicating a differential reactivity of the SH groups and a difference in their involvement in catalysis. The shape of the curve suggests that a set of cysteinyl residues (Cys-I) may be blocked without loss of activity. Two further sets of cysteinyl residues (Cys-II and Cys-III) become unmasked during the carboxymethylation process and it seems that only one of these sets, Cys-II, is required for enzyme activity.

In order to differentiate between the reactivity of these cysteinyl residues, we made use of our previous results. As shown above, binding of inorganic phosphate stabilizes a more compact structure of aldolase. It might be expected

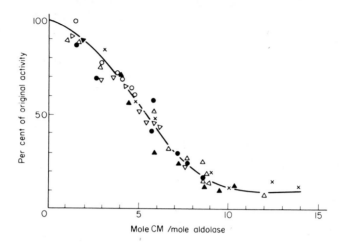

Figure 6. Activity of aldolase as a function of carboxymethyl groups bound to the protein.

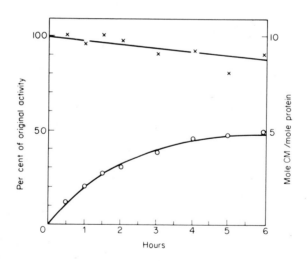

Figure 7. Progress curve for the reaction of aldolase with [^{14}C] bromoacetate in the presence of phosphate at 37°C. Tris buffer as in Fig. 5 and 0.07M phosphate. Protein concentration, 6.4×10^{-5}M; bromoacetate concentration, 1×10^{-2}M. x--x, enzyme activity; o--o, radio-activity.

that in the presence of phosphate the conformational changes and, as a consequence, the unmasking of SH groups, will be slowed down and carboxy-methylation of the cysteinyl residues located at the surface of the protein (Cys-I) will proceed more rapidly than the unmasking of Cys-II. In fact, we find that if carboxymethylation is performed in the presence of 0.07M phosphate, about four carboxymethyl groups can be bound to aldolase, with practically no loss of activity (Fig. 7).

In order to determine which residues were carboxymethylated, samples of inactive 10-CM and active 4-CM aldolase derivatives were digested with trypsin and peptide maps prepared and examined [12, 14]. A tryptic digest of aldolase

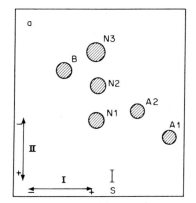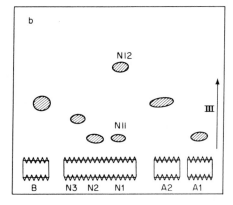

Figure 8. Tracing of autoradiograms of tryptic [^{14}C] CM-Cys peptides of aldolase fully carboxymethylated in the presence of urea. (a) two-dimensional electrophoresis of the digest at pH 6.5 (I) and pH 2.1 (II); (b) the radioactive spots revealed by autoradiography were cut out, sewn to another sheet of paper and chromatographed in iso-amylalcohol-pyridine-water system (III).

which had been fully carboxymethylated in urea was used as a control (Fig. 8). The autoradiogram of the two-dimensional electrophoretic pattern of fully carboxymethylated aldolase shows six spots, all of which contain CM-cysteine. In a second run in the iso-amylalcohol system, peptide N_1 is resolved further into two peptides, showing that a total of seven different CM-cysteinyl peptides are produced. Figure 9 shows the radioactive peptides obtained by trypsin digestion of 10-CM and 4-CM aldolase. In the enzymatically-active 4-CM aldolase only one peptide, peptide B, is highly labelled. The enzymatically-inactive 10-CM aldolase contains two highly radioactive spots, peptides B and N_{11}, and a third peptide, N_{12}, is faintly labelled. It was also shown that, during the carboxymethylation process, peptide B is labelled more rapidly than peptides N_{11} and N_{12}.

Since peptide B is labelled both in the presence and absence of phosphate, its cysteinyl residue must be located at the surface of the protein. Blocking of this

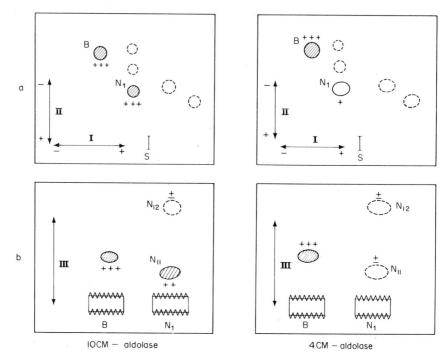

Figure 9. Tracing of autoradiograms of tryptic [^{14}C] CM-Cys peptides of selectively carboxymethylated 10-CM and 4-CM aldolase. The digests were treated as in Fig. 8.

cysteinyl residue does not affect enzyme activity. This cysteinyl residue was designated above as Cys-I. It follows that blocking of the cysteinyl residue of peptide N_{11} (Cys-II) is the cause of the inactivation. The faint labelling found in peptide N_{12} suggests that the cysteinyl residue of this peptide belongs to the set of Cys-III. Furthermore, it seems that peptides B and N_{11} are present in each subunit of aldolase, but it is not yet established whether each subunit contains peptide N_{12}.

The cysteinyl peptides of aldolase have been isolated and the amino acid composition and a partial sequence analysis of peptides B, N_{11} and N_{12} is presented in Table 1. Peptide B is a large peptide and it can be identified with the tryptic peptide which contains the lysine residue involved in Schiff base formation with dihydroxyacetone phosphate [15, 16]. It may be noted that the cysteinyl residue which is near the active lysine in the primary structure can be blocked without loss of enzyme activity. Peptide N_{11}, which contains Cys-II, is a smaller peptide.

If chemical modification of an enzyme results in inactivation, and if, in addition, the modification is slowed down in the presence of substrate, it is often assumed that the substrate binding site of the enzyme has been affected. As this

Table 1. Tryptic [^{14}C]CM-Cys peptides of selectively carboxymethylated aldolase.

Peptide	Partial sequence
N_{11}	Ala-Leu-Ala-CM$\overset{*}{\text{C}}$-Ser (Asn$_1$,Gln$_1$,Gly$_1$,Ala$_1$,Leu$_1$)Lys
B	Ala-Leu-Ser (His$_2$,Asx$_1$,Ile$_1$,Tyr$_1$) (Asx$_1$,Thr$_1$,Glx$_1$,$_*$Pro$_1$,Gly$_1$ Leu$_3$,Lys$_1$) Met-Val-Thr-Pro (His$_1$, CMC$_1$,Thr$_1$,Gln$_1$, Gly$_1$,Ala$_1$) Lys
Active	Ala-Leu-Ser (His$_2$,Asn$_1$,Ile$_1$,Tyr$_1$) (Asn$_1$,Thr$_1$,Gln$_1$,$_*$Pro$_1$,Gly$_1$,
Lys- containing	Leu$_3$,Lys$_1$†) Met-Val-Thr-Pro (His$_1$,CMC$_1$,Thr$_1$,Gln$_1$,
peptide †[16]	Gly$_1$, Ala$_1$) Lys
N_{12}	(CM$\overset{*}{\text{C}}_1$,Pro$_2$,Leu$_2$,Trp$_1$)Lys

†Lys–Schiff's base-forming lysine. The sequence of this peptide has been determined [16], and is written in this form for the sake of comparison with peptide B.

is the case with the carboxymethylation of Cys-II in aldolase, we felt that this assumption should be tested experimentally [14]. We therefore devised an indirect method in the following way. We measured the rate of heat denaturation of samples of 10-CM aldolase in the presence and absence of FDP and inorganic phosphate. If the substrate binding site had been damaged, it might have been expected that the inactive enzyme would not be protected against heat treatment. Since the modified enzyme was inactive, the progress of denaturation had to be followed by measuring the radioactivity and protein content of the supernatants of the heat-treated samples. Unexpectedly, we found that 10-CM aldolase was protected by phosphate and also by FDP (Fig. 10). The modified enzyme is nevertheless more susceptible to heat treatment than native aldolase, and the rate constant of denaturation decreases in the presence of increasing concentrations of phosphate, indicating that 10-CM aldolase can be saturated with phosphate.

However, if we compare the phosphate dependence of protection of 10-CM aldolase with that of native and mercaptidated aldolase, we do not find the intermediate plateau characteristic of the tight phosphate binding site. This suggests that carboxymethylation of Cys-II has affected the tight phosphate binding site of aldolase, while the loose phosphate binding site has not been impaired.

From these investigations the conformation and the active center of aldolase can be depicted in the following way:

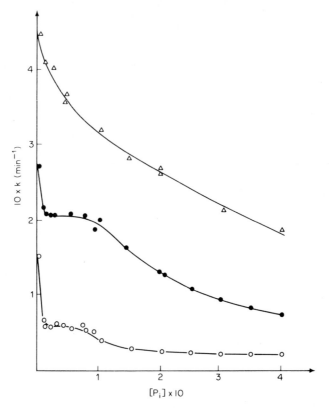

Figure 10. Phosphate dependence of heat denaturation of 10-CM aldolase, aldolase containing ten mercaptide bonds, and native aldolase. Conditions of denaturation as in Fig. 2. The progress of denaturation was followed by measuring the radioactivity and/or protein content of the supernatant of the heat-treated samples. △–△, 10-CM aldolase; ●–●, aldolase containing ten mercaptide bonds; ○–○, native aldolase.

1. At the surface of the protein and in the vicinity of the Schiff's base-forming lysine, a cysteinyl residue is located (Cys-I in peptide B) which is certainly not involved in catalysis.

2. Both PMB treatment and carboxymethylation of this residue alter the steric structure of aldolase and this alteration results in the unmasking of further SH groups. The modified proteins have a less compact structure than native aldolase. However, aldolase which contains ten mercaptide bonds is fully active, while aldolase which contains ten carboxymethylated cysteinyl residues is completely inactive. It follows that, using different reagents, different portions of the protein are altered and different SH groups become unmasked.

3. The conformation of the enzyme-phosphate complex is markedly different from that of the free enzyme. Based on this phenomenon, it could be

shown that the cysteine of peptide N_{11} (Cys-II) becomes unmasked upon carboxymethylation. If this cysteine is blocked, the tight phosphate binding site of aldolase is damaged.

4. It is to be expected that the Schiff's base-forming lysine of peptide B occurs in close proximity to the cysteinyl residue of peptide N_{11} in the three-dimensional structure of native aldolase.

REFERENCES

1. Szabolcsi, G., Biszku, E. and Szörényi, E. T., *Biochim. biophys. Acta* **35** (1959) 237.
2. Szabolcsi, G. and Biszku, E., *Biochim. biophys. Acta* **48** (1961) 355.
3. Szabolcsi, G., Boross, L. and Biszku, E., *Acta physiol. hung.* **25** (1964) 149.
4. Kawahara, K. and Tanford, C., *Biochemistry* **5** (1966) 1578.
5. Závodszky, P. and Biszku, E., *Acta Biochim. biophys. Acad. Sci. Hung.* **2** (1967) 109.
6. Winstead, J. A. and Wold, F., *J. biol. Chem.* **239** (1964) 4212.
7. Morse, D. E., Chan, W. and Horecker, B. L., *Proc. natn. Acad. Sci. U.S.A.* **58** (1967) 628.
8. Swenson, A. D. and Boyer, P. D., *J. Am. chem. Soc.* **79** (1957) 2174.
9. Závodszky, P., Biszku, E., Abaturov, L. B. and Szabolcsi, G., in preparation.
10. Ginsburg, A. and Mehler, A. H., *Biochemistry*, **5** (1966) 2623.
11. Kowal, J., Cremona, T. and Horecker, B. L., *J. biol. Chem.* **240** (1965) 2485.
12. Szajáni, B., Sajgó, M. and Szabolcsi, G., *Acta Biochim. biophys. Acad. Sci. Hung.* in press.
13. Szajáni, B. and Szabolcsi, G., *Acta Biochim. biophys. Acad. Sci. Hung.* in press.
14. Sajgó, M., Szajáni, B., Biszku, E. and Szabolcsi, G., in preparation.
15. Horecker, B. L., Rowley, P. T., Grazi, E., Cheng, T. and Tchola, O., *Biochem. Z.* **338** (1963) 36.
16. Lai, C. Y., Tchola, O., Cheng, T. and Horecker, B. L., *J. biol. Chem.* **240** (1965) 1347.
17. Penhoet, E., Rajkumar, T. and Rutter, W. J., *Proc. natn. Acad. Sci. U.S.A.* **56** (1966) 1275.

FEBS Symposium, Volume 18, 1970, pp. 51-60

Chemical and Crystallographic Studies of Glyceraldehyde 3-Phosphate Dehydrogenase from Lobster Muscle

P. M. WASSARMAN and H. C. WATSON

*Medical Research Council Laboratory of
Molecular Biology, Cambridge, England*

A great deal has been learned about the biological function of glyceraldehyde 3-phosphate dehydrogenase (GPDH) [1] from a knowledge of its amino acid sequence. To gain further insight into the relationship between structure and function, it is now desirable to examine in detail a three-dimensional model of the enzyme. The fact that such a model can provide valuable biochemical information has been shown by recent crystallographic work on other enzymes, such as lysozyme [2], ribonuclease [3, 4], carboxypeptidase [5], and chymotrypsin [6]. The results of these and other studies have already illustrated how a knowledge of the spatial arrangement cf the atoms in the polypeptide chains of an enzyme can facilitate an understanding of its catalytic action.

Examination of the properties of lobster muscle GPDH (L-GPDH) indicates that it is particularly well suited to X-ray crystallographic analysis. In addition to being representative of a large class of enzymes important in metabolic processes, L-GPDH possesses several characteristics which enhance its applicability to study by crystallographic methods:

1. L-GPDH is relatively easy to prepare in large quantities from a readily available source,
2. Large, well-shaped crystals of L-GPDH can be obtained under not too stringent conditions,
3. Crystalline L-GPDH is highly ordered, giving an X-ray diffraction pattern which extends to interatomic distances,
4. L-GPDH has been studied extensively by chemical methods, its primary structure is known, and it contains several prospective binding sites for heavy atoms.

The enzyme used throughout this investigation was prepared according to the procedure outlined by Allison and Kaplan [7] and was characterized by the following criteria: molecular weight, 144,000; extinction coefficient (0.1%, 1 cm, 280 mμ), 1.00 ± 0.05; absorbance 280 mμ: absorbance 260 mμ, 1.07 ± 0.03;

specific activity (µmoles NADH formed/min/mg), 115 ± 5; moles bound NAD$^+$/mole L-GPDH, 4; cysteine residues/mole L-GPDH, 20. Harris and co-workers [8] have shown that L-GPDH is composed of four identical polypeptide chains, each containing a cysteine residue in position 148 which is acylated by substrate during catalysis.

Crystals of native L-GPDH, approximately 1 mm in each dimension, can be grown from 3.0M ammonium sulfate solutions (1 mM EDTA) at pH 5.5 to 6.2 either at 4°C or room temperature (Fig. 1). These crystals are optically biaxial and display the yellow color characteristic of L-GPDH solutions. Watson and Banaszak [9] showed that crystals of L-GPDH are orthorhombic with space group P2$_1$2$_1$2$_1$ and that the unit cell (a = 149Å, b = 140 Å, c = 80.5 Å) contains four molecules of approximately 140,000 molecular weight. An X-ray diffraction photograph of the hk0 reciprocal lattice section, taken at 2.8 Å resolution, is shown in Fig. 2. L-GPDH crystals, grown from ammonium sulfate solutions at 4°C, exhibit enzymatic activity with glyceraldehyde 3-phosphate as substrate, even after prolonged storage.

In order to obtain crystallographic phase information, an attempt has been made to use the method of multiple isomorphous replacement [10]. This method makes use of the small changes which are produced in a crystal's X-ray

Figure 1. Photomicrograph of crystals of native L-GPDH. Each numbered scale division is equivalent to 0.25 mm.

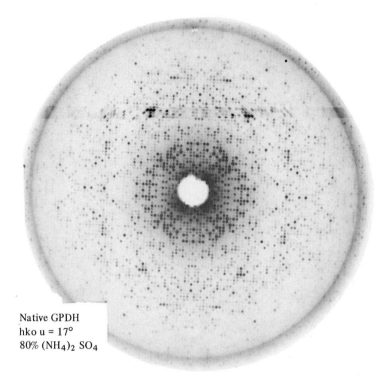

Native GPDH
hko u = 17°
80% $(NH_4)_2 SO_4$

Crystals are orthorhombic with space group $P2_1 2_1 2_1$.
Unit cell dimensions: a = 149 Å, b = 140 Å, c = 80 Å.

Figure 2. An X-ray diffraction photograph of native L-GPDH through the hk0 reciprocal lattice section. The photograph is oriented with the b* axis horizontal and contains reflections to spacings which correspond to 2.8 Å resolution.

diffraction spectrum when a small number of heavy atoms are attached to the protein molecule. The traditional approach, that of soaking crystals in the presence of a large molar excess of heavy atom reagents, has proved unsuccessful with L-GPDH. Reaction between the enzyme and the heavy atom obviously takes place, as judged by changes in the X-ray diffraction patterns; but because the reactions have not taken place with a high degree of specificity the resulting Patterson [11] calculations are difficult to interpret. In view of these difficulties a detailed study of the thiol groups of L-GPDH—a very promising site of attachment for a heavy atom—has been carried out. Chemical and crystallographic methods have been used to examine the reactivity of L-GPDH towards two sulfhydryl reagents in particular, iodoacetic acid and p-hydroxymercuribenzoate.

CARBOXYMETHYLATED L-GPDH

The reaction of iodoacetic acid with L-GPDH has been shown to take place preferentially at "active site" cysteine residue 148 [8, 12]. The carboxymethylation of this particular thiol results in complete inactivation of L-GPDH and in abolition of the enzyme's broad absorption band ("Racker Band") around 360 to 400 mμ [13, 14]. This latter effect has been attributed to an alteration of the interactions between the enzyme and its coenzyme, NAD^+. The precise nature of the alteration, which is mediated by carboxymethylation of the "active site" cysteine residue, has remained obscure. If, as has been suggested [15, 16], the spectrum of the holoenzyme derives from a charge-transfer complex between NAD^+ and an amino acid (most likely tryptophan) [23], then it is possible that the inclusion of a carboxymethyl group in the proximity of the interacting groups could lead to the loss of the spectrum without causing a large conformational change. Crystallographic experiments were carried out on carboxymethylated L-GPDH to determine whether such a conformational change did occur.

Despite complete inactivation and loss of the characteristic absorption band, carboxymethylated L-GPDH crystallized under conditions identical to those used for the native enzyme. Crystals of carboxymethylated L-GPDH grown from ammonium sulfate solutions could be distinguished visibly from crystals of native enzyme *only* by the conspicuous lack of yellow color. These crystals were completely isomorphous with those of native L-GPDH. A comparison at 6 Å resolution of reflections in the hk0 and 0kl sections of X-ray diffraction patterns of native and carboxymethylated L-GPDH revealed *no significant intensity difference* (Fig. 3).

Solutions prepared from thoroughly washed crystals of carboxymethylated L-GPDH exhibited a ratio of absorbance at 280 mμ to that at 260 mμ of 1.07 ± 0.03, identical to that of native L-GPDH, and indicative of the presence of four moles of bound NAD^+ per tetramer [17, 18]. Chromatography of these solutions on Sephadex G-25 resulted in the removal of a significant portion of the bound coenzyme. The ratio of absorbance at 280 mμ to that at 260 mμ increased from 1.07 ± 0.03 to 1.3 ± 0.05 for native L-GPDH, and to 1.7 ± 0.05 for carboxymethylated L-GPDH following Sephadex treatment.

These chemical results, when combined with the crystallographic evidence, suggest that the conformation of L-GPDH is largely preserved following carboxymethylation of the "active site" cysteine residue, although the *detailed* mode of interaction of NAD^+ with protein is altered. The alteration is expressed as a change in the spectral properties of L-GPDH and as an increase in the dissociation constant of the enzyme-coenzyme complex. The absence of larger X-ray intensity differences between native and carboxymethylated L-GPDH strongly suggests that the loss of the broad absorption band upon

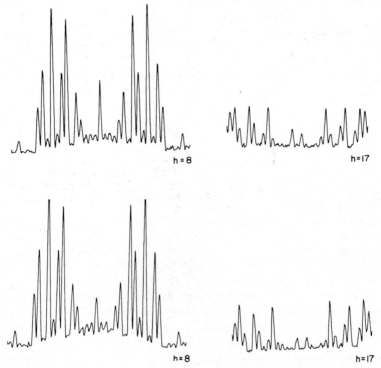

Figure 3. Two pairs of microdensitometer traces chosen at random from the projection set to show the similarity of the X-ray diffraction spectra from *native* (upper traces) and *carboxymethylated* (lower traces) crystals of L-GPDH. Differences between reflections on either side of the center point illustrate the errors involved in the taking or measuring of the X-ray photographs.

carboxymethylation of the enzyme is due to a change in *charge interactions* rather than to a structural change.

ORGANOMERCURY-L-GPDH

The reaction of the thiol groups of L-GPDH with *p*-hydroxymercuribenzoate (PMB) has been followed spectrophotometrically according to the method described by Boyer [19]. In the presence of equivalent molar concentrations of PMB, two thiol groups per L-GPDH subunit react with the mercurial. Titrations of modified forms of the enzyme—carboxymethylated and oxidized L-GPDH—suggest that one of the two sites of mercaptide formation is at cysteine residue 152, the thiol four residues removed from the "active site". The results of these experiments are summarized in Table 1.

The relationship between binding of PMB and inhibition of enzymatic activity appears to be influenced to a great extent by the conditions under which

Table 1. Stoichiometry of binding of PMB to L-GPDH.

Molar† excess PMB	Moles PMB bound per mole enzyme*		
	Native L-GPDH	Carboxymethyl L-GPDH	Oxidized L-GPDH
2	2	2	2
4	4	4	4
6	6	6	~4
8	~8	~8	~4

*Calculated from the A_{250} mμ, using a molar extinction coefficient (E_{250} mμ) of $7.6 \times 10^3 M^{-1} cm^{-1}$ for mercaptide formation.

†The molar excess was calculated on the basis of a molecular weight of 140,000 for L-GPDH.

the reaction is carried out. In the presence of rather low concentrations of L-GPDH, stoichiometric binding of eight moles of PMB per tetramer (two moles of PMB per subunit of L-GPDH) results in complete inactivation. Simultaneously, the enzyme's 360 to 400 mμ absorption band is lost. However, under certain conditions it is possible to react PMB with one thiol per monomer while retaining all of the original activity and the spectral properties characteristic of the native enzyme. This can be achieved in either of two ways: by increasing the concentration of L-GPDH titrated with the mercurial, or by carrying out the reaction in the presence of a large excess of NAD$^+$. These results are summarized in Table 2. Therefore, it is possible to preferentially react a single, non-active center cysteine residue of L-GPDH with an organic mercurial

Table 2. The effect of PMB on the activity of L-GPDH.

Molar† excess PMB	Per cent remaining enzymatic activity*		
	L-GPDH (3.0 μM)	L-GPDH (150 μM)	L-GPDH (3.0 μM) NAD$^+$ (300 μM)
2	75	95–100	100
4	50	90–95	95–100
10	<5	<5	<20

*Activity measurements were made exactly two minutes after the addition of PMB to the reaction mixture.

†The molar excess was calculated on the basis of a molecular weight of 140,000 for L-GPDH.

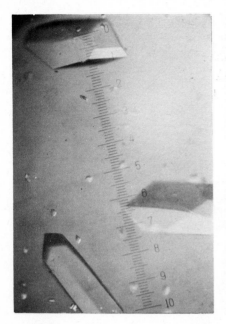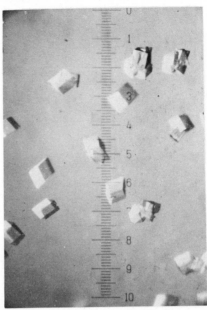

Figure 4. Photomicrograph of crystals of Type A (left) and Type B (right) of PMB-reacted L-GPDH.

(similar results have been obtained with *o*-hydroxymercuribenzoate and *p*-hydroxymercuribenzenesulfonate). Such a derivative, which retains all of the catalytic and physical properties of the native enzyme, should be ideal for phasing X-ray diffraction data.

As a consequence of mercaptide formation at *two different sites* on the enzyme molecule, two crystalline forms of PMB-L-GPDH have been obtained which are termed Type A and Type B crystals (Fig. 4).

Type A crystals can be grown from a solution of L-GPDH which has been incubated with a four-fold molar excess of PMB in the presence of a large excess of NAD^+. These crystals are completely isomorphous with the native enzyme, despite the presence of one mole of PMB per L-GPDH subunit; in addition, they exhibit the characteristic yellow color of native enzyme indicative of bound NAD^+. A solution prepared from thoroughly washed Type A crystals has an absorption spectrum which differs from that of native L-GPDH only in the region below 275 mμ where mercaptides absorb. The specific activity of such a solution is comparable to that of native L-GPDH. These crystals apparently represent enzyme which has been preferentially reacted with PMB at a non-active center cysteine residue (i.e. not at cysteine residues 148 or 152); this heavy atom derivative retains all of the enzymatic and physical properties associated with the native enzyme.

Type B crystals can be grown from a solution of L-GPDH which has been *partially* inhibited upon binding four moles of PMB per mole of tetramer. The crystals appear as rhombohedral plates, invariably have a dividing crack perpendicular to one of the faces, and are colorless. X-ray diffraction photographs show that these crystals are, in fact, disordered, suggesting a certain

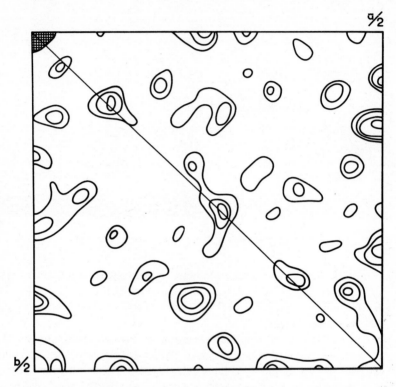

Figure 5. Difference Patterson projections calculated using data collected from native and non-active center PMB-reacted L-GPDH crystals. The symmetry of the Patterson peaks about the diagonal through the origin indicates the presence of at least one two-fold axis through the molecule.

degree of molecular variation. This effect is most probably related to observations [20] that PMB inhibition of GPDH is accompanied by the release of bound NAD^+; in this regard it should be noted that attempts to prepare crystals of the apo-enzyme or of L-GPDH totally inactivated with PMB have always resulted in precipitation of the protein. These results are consistent with those of other investigators [21, 22] which indicate that the NAD^+-free enzyme is quite unstable relative to the holoenzyme.

Three-dimensional X-ray diffraction data have been collected for the native enzyme and for several promising derivatives. As was mentioned earlier, the resulting Patterson syntheses have so far defied solution. Low resolution (6 Å) two-dimensional Patterson syntheses of the non-active center mercurial derivative have been calculated (Fig. 5), from which it is possible to derive a solution which possesses the expected symmetry and also predicts the correct position for the center of the molecule. To eliminate the possibility of the choice of an incorrect homometric structure, high resolution (3 Å) two-dimensional X-ray diffraction data have been collected. If, as expected, the interpretation of the low resolution is confirmed, a full three-dimensional investigation of this derivative will then be undertaken.

ACKNOWLEDGEMENTS

The authors wish to acknowledge the contributions made by Dr. L. Banaszak in the early stages of this work and the excellent technical assistance of Miss Jean Parkinson. One of us (P.M.W.) is a fellow of The Helen Hay Whitney Foundation.

REFERENCES

1. Velick, S. F. and Furfine, C., *in* "The Enzymes", Vol. 7 (edited by P. D. Boyer, H. Lardy and K. Myrbäck), Academic Press, New York, 1963, p. 243.
2. Blake, C. C. F., Koenig, D. F., Mair, G. A., North, A. C. T., Phillips, D. C. and Sarma, V. R., *Nature, Lond.* **206** (1965) 757.
3. Kartha, G., Bello, J. and Harker, D., *Nature, Lond.* **213** (1967) 862.
4. Wyckoff, H. W., Hardman, K. D., Allewell, N. M., Inagami, T., Johnson, L. N. and Richards, F. M., *J. biol. Chem.* **242** (1967) 3984.
5. Lipscomb, W. N., Hartsuck, J. A., Reeke, G. N. Jr., Quiocho, F. A., Bethge, P. H., Ludwig, M. L., Steitz, T. A., Muirhead, H. and Coppola, J. C., *Brookhaven Symp. Biol.* **21** (1968) 24-90.
6. Sigler, P. B., Blow, D. M., Matthews, B. W. and Henderson, R., *J. molec. Biol.* **35** (1968) 143.
7. Allison, W. S. and Kaplan, N. O., *J. biol. Chem.* **239** (1964) 2140.
8. Davidson, B. E., Sajgó, M., Noller, H. F. and Harris, J. I., *Nature, Lond.* **216** (1967) 1181.
9. Watson, H. C. and Banaszak, L. J., *Nature, Lond.* **204** (1964) 918.
10. Holmes, K. C. and Blow, D. M., *in* "Methods of Biochemical Analysis", Vol. 13 (edited by D. Glick), Interscience, New York, 1965, p. 113.
11. Lipson, H. and Cochran, W., *in* "The Determination of Crystal Structures", (edited by L. Bragg), Bell and Sons, London, 1966.
12. Allison, W. S. and Harris, J. I., "Abstracts of the Second FEBS Meeting, Vienna, 1968", Pergamon Press, Oxford, Abstract 205.
13. Colowick, S. P., van Eys, J. and Park, J. H., *in* "Comprehensive Biochemistry", Vol. 14 (edited by M. Florkin and E. Stotz), Elsevier, Amsterdam, 1966, p. 63.
14. Krimsky, I. and Racker, E., *J. biol chem.* **198** (1952) 721.

15. Friedrich, P., *Biochim. biophys. Acta* **99** (1965) 371.
16. Kosower, E. M., *Biochim. biophys. Acta* **56** (1962) 474.
17. Ferdinand, W., *Biochem. J.* **92** (1964) 578.
18. Murdock, A. L. and Koeppe, O. J., *J. biol. Chem.* **239** (1964) 1983.
19. Boyer, P. D., *J. Am. chem. Soc.* **76** (1954) 4331.
20. Velick, S. F., *J. biol. Chem.* **203** (1953) 563.
21. Listowsky, I., Furfine, C. S., Betheil, J. J. and England, S., *J. biol. chem.* **240** (1965) 4253.
22. Furfine, C. S. and Velick, S. F., *J. biol. Chem.* **240** (1965) 844.
23. Shifrin, S., *Biochim. biophys. Acta* **81** (1964) 205.

FEBS Symposium, Volume 18, 1970, pp. 61-81

Activity of some Autolyzed Derivatives of Bovine Trypsin

S. MAROUX and P. DESNUELLE

Centre de Biochimie et de Biologie Moléculaire
CNRS, Marseille, France

Since the catalytic activity of an enzyme arises from a specific three-dimensional arrangement of amino acid residues in limited regions (the "active sites"), it may be expected that the integrity of the whole molecule is not necessary for activity. Degraded and still active forms may exist, provided that the unique structure of the sites is preserved.

Bovine ribonuclease, for instance, remains fully active after hydrolysis of bond Ala_{20}-Ser_{21} or bond Ser_{21}-Ser_{22} by subtilopeptidase BPN' [1, 2]. However, activity of the enzyme is lost when peptide S, containing one of the two histidines (His_{12}) of the catalytic site, is removed. Inactivation also occurs when bond Phe_{120}-Asp_{121} in the immediate vicinity of the second catalytic histidine (His_{119}) is specifically cleaved by trypsin [3]. Similar observations are made with other pancreatic enzymes such as carboxypeptidase and chymotrypsin. Several active forms of bovine carboxypeptidase A, differing from each other by a number of N-terminal residues, are known to exist [4]. On the other hand, the primary chymotrypsin Aπ, arising from bovine chymotrypsinogen A by cleavage of bond Arg_{15}-Ile_{16}, rapidly autolyzes at the level of bond Leu_{13}-Ser_{14} without loss in activity. An additional degradation occuring in an internal region of the sequence is observed when two bonds (Tyr_{146}-Thr_{147} and Asn_{148}-Ala_{149}) in bovine chymotrypsinogen A are cleaved by chymotrypsin [5]. The resulting neochymotrypsinogens may still be activated by trypsin and give rise to chymotrypsins of the α type containing three chains instead of two, as in chymotrypsin Aπ and Aδ. Similar degradations are also observed in an homologous region (bonds Tyr_{146}-Asn_{147} and Leu_{149}-Lys_{150}) of bovine chymotrypsinogen B when this zymogen is incubated with chymotrypsin A or B [6]. Therefore, two kinds of highly specific cleavages can occur in different regions of the chymotrypsinogen molecule, one activating cleavage induced by trypsin and two non-deactivating cleavages induced by chymotrypsin.

Abbreviations used: BAEE, benzoyl-L-arginine ethylester; DFP, diisopropylfluorophosphate; TLCK, tosyl-L-lysine chloroketone; DNP, dinitrophenyl; AE, aminoethyl.

The case of bovine trypsin (EC 3.4.4.4) is most interesting because of the ability of the enzyme to autolyze rapidly under a variety of conditions. This autolysis, which is presumably specific for basic bonds, is attributed [7] to the digestion of reversibly-denatured inactive molecules (T_i) by active ones (T_a). The molecules designated T_i have recently been shown to consist of several species with different conformations [8]. Since these species are in a dynamic equilibrium, depending upon a number of parameters (pH, temperature, presence of Ca^{2+}, etc.), it may be anticipated that the "accessibility" of basic bonds to tryptic attack can be modified by varying the experimental conditions under which autolysis is performed.

Despite these unique features, little information has been gained in recent years about the chemical aspects of trypsin autolysis and its effect on the activity of the enzyme. In fact, most investigations have been performed at the kinetic level and an absolute correlation has been taken for granted between autolysis and inactivation. If, as suggested below, some autolyzed derivatives of trypsin still retain activity, these studies would not be fully reliable. In addition, they cannot give any information about the chemical nature and the position of the bonds cleaved by autolysis.

The arrangement of the 229 amino acid residues in bovine trypsinogen is now known [9, 10]. Autocatalytic activation in the presence of Ca^{2+} is brought about by the cleavage of bond Lys_6-Ile_7, which liberates the N-terminal hexapeptide Val(Asp)$_4$Lys [11, 12]. Consequently, active trypsin is composed of a single chain with 223 residues*. On the other hand, residues Ser_{183} and His_{46} are known to be involved in the catalytic site of the enzyme [13, 14].

The purpose of this study is to show that incubation at pH 6.0 of dilute solutions of pure bovine trypsin in the presence of Ca^{2+} induces almost exclusively the cleavage of bond Arg_{105}-Val_{106}. Moreover, three bonds (Arg_{105}-Val_{106}, Lys_{131}-Ser_{132}, and Lys_{49}-Ser_{50}) are substantially autolyzed in commercial samples of crystalline trypsin. Results obtained are consistent with the view that cleavage of the first two bonds (Arg_{105}-Val_{106} and Lys_{131}-Ser_{132}) does not inactivate trypsin. Since these bonds are separated by disulfide bridges, no part of the molecule is actually split off. But active trypsins with one, two or three "open" chains are likely to exist. Some of these results have already been published [15].

TECHNIQUES

All assays were carried out at 2-4°C unless otherwise stated.

---

*For the sake of uniformity, trypsin residues will be numbered as in trypsinogen. The N-terminal Ile and the C-terminal Asn will be referred to as the 7th and 229th residues, respectively.

1. Trypsinogen

Samples of bovine trypsinogen purified from the mother liquors of the crystallization of chymotrypsinogen A [7] were used for the first series of assays reported below. The filter cake obtained by precipitation in 0.8 saturated ammonium sulfate was dialyzed against 1mM HCl and lyophilized; 3 g dry powder were dissolved in 90 ml 0.08M ammonium acetate buffer (pH 6.0) and charged onto a 3 x 50 cm CM-cellulose column equilibrated with the same buffer. Results obtained were similar to those already described some years ago [16]. After the emergence of two small peaks with potential activity against the chymotryptic substrate acetyl-L-tyrosine ethylester, trypsinogen was eluted by raising the buffer molarity to 0.15. The fractions under the trypsinogen peak were pooled, acidified to pH 3, dialyzed against 1 mM HCl and lyophilized; 800 mg dry powder (yield, 75%) were obtained with no direct activity against the substrate BAEE, and a potential activity between 47–54. The preparations were found to be homogeneous by disc electrophoresis at pH 5.0 (15% crosslinked gel) and to contain no other N-terminal residue than valine (one residue per mole). They could be kept at −15°C for several weeks without any detectable activation or loss of potential activity.

A second series of experiments was performed with commercial crystalline trypsinogen (Worthington 1 x crystallized).

2. Trypsin and DFP-inhibited trypsin

Trypsin was either prepared by autoactivation of chromatographically-purified or commercial trypsinogen, or obtained from commercial sources. Full activation of trypsinogen was obtained by a 2 h incubation of 10 mg/ml solutions in a Tris-HCl buffer (pH 7.8), 50 mM in Tris and 50 mM in $CaCl_2$ with trypsin (weight ratio, 1%). Some of these preparations were inhibited by treatment with 20 moles/mole DFP for 2 h. The solutions were dialyzed against 1 mM HCl and lyophilized. Three commercial samples (TW_1, TW_2, TW_3) of trypsin (Worthington 2 x crystallized) were also used and inactivated by DFP as indicated above.

3. Determination of trypsin activity

Trypsin activity was measured titrimetrically at pH 7.9 and 25°C in a Radiometer pH-stat Model TTT1, using a 10 mM solution of the substrate BAEE. The flow rate of 0.1M NaOH was recorded for 3 min. Specific activities were expressed in μmoles substrate hydrolyzed per min and per mg protein ($E_{1\%}^{1cm}$ of pure trypsin at 280 mμ, 14.4).

4. N-terminal residues

Solutions of proteins or peptides in bicarbonate were condensed with 1-fluoro-2,4-dinitrobenzene and the resulting DNP-derivatives were hydrolyzed at 115°C for 18 h in tridistilled 6N HCl. The ether-soluble DNP-amino acids were separated, identified and quantitated by the technique described by Murachi and Neurath [17]. DNP-leucine and DNP-isoleucine, which were not separated by this technique, were hydrolyzed with concentrated ammonia and the corresponding amino acids were identified in an automatic analyzer.

5. Fingerprints

Peptide maps were obtained by electrophoresis-chromatography on Whatman 3 MM paper. Electrophoresis was run for 90 min at 36 V/cm in a pyridine-acetic acid-water buffer (1:10:289 by volume) at pH 3.5. The solvent system butanol-acetic acid-pyridine-water (15:3:10:12 by volume) was used for chromatography.

6. Radioactivity

Aliquots of intact or autolyzed trypsin which had been labelled with [^{32}P] DFP (Amersham, England) were dissolved in 10 ml Bray's mixture and counted in a Packard Tricarb Spectrometer. Specific radioactivities were expressed in c.p.m. per mg trypsin, using the same absorbance coefficient as above.

RESULTS

I. Chromatography of active trypsin on CM-cellulose

It was found in this laboratory that some commercial crystalline trypsin samples could be freed of inactive or chymotrypsin-like impurities by chromatography on CM-cellulose in a citrate buffer at pH 6.0 [16]. The choice of a relatively high pH value for this chromatography was dictated by considerations of ion-exchange efficiency. Indeed, trypsin preparations could be obtained in this way with a specific BAEE-splitting activity 20% higher than the original crystalline material. Results observed when activated trypsinogen, instead of commercial trypsin, was submitted to chromatography on CM-cellulose under similar conditions, are indicated by Fig. 1.

Fig. 1(a) shows that, when trypsinogen was activated in the presence of 50 mM CaCl$_2$, a small inactive peak amounting to about 6% of the total proteins was eluted ahead of trypsin. Since this inactive peak was considerably larger (60% of total proteins, Fig. 1(b)) when activation was performed in the absence of added Ca^{2+}, it could be assumed to contain the "inert proteins" first described by Kunitz et al. [7]. The formation of these compounds was depressed, but not entirely suppressed, by 50 mM CaCl$_2$. On the other hand, it was noteworthy that most fractions of the trypsin peak in Fig. 1(b) had nearly the same specific activity as in Fig. 1(a) and that this activity (46-54) was very close to the potential specific activity of pure trypsinogen. Chromatography on CM-cellulose at pH 5.5, therefore, appeared to be a quite efficient technique for the purification of trypsin.

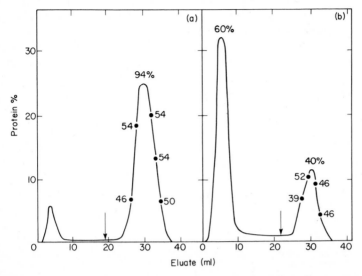

Figure 1. Chromatography of activated trypsinogen on CM-cellulose. Chromatographically-purified trypsinogen (potential specific activity, 54) was activated in the presence of 50 mM $CaCl_2$ (a), or without added Ca^{2+} (b). The activated solutions were brought to pH 3, dialyzed against 1 mM HCl and lyophilized. 20 mg dry powder dissolved in 2 ml sodium citrate buffer (pH 5.5) 30 mM in citrate and 13 mM in $CaCl_2$ were charged onto a 0.9 x 16 cm CM-cellulose (Biorad) column equilibrated with the same buffer. Arrow indicates the point where the citrate molarity of the buffer was raised to 0.1M. Fraction volume, 1.4 ml. Ordinates, protein content of the fractions in % of the total proteins of the charge. Numbers along the second peak give the specific activity of the fractions against BAEE (trypsin activity).

However, condensation with fluorodinitrobenzene of the compounds under the trypsin peak in Fig. 1(a) revealed the presence, not only of the N-terminal isoleucine residue characteristic for undegraded trypsin, but also of N-terminal valine (0.29 mole/mole), serine (0.20 mole/mole) and aspartic acid or asparagine (0.13 mole/mole). In contrast, no DNP-amino acid other than DNP-isoleucine was identified when DFP-inhibited trypsin, rather than the active enzyme, was submitted to chromatography. Hence, it was clear that active trypsin underwent a limited and relatively specific autolysis during chromatography at pH 5.5 in the presence of 13 mM Ca^{2+}.

II. First series of assays on trypsin autolysis

Since valine was predominant among the N-terminal residues set free by autolysis, it was interesting to see which one of the two basic bonds (Arg_{105}-Val_{106} and Lys_{214}-Val_{215}) involving this residue in trypsin sequence [9, 10] was actually cleaved during the process. Table 1 indicates how conditions for maximal production of N-terminal valine were found.

Results reported in Table 1 at first confirmed the well-known stabilizing effect of Ca^{2+} on trypsin. A 1% solution at pH 7.8 in 50 mM $CaCl_2$ could be kept for 20 h with very little degradation. The same observation was made at pH 6.0. But a new fact was that more dilute trypsin solutions were unstable at pH 6.0, even

Table 1. N-terminal residues appearing during trypsin autolysis.

Conditions of incubation					Trypsin			
					N-terminal residues (mole/mole, uncorrected)			
Trypsin (%)	Ca^{2+} (mM)	pH	Time (h)	Specific activity	Ile	Val	Ser	Asp
1.0	50	7.8	20	48	0.72	0.07	0.03	0.07
1.0	50	6.0	20	51	0.61	0.04	0.04	0.02
0.1	50	6.0	20	51	0.82	0.17	0.06	0.04
0.2	10	6.0	20	47	0.76	0.26	0.04	0.05
0.1	10	7.2	1.5	42	0.75	0.48	0.58	0.14
0.1	0	7.9	2	38	0.71	0.28	0.17	0.18

in the presence of 50 mM Ca^{2+}. Formation of N-terminal valine was somewhat enhanced by lowering the Ca^{2+} concentration to 10 mM. For higher pHs, and in the complete absence of added Ca^{2+}, autolysis was very rapid and largely non-specific. In addition, the activity of the solutions, which remained nearly constant during the first assays, markedly decreased during the final ones. Therefore, conditions indicated on the fourth line in Table 1 (incubation of a 0.2% trypsin solution for 20 h in presence of 10 mM Ca^{2+}) were adopted for subsequent experiments.

If the bond cleaved during autolysis was the one linking Lys_{214} to Val_{215}, a pentadecapeptide (residues 215-229) should be liberated. This peptide was never detected, even after complete denaturation of the molecules by reduction-carboxymethylation [18]. Direct evidence for the cleavage of the other bond was obtained as follows: A sample of autolyzed trypsin (N-terminal isoleucine, valine, serine and aspartic acid: 0.75, 0.26, 0.06 and 0.13 mole/mole, respectively) was treated with DFP and the disulfide bridges were split by reduction-carboxymethylation. After condensation with fluorodinitrobenzene, 23 mg of the DNP-derivatives suspended in 3 ml water were digested at pH 7.9 for 15 h at 30°C by 0.15 mg subtilopeptidase. The ether-soluble fraction contained two α-N-DNP-peptides: DNP-Val (Ser, Ala) and DNP-Val (Ser, Ala, Ile), which were purified by two-dimensional paper chromatography in 1.5M phosphate, pH 6.0, and *tert* amyl alcohol-2M ammonia (4:1 by volume). The ethyl acetate extract contained a number of ϵ-N-DNP-lysine and O-DNP-tyrosine peptides, but no other α-DNP-derivatives. Since residue Val_{106} is followed in the trypsin sequence by Ala_{107}, Ser_{108} and Ile_{109}, whereas the residues situated after Val_{215} are different, the identification of the two DNP-peptides mentioned above conclusively proved that bond Arg_{105}-Val_{106} was autolyzed. In contrast, bond Lys_{214}-Val_{215} was probably intact.

Another sample of autolyzed trypsin (N-terminal isoleucine, valine, serine and aspartic acid: 0.72, 0.28, 0.18 and 0.09 mole/mole, respectively) was inactivated by $[^{32}P]$DFP prior to chromatography. The fractions under the DFP-inhibited trypsin peak were dialyzed and lyophilized. A solution of the powder in a Tris buffer (pH 8.6) 8M in urea was reduced and carboxymethylated. The chains were dialyzed, lyophilized, dissolved in 10 mM phosphate buffer (pH 8.0) 6M in urea and percolated through Sephadex G-100. Figure 2(a) shows that two peaks designated A and B were separated. Amino acid analysis indicated that peak A was composed of the long chain of undegraded trypsin with 223 residues. Peak B emerging later could be assumed to contain the shorter chains resulting from autolysis. These chains were further fractionated on DEAE-cellulose as indicated in Fig. 2(b). Four peaks designated (1), (2), (3) and (4) were obtained in the experiment related to Fig. 2. In other experiments, performed under apparently identical conditions, only two peaks were separated on DEAE-cellulose: (1') emerging at the same position as peaks (1) and (2), and (2') eluted later by the gradient, like peaks (3) and (4).

Results recorded in Table 2 suggested that peaks (1) and (4) were composed, respectively, of the N-terminal part of the trypsin chain extending from Ile_7 to Arg_{105}, and of the C-terminal part of the same chain from Val_{106} to Asn_{229}. Identification of this latter was facilitated by the complete absence of arginine and histidine. Therefore full support was given to the above assumption, according to which bond Arg_{105}-Val_{106} was autolyzed during incubation of trypsin at pH 6. In addition, the amino acid composition of peaks (2) and (3) was somewhat reminiscent of sequences 106–176 and 177–229, respectively. But, for reasons given later, a more precise study of these fragments was not undertaken. When only two peaks (1' and 2') were separated on DEAE-cellulose, these peaks also had the same amino acid composition as sequences 7-105 and 106-229, respectively.

Another point of interest was that peak B in Fig. 2 had nearly the same specific radioactivity as peak A. After DEAE-cellulose chromatography, radioactivity was found under peaks (3) and (4), whereas peaks (1) and (2) were not radioactive. Since peak (4), and presumably also peak (3), could be assumed to contain residue Ser_{183} binding DFP, this fact appeared to confirm that a functional catalytic site was still present in autolyzed trypsin molecules.

However, this confirmation would not be valid if some intact molecules were degraded after labelling by radioactive DFP. To check this point, trypsin

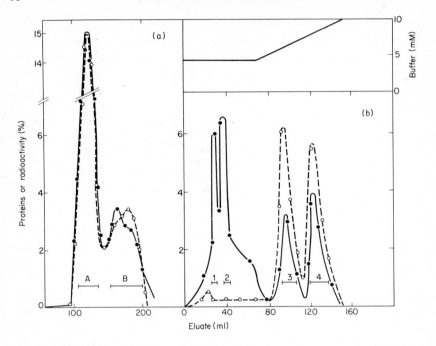

Figure 2. Typical separation of S-carboxymethylated chains. (a) The chains (90 mg) in 5 ml 10 mM phosphate buffer (pH 8.0) 6M in urea, were passed through a 2 x 145 cm Sephadex G-100 column equilibrated and eluted with the same buffer. Fraction volume, 5 ml. (b) Fractionation of peak (B) in (a) on a 1.4 x 25 cm DEAE-cellulose column equilibrated with a 4.5 mM phosphate buffer (pH 8.0) 8M in urea. Peaks (3) and (4) were eluted by a linear gradient of the buffer concentration. Solid and dashed lines indicate, respectively, the protein content and the radioactivity of the fractions in % of the total introduced onto the column.

prepared from pure trypsinogen was inactivated at once by DFP, reduced and carboxymethylated. The resulting product was percolated as above through Sephadex G-100. A single fast migrating peak (peak A) containing the intact chain should normallly have been obtained. In fact, peak B was small in some of these blank assays, but in others it amounted to 30% of the total protein. Moreover, as already pointed out before, four or only two peaks were separated on DEAE-cellulose in a series of assays performed with trypsin samples that were autolyzed to the same extent. Hence, it was obvious that uncontrolled degradations sometimes occured during the preparation of S-carboxymethylated chains.

To sum up, this first series of assays demonstrated that bond Arg_{105}-Val_{106} was highly sensitive in trypsin. On the other hand, cleavage of this bond did not appear to inactivate the enzyme since the specific activity and the number of functional sites remained essentially unaltered. But, owing to uncontrolled

Table 2. Amino acid composition of peaks (1) and (4) in Fig. 2.

| | Peak (1) | Peak (4) | Sequences* | |
			7–105	106–229
Ala	7.0	5.8	6	8
Arg	0.7	0.0	2	0
Asx	12.5	8.0	13	9
Cys†	3.6	8.4	3	9
Glx	7.9	7.7	7	7
Gly	9.3	15.0	9	16
His	2.5	0.0	3	0
Ile	7.4	5.3	8	7
Leu	7.0	6.2	7	7
Lys	4.4	8.9	4	10
Met	0.9	1.0	1	1
Phe	1.8	1.4	2	1
Pro	2.9	5.6	3	6
Ser‡	13.5	14.6	13	20
Thr	3.4	3.9	3	7
Tyr	3.5	4.5	5	5
Val	8.7	6.9	9	8

*According to Walsh and Neurath [9] and Mikes et al. [10].
†As S-carboxymethyl-cysteine.
‡Corrected for a 17% loss during hydrolysis (48 h).

degradations occurring after autolysis, identification of labelled fragments was not absolutely conclusive. These degradations seem to affect mainly bond Lys_{176}-Asn_{177}. But the presence of N-terminal valine in peak B, corresponding to blank experiments, suggests that they also affect bond Arg_{105}-Val_{106}.

III. Second series of assays with commercial trypsin
Commercial, crystalline trypsin is known to contain a number of N-terminal residues other than the isoleucine of the intact chain. These residues probably arise from autolysis during crystallization at alkaline pHs, and storage.

1. Preliminary assays
Three commercial samples (Worthington) designated TW_1, TW_2, TW_3 (specific activities, 47, 36 and 46, respectively), and a sample T_0, prepared as described earlier by autoactivation of commercial trypsinogen, were used for these assays. The sample T_0 had a specific activity of 50 and served as reference. The samples were freed of inactive products by chromatography on CM-cellulose with the results indicated in Fig. 3 for T_0 and TW_2. As expected, the inactive peak related

to T_0 was small. It was larger with TW_2 and two distinct regions (a) and (b) were visible. In both cases, however, the fractions forming the trypsin peak had a high specific activity (49–53 for T_0 and 49–50 for TW_2) and this activity was nearly the same as the potential activity of pure trypsinogen.

Another aliquot of T_0 was incubated with $[^{32}P]DFP$ immediately after activation. TW_2 was labelled in the same way and the two solutions were chromatographed on CM-cellulose. Figure 3 shows that specific radioactivities of the fractions under the DFP-inhibited trypsin peak were not significantly lower for TW_2 than for T_0. A similar observation could be made with TW_1 and TW_3.

In other experiments, the four trypsin samples were inhibited with cold DFP, purified by chromatography on CM-cellulose and then condensed with fluoro-dinitrobenzene. Table 3 indicates that isoleucine was the single N-terminal residue in T_0, whereas N-terminal valine and serine were also present in TW_1, TW_2 and TW_3. If losses occurring during acid hydrolysis of the DNP-proteins were assumed to be the same for these three DNP-amino acids, TW_1, TW_2 and TW_3 would contain about 0.3 mole N-terminal valine for one mole isoleucine. The corresponding content in N-terminal serine would be about 0.2 mole for TW_1 and TW_2, and 0.13 mole for TW_3. These latter values are probably a minimum, because of the known lability of DNP-serine in acid.

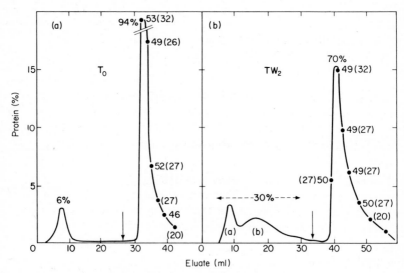

Figure 3. Chromatography of T_0 and TW_2 on CM-cellulose. The conditions were essentially the same as in Fig. 1. (a) Reference sample (T_0) prepared from trypsinogen. (b) Commercial crystalline sample TW_2 (Worthington). Fraction volume, 1.7 ml. Ordinates, protein content of the fractions in % of total proteins. Numbers along the trypsin peak indicate the specific activity of the fractions against BAEE. Numbers in parentheses indicate the specific radioactivity of the fractions when trypsin was labelled with $[^{32}P]DFP$ prior to chromatography.

2. Autolytic cleavages in trypsins TW_1, TW_2 and TW_3

a. Dialysis of the aminoethylated chains

The uncontrolled degradations of the S-carboxymethylated chains were probably caused by their lack of solubility and further treatment at alkaline pHs in the presence of high concentrations of urea. Hence, acid-soluble S-aminoethylated chains were prepared [19] for subsequent assays. Immediately after preparation, these chains were dissolved in acid and all further treatments were conducted in acid medium. Another technical improvement over the studies described earlier was to incubate DFP or TLCK-inhibited trypsin preparations at pH 3.0 for 15 min in 8M urea before reduction-aminoethylation. In this way, any traces of active molecules were fully denatured.

Table 3. N-terminal residues in DFP-inhibited trypsin samples after purification on CM-cellulose.

N-terminal residues (mole/mole, uncorrected)	DFP-inhibited trypsin				DFP-inhibited TW_2		
	T_0	TW_1	TW_2	TW_3	Unpurified sample	Inactive peaks (a)	(b)
Ile	0.58	0.66	0.63	0.63	—	—	—
Ile + Leu	—	—	—	—	1.02	1.87	1.77
Val	0.05	0.20	0.19	0.19	1.03	2.62	0.21
Ser	0.05	0.12	0.14	0.08	0.28	0.79	0.14
Asx	0.01	0.03	0.02	0.02	0.25	0.89	0.15
Ala	0.00	0.00	0.00	0.00	0.21	0.85	0.07

The chromatographically-purified DFP-treated samples were reduced and aminoethylated. The solutions containing the chains were brought to pH 2, desalted on Sephadex G-25 (coarse), equilibrated with 10 mM HCl, and lyophilized; 200 mg of the dry powder in 12 ml 10 mM HCl were dialyzed 5 times against 30 ml 10 mM HCl. No dialyzable fragments could be detected in experiments carried out with the reference sample T_0. But dialyzable fractions, designated DW_1, DW_2 and DW_3, were consistently obtained with the autolyzed samples TW_1, TW_2 and TW_3.

b. The dialyzable fraction D

Figure 4 gives an example of the technique used for the purification of the dialyzable fraction by passage through Sephadex G-50 (fine) equilibrated with 10 mM HCl. As indicated in Table 4, the main peak thus obtained was found to contain high amounts of N-terminal isoleucine and valine, but no N-terminal serine. Identification of the dipeptide DNP-Ile-Val in the hydrolyzate of the DNP-derivatives further showed that isoleucine originated from the N-terminus of trypsin. Hence it appeared likely that the dialyzable fraction was composed of two short chains, one corresponding to the N-terminal sequence of trypsin, the other to a sequence beginning with the Val_{106} residue earlier identified.

This assumption was confirmed by the amino acid composition of DW_2 given in column (1) of Table 5. Owing to the complete absence of arginine in the N-terminal sequence extending from residue 7 to 49 in trypsin, and to the exclusive presence in this sequence of 1, 2 and 4 residues of phenylalanine, histidine and tyrosine, its amount in the aliquot submitted to analysis was seen to be 0.025 μmole. Next, its contribution to the amino acid content of the analyzed mixture was calculated (column (3)) and the composition of the second chain was obtained by difference (columns (4) and (5)). The composition of the sequence extending from residue 106 to 131 in trypsin is given in column (6) for the purpose of comparison. Taking into account the relatively large errors inherent in difference methods, and the fact that normal values were obtained for valine and histidine with samples DW_1 and DW_3, the agreement between columns (5) and (6) appeared to be quite satisfactory and to warrant identification of the second dialyzable chain with sequence 106–131. In conclusion, amino acid analysis indicated that the dialyzable fraction was composed of two chains originating from sequences 7–49 and 106–131. For one mole of chain

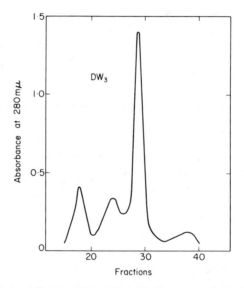

Figure 4. Typical purification of the dialyzable fraction on Sephadex G-50. The 1 x 200 cm column of Sephadex G-50 (fine) was equilibrated and eluted with 10 mM HCl. Fraction volume, 2.9 ml.

Table 4. N-terminal residues in aminoethylated chains.

| | N-terminal residues* | | |
	Ile	Val	Ser
AW_1	0.59	0.00	0.02
DW_2	0.37	0.23	0.00
BW_1	0.54	0.07	0.31
BW_2	0.57	0.21	0.32
BW_3	0.56	0.08	0.56

*Molar ratios for fractions A and B. The amount of fraction DW_2 submitted to analysis was unknown.

106–131, 0.45, 1.5 and 0.11 moles of chain 7–49 were present in DW_1, DW_2 and DW_3, respectively.

Further identification of the dialyzable chains was obtained by fingerprinting peptic digests of fraction D and intact trypsin (fraction A, see later). All the peptides arising from fraction D during this digestion must be present in intact trypsin, with the exception of the two related to the C-terminus of the chains resulting from tryptic attack. Two peptides not found in digests of fraction A could actually be characterized after Sephadex G-25 filtration and fingerprinting of digests of fraction D. The first (Ala-Ala-His-AE.Cys-Tyr-Lys) corresponded to the C-terminal part of sequence 7–49, and the second (Try-Gly-Asn-Thr-Lys) to the C-terminal part of sequence 106–131. Consequently, substantial autolysis of bonds Lys_{49}-Ser_{50}, Arg_{105}-Val_{106} and Lys_{131}-Ser_{132} was demonstrated in the three crystalline samples TW_1, TW_2 and TW_3.

c. The non-dialyzable fraction

The non-dialyzable fraction prepared from the reference sample T_0 and from commercial TW_3 was passed through Sephadex G-75 with the results indicated in Fig. 5.

Fig. 5 is similar in a number of respects to Fig. 2(a). Peak A could be shown to contain exclusively an N-terminal isoleucine (see AW_1 in Table 4), to have the same amino acid composition as trypsin (see column 7 in Table 5) and to give a single band by disc electrophoresis at the same position as the aminoethylated enzyme. Therefore, like the corresponding peak A in Fig. 2, it was composed of the complete chain of trypsin originating from non-degraded molecules.

The slower moving peak B (BW_1, BW_2 and BW_3) could be expected to be formed by the non-dialyzable large fragments resulting from autolysis. As shown

Table 5. Amino acid composition of fractions D and A.

	Sequence 7–49			Second chain		
	Fraction DW$_2$	Composition	In 0.025 μmoles	Difference (1) – (3)	Composition	Sequence 106–131	Fraction AW$_2$
	(1)	(2)	(3)	(4)	(5)	(6)	(7)
Ala	0.128	3	0.075	0.053	3.1	3	14.2
Arg	0.005	0	0.000	0.005	0.0	0	2.3
Asx	0.036	2	0.050	–	–	1	21.8
Cys*	0.097	3	0.075	0.022	1.3	2	11.0
Glx	0.067	2	0.050	0.017	1.0	1	15.5
Gly	0.205	6	0.150	0.055	3.2	3	25.0
His	*0.049*	2	*0.050*	0.000	0.0	0	3.0
Ile	0.085	2	0.050	0.035	2.0	2	14.8
Leu	0.102	2	0.050	0.052	3.0	2	14.4
Lys	0.062	1	0.025	0.037	2.2	1	15.2
Met	0.000	0	0.000	0.000	0.0	0	1.1
Phe	*0.025*	*1*	*0.025*	0.000	0.0	0	3.2
Pro	0.052	1	0.025	0.027	1.6	1	7.8
Ser†	0.211	5	0.125	0.086	5.1	5	33.2
Thr	0.092	2	0.050	0.042	2.5	3	9.4
Trp‡	0.038	1	0.025	0.013	0.8	1	–
Tyr	*0.102*	*4*	*0.100*	0.002	0.0	0	10.0
Val	0.130	5	0.125	0.005	0.0	1	16.0

*As aminoethylcysteine.
†Corrected for a 17% loss during hydrolysis (48 h).
‡By spectrophotometry [22].

by Table 4, most of the N-terminal serine was found in this peak, whereas most of the N-terminal valine was in the dialyzable fraction. This distribution is perfectly consistent with the degradation scheme put forward earlier, following study of the dialyzable fraction.

Another point of interest was that peak B corresponded to 12% of the total proteins existing in chromatographically-purified T$_0$. Hence, the S-aminoethylated derivatives arising from the reference sample were not completely free of degraded fragments. In contrast, however, to what was observed earlier with S-carboxymethylated chains, the size of peak B was found constant in a number of blank experiments and to be always much smaller than that observed in assays starting from autolyzed trypsin. The proportions of peaks A and B were 80-12% for T$_0$, 53-37% for TW$_1$, 52-40% for TW$_2$ and 45-43% for TW$_3$.

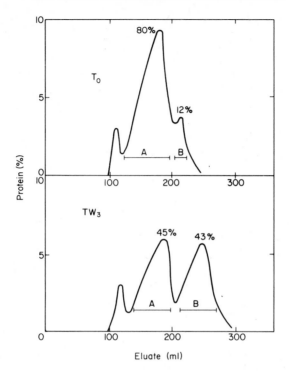

Figure 5. Fractionation of the undialyzable compounds on Sephadex G-75. The column (1.4 x 200 cm) of Sephadex G-75 was equilibrated and eluted with 10 mM HCl. Fraction volume, 6 ml. Ordinates, protein content of the fractions in % of total proteins. Upper and lower diagrams: aminoethylated chains from intact (T_0) and autolyzed (TW_3) trypsins, respectively.

3. Inhibition by TLCK

Since uncontrolled degradations of the chain were now reduced to a minimum, it was of interest to see whether or not labelled fragments could be isolated after treatment of autolyzed trypsin with specific inhibitors. TLCK was used here rather than DFP because of the more favorable position of residue His_{46} with respect to the autolyzed bonds. This residue in the dialyzable fraction originated exclusively from molecules degraded at the level of bond Lys_{49}-Ser_{50}. The same residue in fraction B could be considered as a good "reporter" for the activity of molecules in which bonds Arg_{105}-Val_{106} and Lys_{131}-Ser_{132} were cleaved. Finally, His_{46} in fraction A, originating from intact trypsin, served as internal standard.

100 mg TW_3 were dissolved in 10 ml Tris-HCl buffer (pH 7.0) 50 mM in Tris and $CaCl_2$. After 18 h incubation with 4.3 mg TLCK (Mann Research Laboratories), activity had dropped to 1% of the original value. Fractions A, B

and D were prepared in the usual way, oxidized by performic acid and hydrolyzed. The carboxymethylhistidine formed at the expense of the labelled histidine residue during oxidation [20] was quantitated in an automatic analyzer. Table 6 shows that the number of active histidine residues was the same in mixture A + B as in intact trypsin (fraction A). It was only slightly lower in isolated fraction B, whereas no carboxymethyl-histidine could be detected in the dialyzable fraction D.

These results definitely confirm that trypsin activity is not essentially altered by autolysis of bonds Arg_{105}-Val_{106} and Lys_{131}-Ser_{132}. They also suggest that the cleavage of bond Lys_{49}-Ser_{50} inactivates the enzyme. But, for reasons given in the following section, this latter conclusion is still tentative.

Table 6. Carboxymethyl-histidine (CM-His) in oxidized fractions from TLCK-treated trypsin.

Fraction	Experimental results (μmoles)		In 1 mole trypsin	
	CM-His*	Phe†	CM-His*	Phe†
A	0.065	0.493	0.43	3.0
A + B	0.075	0.498	0.45	3.0
B	0.045	0.397	0.34	3.0
D	0.000	0.094	0.00	1.0

*The color value of glycine was used for CM-His.
†Calculation is made by reference to three, three and one phenylalanine residues, respectively, in fractions A, B and D.

DISCUSSION

The property of bovine trypsin to autolyze rapidly may be related in some way to the specificity of the enzyme towards basic bonds. Owing to their polar character, these bonds can be expected to be at, or near, the surface of the native molecules and, consequently, to be more "accessible" than hydrophobic bonds for which chymotrypsin and other stable proteases are specific. However, it must be emphasized that trypsin is not subject to self-digestion unless its configuration has been modified by denaturation.

The main purpose of this work was to see whether some autolyzed trypsin derivatives were enzymatically active. The most degraded species so far known in the chymotrypsin family is chymotrypsin Aα with three "open" chains. Activity of this species is easy to demonstrate after purification. In contrast, autolyzed trypsin derivatives could not be separated either from each other, or from the intact molecules remaining in the incubated or crystalline preparations. But the shorter chains arising from self-digestion were fully identified after splitting of the disulfide bridges. Specific activity determinations and titrations of catalytic sites by DFP or TLCK unequivocally proved that at least some autolyzed derivatives were still active. The main difficulty encountered was the instability of S-carboxymethylated chains in alkaline medium in the presence of urea. Better results were obtained with S-aminoethylated chains which were found to be soluble in acid and could be fractionated by dialysis and chromatography in the absence of urea. This observation may be of practical interest in other cases.

Among all the basic bonds of trypsin, the one linking residue Arg_{105} to Val_{106} appeared to be most labile. It was cleaved first during incubation of the active enzyme at pH 6.0 in the presence of Ca^{2+}. Appreciable autolysis of the same bond was also observed in the crystalline enzyme, despite the widely different conditions under which incubation and crystallization were performed. The reason for this remarkable lability is still unknown. It may be postulated that, in addition to the usual factors controlling the specificity of a proteolytic attack ("accessibility" of the bond, nature of the adjacent residues) a highly specific cleavage in a native protein requires a special fit between the conformation of the substrate in the neighbourhood of the bond and the structure of the active site in the enzyme.

In addition to the Arg_{105}-Val_{106} bond, two other basic bonds involving a serine residue (Lys_{131}-Ser_{132} and Lys_{49}-Ser_{50}) were found to be autolyzed in crystalline trypsin. The percentages of the three cleavages were roughly calculated, using the amounts of stable DNP-isoleucine and DNP-valine recovered in the autolyzed fragments. The percentage for bond Arg_{105}-Val_{106} was directly derived from the amount of DNP-valine in chromatographically-purified TW_1, TW_2 and TW_3 (Table 3). Moreover, assuming that the passage of N-terminal Val_{106} into the dialyzable fraction was mainly caused by the cleavage of bond Lys_{131}-Ser_{132}, the value for this latter was calculated from the difference between the amount of DNP-valine in the original sample (Table 3) and in the corresponding fraction B (Table 4). Finally, the percentage for bond Lys_{49}-Ser_{50} was deduced from the ratio of the two short chains in the dialyzable fraction (Table 5). Results obtained in this way are given by Table 7.

Table 7 shows that bond Arg_{105}-Val_{106} was cleaved in approximately one molecule out of three. The percentage was slightly lower for bond Lys_{131}-Ser_{132} and it was quite variable for bond Lys_{49}-Ser_{50}. Since more than half of the molecules were still intact in the autolyzed samples, cleavage of bond Arg_{105}-Val_{106} appeared to be rate-limiting and to favor the further splitting of the other two bonds. As a consequence, autolyzed trypsin was essentially composed of molecules either intact or degraded at several points. Indeed, the presence of substantial amounts of chain 106–131 in the dialyzable fraction proved that bonds Arg_{105}-Val_{106} and Lys_{131}-Ser_{132} were simultaneously cleaved in a number of autolyzed molecules.

Another point of considerable interest was that some of the cleavages mentioned above did not inactivate trypsin. An initial observation supporting

Table 7. Autolytic cleavages (%) in samples TW_1, TW_2 and TW_3.

	Autolyzed molecules (% of total)	Number of autolyzed bonds (in 100 moles)		
		105–106	131–132	49–50
TW_1	37	30	26	12
TW_2	40	30	17	26
TW_3	45	30	24	3

this view was that the specific activity of the enzyme and the specific radioactivity of its $[^{32}P]$ DFP-inhibited derivative were never found to be markedly lower for autolyzed, as compared to intact, preparations. Taken in conjunction with the results in Table 7, and with the fact that the accuracy of activity and radioactivity determinations is certainly better than 20%, this observation strongly suggests that the cleavage of bonds Arg_{105}-Val_{106} and Lys_{131}-Ser_{132} is not de-activating, even if both occur in the same molecule. Final proof for the existence of an active trypsin with three "open" chains (7-105, 106-131 and 132-229) was provided by the identification of large amounts of carboxymethyl-histidine in the hydrolyzates of performic acid-oxidized fraction B originating from TLCK-treated TW_3.

As already pointed out above, the case of bond Lys_{49}-Ser_{50} is still unsettled. Absence of carboxymethyl-histidine in the hydrolyzates of fraction D is consistent with the view that cleavage of this bond inactivates trypsin. However, the corresponding experiment was performed only once with a sample (TW_3) in which autolysis of the bond was slight. Consequently, the amount of chain 7-49 submitted to analysis was barely above the lower limit for detection of carboxymethyl-histidine. In addition, if cleavage of bond Lys_{49}-Ser_{50} really inactivated trypsin, the specific activity and radioactivity of sample TW_2 (cleavage, 26%; see Table 7) should have been, in contrast to the experimental results, lower than normal. Since bond Lys_{49}-Ser_{50} is close to residue His_{46}, the effect of its cleavage on trypsin activity must be carefully re-investigated.

The position of the autolyzed bonds with respect to the disulfide bridges in the trypsin molecule is also of interest. Four of the six disulfide bridges in trypsin are homologous with four of the five bridges in chymotrypsin. As shown by Fig. 6, the other two (bridges 13–143 and 115–216) are situated on both sides of bond Arg_{105}-Val_{106}. Bridge 13–143 is relatively far, but it is the only one binding chain 7–105 to the rest of the molecule. If the same cleavage had occurred in chymotrypsin, the corresponding chain would not be covalently bound. Moreover, bridge 115–216 is closer and it may play a role in the stabilization of the three-dimensional configuration after autolysis of bond Arg_{105}-Val_{106}. This bridge also provides an additional point of attachment for the short chain 106–131. Indeed, this short chain is very firmly held by two bridges (115–216 and 122–189), the latter being homologous with the one connecting chains B and C in bovine chymotrypsin A.

Finally, Table 8 shows that bond Lys_{131}-Ser_{132} is nearly homologous with bonds Tyr_{146}-Thr_{147} or Tyr_{146}-Asn_{147} which are specifically split by chymotrypsin in bovine chymotrypsinogen A or B [5, 6]. This indicates not only that bond cleavages are possible in the same regions of trypsin and chymotrypsin without any serious distortion of the active site, but also that conformational homology should exist in this region, permitting highly specific attack by both enzymes. By contrast, no counterpart of bond Arg_{105}-Val_{106} has so far been identified in chymotrypsin.

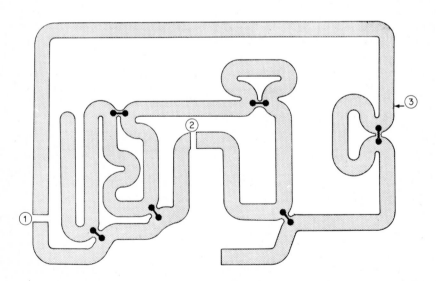

Figure 6. Position of autolytic cleavages in trypsin with respect to the disulfide bridges. The two-dimensional representation of the peptide chain is given according to Holeysovsky et al. [21]. This chain is interrupted at the level of bond Arg_{105}-Val_{106} (1) and bond Lys_{131}-Ser_{132} (2). Arrow (3) indicates the position of bond Lys_{49}-Ser_{50}. Disulfide bridges are represented by short bars connecting different regions of the chain.

Table 8. Homologous cleavages in bovine chymotrypsinogens A and B (ChTg A and B) and in bovine trypsin (T).

ChTg A   Gly - Trp - Gly - Leu - Thr - Arg - Tyr - Thr - Asn - Ala - Asn - Thr - Pro
          140   141   142   143   144   145   146   147   148   149   150   151   152

ChTg B   Gly - Trp - Gly - Lys - Thr - Lys - Tyr - Asn - Ala - Leu - Lys - Thr - Pro

T   Gly - Trp - Gly - Asn - Thr - Lys - Ser - Ser - Gly - Thr - Ser - Tyr - Pro
          126   127   128   129   130   131   132   133   134   135   136   137   138

SUMMARY

Autolysis of bovine trypsin has been investigated by chemical techniques (end group determinations and identification of the chains after splitting of the disulfide bridges) in order to localize the autolytic cleavages in the enzyme molecule and to evaluate their effect on activity.

Whereas 1% solutions of trypsin were stable for 20 h when kept at pH 6 and 4°C in the presence of Ca^{2+}. incubation of more dilute solutions (0.1–0.2%) under the same conditions induced the specific cleavage of bond Arg_{105}-Val_{106} in about one molecule out of three. This bond was also found to be autolyzed in several crystalline samples of the enzyme.

Two other bonds (Lys_{131}-Ser_{132} and Lys_{49}-Ser_{50}) were also partially autolyzed in the crystalline samples. Previous cleavage of bond Arg_{105}-Val_{106} appeared to be necessary for further autolysis, so that autolyzed preparations were composed mainly of molecules either intact or degraded at several points. Among these latter, the predominant species included molecules in which bonds Arg_{105}-Val_{106} and Lys_{131}-Ser_{132} were simultaneously cleaved.

Specific activity determinations and titrations with [32P]DFP and TLCK of residues Ser_{183} and His_{46}, involved in the catalytic site of trypsin, conclusively showed that the simultaneous cleavage of bonds Arg_{105}-Val_{106} and Lys_{131}-Ser_{132} did not inactivate the enzyme. This showed that active trypsin molecules contain three "open" chains like chymotrypsin A. The effect of the cleavage of bond Lys_{49}-Ser_{50} on the activity of trypsin could not be definitely established.

ACKNOWLEDGMENTS

Stimulating discussions with Drs M. Rovery, M. Lazdunski and M. Delaage are gratefully acknowledged. Mrs. A. Guidoni performed the amino acid analyses recorded in this work. Our thanks are also due to Centre National de la Recherche Scientifique (CNRS) and Délégation Générale à la Recherche Scientifique et Technique (DGRST) for financial support.

Added Note. After this manuscript was completed, a paper by D. D. Schroeder and E. Shaw (*J. biol Chem.* **243** (1968) 2943) came to our attention. The authors announced the separation by chromatography on sulfoethyl Sephadex at pH 7.1 of an active "α-trypsin" from crystalline preparations. Bond Lys_{131}-Ser_{132} was shown to be autolyzed in this derivative.

REFERENCES

1. Richards, F. M. and Vithayathil, P. J., *J. biol. Chem.* **234** (1959) 1459.
2. Doscher, M. S. and Hirs, C. H. W., *Biochemistry* **6** (1967) 304.
3. Fujioka, H. and Scheraga, H. A., *Biochemistry* **4** (1965) 2197.
4. Sampath Kumar, K. S. V., Clegg, J. B. and Walsh, K. A., *Biochemistry* **3** (1964) 1728.
5. Rovery, M., Poilroux, M., Yoshida, A. and Desnuelle, P., *Biochim. biophys. Acta* **23** (1957) 608.
6. Guy, O., Rovery, M. and Desnuelle, P., *Biochim biophys. Acta* **124** (1966) 402.
7. Northrop, J. H., Kunitz, M. and Herriott, R., "Crystalline Enzymes", 2nd Edition, Columbia University Press, New York, 1948.
8. Delaage, M. and Lazdunski, M., *Biochim. biophys. Acta* **105** (1965) 608.
9. Walsh, K. A. and Neurath, H., *Proc. natn. Acad. Sci. U.S.A.* **52** (1964) 884.
10. Mikes, O., Tomasek, V., Holeysovsky, H. and Sorm, F., *Biochim. biophys. Acta* **117** (1966) 281.
11. Davies, E. W. and Neurath, H., *J. biol. Chem.* **212** (1955) 515.
12. Desnuelle, P. and Fabre, C., *Biochim. biophys. Acta* **18** (1955) 49.
13. Dixon, G. H., Kauffman, D. L. and Neurath, H., *J. biol. Chem.* **233** (1958) 1373.
14. Petra, P. H., Cohen, W. and Shaw, E. N., *Biochem. biophys. Res. Commun.* **21** (1965) 612.
15. Maroux, S., Rovery, M. and Desnuelle, P., *Biochim. biophys. Acta* **140** (1967) 377.
16. Maroux, S., Rovery, M. and Desnuelle, P., *Biochim. biophys. Acta* **56** (1962) 202.
17. Murachi, T. and Neurath, H., *J. biol. Chem.* **235** (1960) 99.
18. Crestfield, A. M., Moore, S. and Stein, W. H., *J. biol. Chem.* **238** (1963) 622.
19. Raftery, M. A. and Cole, R. D., *J. biol. Chem.* **241** (1966) 3457.
20. Stevenson, K. J. and Smillie, L. B., *J. molec. Biol.* **12** (1965) 937.
21. Holeysovsky, V., Mesrob, B., Tomasek, V., Mikes, O. and Sorm, F., *Colln Czech. chem. Commun.* **33** (1968) 441.
22. Spies, J. R. and Chambers, D. C., *Analyt. Chem.* **21** (1949) 1249.

FEBS Symposium, Volume 18, 1970, pp. 83-89

Tryptic Activity in Invertebrates and Molecular Evolution

R. ZWILLING and G. PFLEIDERER

*Department of Chemistry, Ruhr-University Bochum,
463 Bochum, Germany*

For our studies on the molecular evolution of proteins, we have chosen the endopeptidases for several reasons. Apart from the hemoglobins, they occur in almost all living organisms and organs, for protein synthesis—a basic function of life itself—implies protein disintegration; and, in contrast to the astonishing constancy of cytochrome *c* during evolution, proteases promise to be a most variable type of protein, coming, in the digestive tract, into direct contact with the exterior environment and its manifold demands for adaptation. The proteases, being enzymatically-active proteins, can be assayed very specifically and their differentiated cleavage specificity, for example with the B-chain of insulin, is a very good starting point for comparative studies. They can be distinguished with the aid of numerous well-elaborated natural and synthetic inhibitors. And, by no means the least of their advantages: bovine trypsin and chymotrypsin are among the most extensively studied and best known of all proteins.

However, while the past few years have resulted in the elucidation of the complete covalent structures of bovine trypsinogen and chymotrypsinogen, little is known about proteases occurring in invertebrates. One of the reasons for this is apparently the difficulty or impossibility of obtaining the raw material in quantities sufficient for purification procedures, for invertebrates are small animals. Nevertheless, they represent eight of the nine major groups of the animal kingdom and are consequently of considerable importance for evolutionary studies at the molecular level.

The decapode crayfish *A. fluviatilis* and *A. leptodactylus* provide an opportunity for accumulation of large quantities of digestive fluid by a favorable circumstance. A curved pipette can be introduced through the oesophagus into the cardia, and each time about 3 ml of the fluid from this stomach-like organ can be collected. This procedure can be repeated every two days, interrupted only by feeding periods. Thus we have been able to accumulate, from 700 crayfish, more than two litres of cardia fluid. The decapode crayfish possess, amongst others, an enzyme that hydrolyzes the B-chain of reduced and carboxymethylated insulin, exactly as does bovine trypsin, only at the bonds Arg_{22}-Gly_{23} and Lys_{29}-Ala_{30} [1, 2].

Figure 1A demonstrates the three electrophoretic bands produced by the action of bovine trypsin on the B-chain of insulin. These bands represent, from the anode to the cathode, the oligopeptide comprising the amino acid residues from positions 1–22, the heptapeptide Gly_{23} to Lys_{29}, and free alanine. The

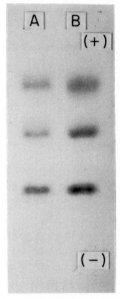

Figure 1. Cleavage specificity of the trypsin-like protease: Electrophoresis of peptides resulting from incubation of insulin B-chain with bovine trypsin and trypsin-like crayfish protease.
  A: bovine trypsin + B-chain
  B: trypsin-like crayfish protease + B-chain.
Electrophoresis in formic acid/acetic acid-buffer, pH 1.9, 1800 V, 2.5 mA/cm, 80 min, stained with ninhydrin.

action of the trypsin-like crayfish protease results in exactly the same peptides (Fig. 1B). An ultimate proof for the identity of the compared bands came from the qualitative amino acid composition analysis of the eluted bands. The strictly basic specificity of the trypsin-like protease was also confirmed with the octapeptide-substrate valyl[5]-angiotensin II, which was split only at its arginine bond.

Recently [1] we reported on a slight additional activity, producing two additional bands from the B-chain upon prolonged incubation. We now know that these were derived from small impurities of a carboxypeptidase B-like activity, reported previously by DeVillez [3] and Kleine [4], which is active against hippurylarginine and is present in the crude extract in high concentration. It can be inhibited completely by ethylenediaminetetraacetic acid, which does not affect trypsin-like activity.

Many interesting questions arise from the fact that two different proteases, from organisms having had no genepool for at least 500 million years, i.e. no interchange of molecular information, do hydrolyze the same two bonds out of twenty-nine possible peptide linkages in the B-chain.

As we reported earlier, the crayfish enzyme is trypsin-like in many additional respects: it hydrolyzes the trypsin-specific amino acid ester substrates like BAEE, TAME, as well as benzoyl-arginine-naphthyl-amide and benzoyl-arginine-*p*-nitranilid; but it does not hydrolyze the chymotrypsin-specific ester substrates, such as acetyltyrosyl-ethylester. Like bovine trypsin, and in contrast to three other protease fractions occurring in the cardia fluid, it is inhibited by bovine pancreatic trypsin inhibitor, soya-bean inhibitor, lima-bean inhibitor, ovomucoid and α_1-anti-trypsin of human serum. It is also inhibited by the trypsin-specific porcine pancreatic inhibitor, isolated by Werle *et al.* [5], which does not inhibit bovine chymotrypsin. This means that there is more resemblance in this respect between the trypsin-like crayfish protease and bovine trypsin than there is between bovine trypsin and bovine chymotrypsin, which do have a part of their primary structure in common. Likewise, the crayfish enzyme is inhibited by the trypsin-specific tosyl-L-lysyl-chloromethane, but not by the chymotrypsin-specific tosyl-L-phenylalanine-chloromethane. We also noticed strong inhibition by phenyl-methyl-sulfonyl-fluoride, which reacts with an active serine residue in the catalytic group of mammalian trypsin. Iodoacetic acid and, as already mentioned, EDTA, do not have any inhibitory effect. From these data it can be inferred that the crayfish protease, like bovine trypsin, is of the "serine-type", having an essential serine and an essential histidine residue in its active site.

This enumeration of similarities could lead to the conclusion that the crayfish enzyme is apparently a mere invertebrate homolog of mammalian trypsin. But the differences are as obvious as the similarities.

A prerequisite for further studies on the structure and catalytic properties was the large-scale preparation of the crayfish enzyme in high purity. The yield by an improved method [2] of salt-free lyophilized trypsin-like protease from 1200 ml of cardia fluid was about 1.5 g, a quantity sufficient to attempt sequence studies.

In a final step of DEAE-anion exchange chromatography, prior to desalting, we obtained a chromatographically-pure peak of activity (Fig. 2B), preceded by a small peak of inactive material (Fig. 2A). The activity of the preparation at pH 8.0 against benzoyl-arginine-ethylester was 56% that of crystalline Worthington trypsin, and the specific activity was 18.2 μmoles BAEE hydrolyzed per min per mg trypsin-like protease. The pure preparation did not contain any detectable trace of hippurylarginine-splitting activity, which disturbed our earlier specificity studies. Incubation of a BAEE-hydrolyzing activity of 3.2 μmoles/min/ml for 60 h at 30°C with the substrate hippurylarginine resulted in a negative ninhydrin reaction.

By disc electrophoresis and amido black staining we then detected two closely related bands in our preparation (Fig. 3B). The electrophoretic migration is towards the anode and, in contrast to the very alkaline bovine trypsin, the crayfish enzyme has a rather low isoelectric point.

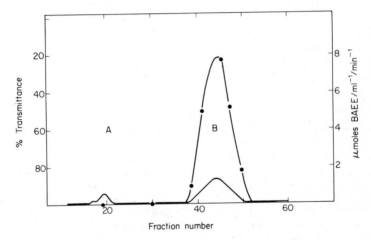

Figure 2. Chromatographically-pure preparation of trypsin-like protease. Column: 1.2 x 40 cm, DEAE-Sephadex A-50 in Tris/HCl-buffer.
  A: inactive material;
  B: trypsin-like protease.

We then developed a method to demonstrate trypsin-like activity directly in disc electrophoresis, using the substrate benzoyl-arginine-naphthylamide and the reaction of free naphthylamide with diazo blue B [2].

In Fig. 3A, one sees from this (orange) colour reaction that both bands are highly active, whereas diazo blue without the substrate does not give this staining effect, so that staining of inactive protein bonds could be excluded (Fig. 3C).

Our studies comprised the three decapode crayfish *A. fluviatilis*, *A. lepto-dactylus* and *Carcinus maenas*. Electrophoresis on acetate membrane foils of the crude extracts revealed that their sub-bands of trypsin-like activity are not identical—with the possible exception of two of them—and with *A. fluviatilis* one finds even three (Fig. 4). We interpret this as due to structural differences. But, if even between these closely related crayfish, evolution is perceptible in this

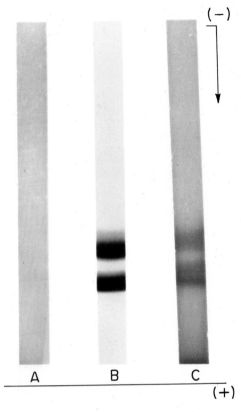

Figure 3. Disc electrophoresis of trypsin-like protease, demonstrating the two active components.

    A: diazo blue without the substrate;

    B: stained with amido black;

    C: after incubation with N-α-benzoyl-L-arginine-β-naphthylamide, stained with diazo blue.

(Facing p. 86)

way, the extent of alterations should be far greater in relation to the systematically-distant mammals.

It should be pointed out in this connection that the crayfish possess another interesting endopeptidase, distinguished from the trypsin-like protease in all of its characteristics, and remarkable for its very low molecular weight of only 11,000 [1].

We have also used disc electrophoresis for the separation and amino acid analysis of the two bands of trypsin-like activity. By methylene blue staining they can be made visible without denaturation, cut out very accurately and eluted for enzymatic tests or amino acid analysis. The separate amino acid analyses of the two bands revealed a complete coincidence between them, but it may be possible that some amino acid residues are not identical.

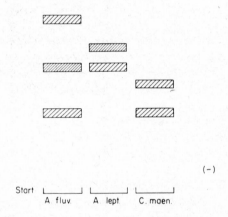

Figure 4. Trypsin-like activity in various crayfish species: Electrophoretic bands of N-α-benzoyl-L-arginine-naphthylamide-hydrolyzing activity from the cardia fluid of *Astacus fluviatilis, A. leptodactylus* and *Carcinus maenas.* Cellulose acetate membrane, pH 8.6.

Following submission of this text, the amino acid compositions of bovine trypsin and the crayfish enzyme were compared [6]. Although there are quantitative differences for almost all amino acids, we should like to draw attention to some major divergencies: serine, the most frequent residue in bovine trypsin, is not the most frequent in the crayfish enzyme, where it is largely replaced by asparagine and glycine. The number of serine residues in the crayfish enzyme is approximately half that in bovine trypsin. In bovine trypsin there is one isoleucine residue more than there are leucine residues, but with the trypsin-like protease the isoleucine content is only about two-thirds of the leucine content. The high

4

lysine content of bovine trypsin is not shared by its invertebrate counterpart, which possesses only about one-third of it.

In addition to the differences from bovine trypsin already mentioned, there are also the following divergences: The invertebrate protease is for at least several days completely stable upon incubation at pH 8 and 30°C; $CaCl_2$ has no stabilizing effect and at pH 3 it is irreversibly denatured, all this in contrast to bovine trypsin. The inactivation of the crayfish protease in acidic medium appears to be a physico-chemical effect, and the inactivation of bovine trypsin is caused by self-digestion. The crayfish protease exhibits against benzoyl-L-lysine methyl ester only two-thirds, and against benzoyl-L-arginine naphthylamide only one-third, of the bovine trypsin activity, when both proteases are compared on the basis of equal activity against BAEE. During our immunological studies in collaboration with Dr. Linke we could not detect an immunological relationship between the two proteases [7]. The crayfish enzyme does attack native proteins, such as ribonuclease and lactic dehydrogenase, which are hardly hydrolyzed by bovine trypsin.

Our findings at least make evident that, in spite of the identical cleavage specificity, there might be considerable structural differences between the two proteases compared. It would therefore be extremely interesting to examine how the catalytic site of the trypsin-like crayfish protease is organized. At present, the question as to whether these two proteases have had a divergent or a convergent molecular evolution is quite open, and it will not be answered until considerably more structural work on the crayfish enzyme is carried out.

As regards trypsin-like activity in invertebrates, it is pertinent to mention that we have isolated from the gut of the beetle *Tenebrio molitor* a trypsin-like endopeptidase characterized by an unusual molecular weight of 60,000 [8].

In gel filtration it is eluted shortly after bovine serum albumin, which has a molecular weight of 67,000. We have called this enzyme with trypsin-like catalytic properties "β-protease from *Tenebrio*", and another protease of the beetle gut, with a molecular weight of about 24,000, but no trypsin-like properties, we have called "α-protease from *Tenebrio*". The β-protease hydrolyzes the trypsin-specific amino acid ester substrates, but not the chymotrypsin-specific ester substrates. Though moving towards the cathode in electrophoresis at pH 8, β-protease has a lower isoelectric point than bovine trypsin.

β-protease, in contrast to α-protease, is inhibited by all natural and synthetic trypsin-inhibitors tested, but the chymotrypsin-specific tosyl-phenylalanyl-chloromethane is again without effect. However, we do not know whether the β-protease also possesses a strictly tryptic cleavage specificity when assayed against the B-chain of insulin or other peptide substrates, but these experiments will be undertaken as soon as sufficient material is available.

Our co-workers in these investigations were Dr. Sonneborn, Mr. Herbold and Mr. Kraft.

ACKNOWLEDGEMENTS

We wish to express our gratitude to Prof. Šorm and Prof. Keil of the Czechoslovak Academy of Science for their generous gift of 700 crayfish, which made much of our work possible.

REFERENCES

1. Pfleiderer, G., Zwilling, R. and Sonneborn, H.-H., *Hoppe-Seyler's Z. physiol. Chem.* **348** (1967) 1319.
2. Zwilling, R., Pfleiderer, G., Sonneborn, H.-H., Kraft, V. and Stucky, I., *Comp. Biochem. Physiol.* **28** (1969) 1275.
3. DeVillez, E. J., *Comp. Biochem. Physiol.* **21** (1967) 541.
4. Kleine, R., *Z. vergl. Physiol.* **56** (1967) 142.
5. Vogel, R., Trautschold, I. and Werle, E., "Natürliche Proteinasen-Inhibitoren", G. Thieme-Verlag, Stuttgart, 1966.
6. Zwilling, R., paper in preparation.
7. Linke, R., Zwilling, R., Herbold, D. and Pfleiderer, G., *Hoppe-Seyler's Z. physiol. Chem.* **350** (1969) 877.
8. Zwilling, R., *Hoppe-Seyler's Z. physiol. Chem.* **349** (1968) 326.

FEBS Symposium, Volume 18, 1970, pp. 91-100

Sequence Studies on Hog Pepsin

V. KOSTKA, L. MORÁVEK, V. A. TRUFANOV,
B. KEIL and F. ŠORM

*Institute of Organic Chemistry and Biochemistry,
Czechoslovak Academy of Sciences, Prague,
Czechoslovakia*

Unlike the other two principal proteases—chymotrypsin and trypsin—involved in the process of digestion, pepsin has for many years received very little consideration from workers engaged in amino acid sequence studies. There is no doubt that the main reason for the little appeal of pepsin to people in the sequence "business" was its non-homogeneity (cf. e.g. [1-3]). The last few years, however, have witnessed an increased interest in this protein and it has become possible to solve or, at least, to circumvent the problem of its non-homogeneity The classical approach to this problem represents the work of Rajagopalan *et al.* [4] who developed a procedure for the preparation of pepsin from pepsinogen. Unfortunately, due to the high price of pepsinogen, this method is not likely to provide a solution in cases where gram quantities of the enzyme are required. The other, less orthodox, approach to the problem of pepsin homogeneity has come about through studies on the purification or fractionation of inactive or chemically-modified proteins [5, 6].

We were faced with the same problem when we started our work on pepsin in 1966, and decided to start with a pepsin derivative purified in inactive form. From amongst several possibilities for chemical modification we chose *S*-sulfonation, mainly because *S*-sulfo derivatives can be reduced and, if necessary, modified again by other methods. We started with twice-crystallized and lyophilized Worthington pepsin, which we converted into its *S*-sulfo derivative by the slightly modified method of Pechère and co-workers [7]. Unlike these authors, we did not dialyze the reaction mixture but removed the low-molecular weight compounds by passage through a column of Sephadex G-25 equilibrated with 0.2M ammonium carbonate. The protein-containing material was desalted by additional gel-filtration on Sephadex G-25 and lyophilized. This product was non-homogeneous when tested by disc electrophoresis [8] at pH 8.6, and end-group analysis, and obviously was contaminated with low-molecular weight

91

products of autolysis. To remove these contaminants, we submitted S-sulfo-pepsin to gel filtration on Sephadex G-100, equilibrated with 0.3M ammonium acetate in 8M urea at pH 6.0 in order to minimize the formation of cyanate [9]. The high-molecular weight material was desalted and lyophilized. S-sulfo-pepsin purified by this operation gave a single zone when subjected to disc electrophoresis in Tris-glycine buffer at pH 8.6. Our derivative was also homogeneous with respect to its end groups. The N-terminal end group was determined by two independent techniques. The conventional dinitrophenylation technique [10]

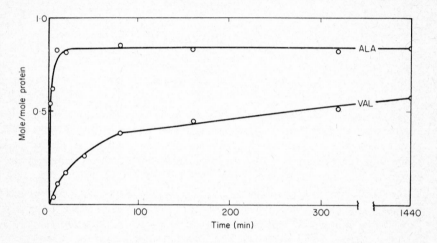

Figure 1. Time dependence of liberation of amino acids from S-sulfo-pepsin by carboxypeptidase A. The digestion was carried out in 0.05M Tris-HCl buffer, pH 8.5 at 37°C. The enzyme to substrate ratio was 1:50.

indicated the presence of N-terminal leucine or isoleucine. When S-sulfo-pepsin was subjected to Edman degradation [11] (this experiment was performed by Dr. Kluh of our Laboratory) only isoleucine was found. The identity of the C-terminal amino acid was examined in kinetic experiments with carboxypeptidase A. S-sulfo-pepsin was digested in 0.05M Tris-HCl buffer at pH 8.5 at 37°C, and a molar enzyme to substrate ratio of 1:50. The liberated amino acids were determined in the amino acid analyzer [12]. The results of the kinetic experiment are shown in Fig. 1.

As can be seen, carboxypeptidase A readily liberates from S-sulfo-pepsin 0.62 mole alanine per mole protein after 5-min digestion. The penultimate amino acid is valine, which is liberated at a considerably slower rate, and levels off at 0.57 mole per mole protein after 24 h. Since no other amino acids are liberated (when we disregard the very small quantities of serine) we are able to conclude that, (a) the C-terminal sequence of hog pepsin is Alanyl-Valyl, and (b) that valine is preceded by proline and the digestion cannot therefore proceed further. The results of our kinetic experiments deserve special interest with respect to the data reported by Rajagopalan and co-workers [4]. In their study on the carboxypeptidase A digestion of native homogeneous pepsin, they observed that only alanine was liberated; the small amount of valine which was also present in the digest was considered by them as an impurity. Our data indicated that this is not the case and that valine is the penultimate amino acid residue of pepsin. The discrepancy between our data and that reported by Rajagopalan and co-workers can be accounted for, for example, by the fact that they carried out the digestion at 25°C, i.e. the reaction rate was about half that observed in our experiments at 37°C.

Having established the homogeneity of our derivative, we decided to hydrolyze it to the smallest possible number of specific fragments. For some time we entertained the hope that this task might be accomplished by tryptic digestion. However, having found in preliminary experiments that the two arginines and one lysine are located in a very small region of the molecule, we turned our attention to hydrolysis with cyanogen bromide [13]. Since S-sulfo-pepsin is practically insoluble in aqueous solutions of mineral acids, we employed 80% formic acid [14], although we were well aware of the damage which this medium may cause to certain amino acids, such as tyrosine or serine. In a typical preparative experiment, 0.29 mmole of S-sulfo-pepsin was dissolved in 20 ml of 80% formic acid and hydrolyzed with 17 mmoles of cyanogen bromide for 20 h at 37°C. The solvent and excess reagent were removed by lyophilization.

For the fractionation of the hydrolysate we used the same system of gel filtration which had been employed for the purification of S-sulfo-pepsin. The hydrolysate was dissolved in 0.3M ammonium acetate in 8M urea at pH 6.0 and applied as a 1% solution to a column of Sephadex G-100 equilibrated with the same buffer. The ratio of sample volume to the volume of the column varied between 1:25 and 1:35. The flow rate was 2 ml per cm^2. The elution profile obtained in a typical preparative-scale experiment is presented in Fig. 2.

The pattern appears to suggest that the cyanogen bromide hydrolysis of S-sulfo-pepsin gave rise to the five fragments which may be theoretically expected. Since we have now analyzed in more detail only two of these fragments, and our information on the amino acid composition and end groups of the remaining fragments is incomplete, we cannot make any final conclusions as to the specificity of cyanogen bromide hydrolysis of S-sulfo-pepsin under the foregoing conditions. The hydrolysate, however, did not contain methionine.

We turned our attention, as might be expected, to the material contained in peak CB-1 which, we assumed, should represent the largest high-molecular weight fragment. This material was subjected to gel filtration on Sephadex G-100 under exactly the same conditions as the parent cyanogen bromide hydrolysate. By one single operation a material was obtained which was

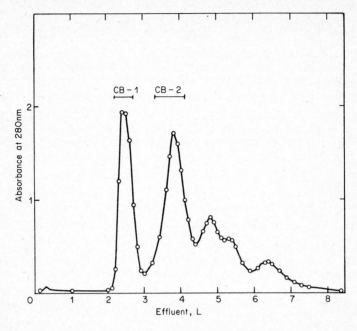

Figure 2. Gel filtration of the cyanogen bromide hydrolysate of *S*-sulfo-pepsin on Sephadex G-100. The column (11 x 80 cm) was equilibrated with 0.3M CH_3COONH_4, 8M in urea, pH 6.0. The same buffer was used as eluent. A flow rate of 2 ml per cm^2 per h was maintained. The fractions indicated by bars were pooled, desalted, and lyophilized.

homogeneous on disc electrophoresis. N-terminal end-group analysis revealed the presence of aspartic acid. When peptide CB-11 was digested with carboxypeptidase A under the same conditions as *S*-sulfo-pepsin, alanine and valine were liberated in the same ratio as from *S*-sulfo-pepsin. This result, together with the absence of homoserine in peptide CB-1, strongly suggested that this peptide represents the C-terminal portion of the molecule of pepsin. The amino acid analysis of peptide CB-11 was repeated several times and yielded intriguing data. A comparison of the nitrogen values, determined analytically, with the values calculated from the amino acid analyses, showed that the peptide sample contained a combustible organic impurity. In our opinion the peptide most likely contains dextran eluted from Sephadex. The presence of dextran may be the cause of variations in values found for certain amino acids. Therefore we precipitated a sample of peptide CB-11 with trichloroacetic acid and performed the amino acid analysis with this sample. The results are shown in Table 1.

A comparison of the nitrogen values showed that the peptide had been freed of dextran by this treatment. The peptide contains three of the four basic amino acid residues of pepsin and is free of cysteine. The tryptophan value was

Table 1. Amino acid composition of C-terminal peptide of hog pepsin.

Amino acid	Found	Nearest integer
Lysine	1.0*	1
Arginine	2.2	2
Aspartic acid	6.1	6
Threonine†	2.8	3
Serine†	3.4	4
Glutamic acid	2.8	3
Proline	2.6	3
Glycine	4.1	4
Alanine	3.3	3
Valine‡	5.9	6
Isoleucine‡	3.0	3
Leucine	4.0	4
Tyrosine	2.4	3
Phenylalanine	2.7	3
Tryptophan§	(1.0)	1
Ammonia¶	(3.0)	(3)
Total		49
Molecular weight		5444

Unless otherwise stated, the values found are averages obtained by duplicate analyses of 20-h and 70-h hydrolysates. These analyses always agreed to ± 3%.

* For the calculation of the molar ratios, the average values of lysine and arginine in μmoles were assumed to represent one and two residues per molecule of peptide CB-11.
† Corrected for destruction during acid hydrolysis.
‡ The average 70-h values were taken for valine and isoleucine.
§ See text.
¶ Found in peptides.

determined by two independent methods.* The possibility of error cannot be excluded, due to prolonged exposure of the material to formic acid. We therefore consider this value merely as informative until we obtain the ultimate proof, i.e. until the complete amino acid sequence is determined. The results of studies in this respect are shown in Fig. 3.

Two principal lines of approach were followed. The amino acid sequence of the C-terminal 27-residue moiety of the peptide was determined [16] by conventional methods, i.e. from analyses of smaller peptides derived from

* The tryptophan content of pepsin and pepsinogen was reinvestigated recently by Neradová and Kostka [15]. Contrary to earlier findings, the presence of five residues was revealed.

280
H-Asp-Val-Pro-Thr-Ser-Ser-Gly-Glu-Leu-Trp-Ile-(Asp,

(290)   300
Thr,Ser,Ser,Glu,Pro,Gly,Val,Leu,Tyr,Phe)-Ile-Leu -Gly-Asp-Val-Phe-Ile-Arg-Gln-

310   320
Tyr-Tyr-Thr-Val-Phe-Asp-Arg-Ala-Asn-Asn-Lys-Val-Gly-Leu-Ala-Pro-Val-Ala-OH

Figure 3. Amino acid sequence of C-terminal peptide of hog pepsin. The numbering system is based on the assumed total number of 321 residues [4] in pepsin.

different enzymatic digests. The N-terminal moiety of the peptide comprising eleven residues was determined by Dr. Kluh [17] in an automatic amino acid sequenator [18] built in our laboratory. There is a gap between the two complete sequences, which is at present under study. Our data are partly in agreement with the results reported by other authors [19, 20], and by Stepanov and his group at this Meeting [21].

The second fragment, CB-2 (Fig. 2) was approached in a similar way, i.e. it was first purified by gel filtration on Sephadex G-100. Since after this treatment the peptide was still slightly non-homogeneous, ion-exchange chromatography was used for its purification [22]. The elution profile obtained is shown in Fig. 4. So

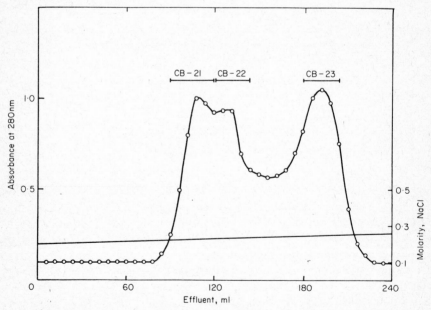

Figure 4. Ion-exchange chromatography of peptide CB-2. A column of DEAE-Sephadex A-25 equilibrated with 0.1M Tris-HCl buffer, pH 7.5, 8M in urea and 0.2M in NaCl was used. An elution gradient of increasing NaCl concentration was employed as shown. The fractions indicated by bars were pooled, desalted, and lyophilized.

Table 2. Amino acid composition of N-terminal peptides of hog pepsin.

Amino acid	Found				Nearest integer			
	CB-2	CB-23	CB-23-AE	CB-23-AE-T3	CB-2	CB-23	CB-23-AE	CB-23-AE-T3
S-(β-aminoethyl)-cysteine			1.8*	1.0†			2	1
Histidine	0.7	0.8	1.2		1	1	1	
Cysteic acid	2.0†	2.0†			2	2		
Aspartic acid	20.0	19.8	21.2	14.6	20	20	21	15
Threonine‡	12.0	13.0	12.2	10.2‡	12	13	12	10
Serine‡	17.9	18.6	21.6	9.2‡	18	19	22	9
Glutamic acid	12.8	13.6	13.6	8.8	13	14	14	9
Proline	7.5	7.9	8.4	5.2	8	8	8	5
Glycine	18.4	18.3	18.3	12.0	18	18	18	12
Alanine	6.6	7.2	7.0	3.0	7	7	7	3
Valine	10.3	10.0	8.3	7.2	10	10	8	7
Isoleucine	10.7	11.0	10.0	8.0	11	11	10	8
Leucine	15.0	15.0	15.1	9.6	15	15	15	10
Tyrosine	9.1	8.4	8.7	7.0	9	9	9	7
Phenylalanine	8.5	8.5	8.7	7.2	9	9	9	7
Tryptophan§					2	2	2	2
Homoserine¶					1	1	1	
Total					156	159	159	105

Unless otherwise stated, the values found are averages obtained by duplicate analyses on 20-h and 70-h hydrolysates. These analyses always agreed to ± 3%.

* For the calculation, the sum of average μmoles of S-(β-aminoethyl)-cysteine plus histidine was taken to represent three residues.
† The calculation of molar amino acid ratios was based on the assumed presence of two cysteic acid residues in peptides CB-2 and CB-23 and of one S-(β-aminoethyl)-cysteine residue in peptide CB-23-AE-T3.
‡ Values not extrapolated to zero time of hydrolysis.
§ The results of the analyses were not unambiguous. The presence of two tryptophan residues seems to be strongly suggested by the results of preliminary experiments with the chymotryptic digest of peptides CB-2, CB-23, and CB-23-AE-T3.
¶ Not determined analytically.

far we have studied in more detail only peptide CB-23, which was homogeneous. The presence of N-terminal isoleucine, and the finding of the N-terminal sequence Ile-Gly-Asp in informative experiments [23], suggested the possibility that the peptide might be derived from the N-terminal portion of the molecule of pepsin. The peptide contains 159 amino acid residues (Table 2), including one histidine and two half-cysteines. The presence of two half-cysteines offered the possibility of additional specific fragmentation of the peptide. Peptide CB-2 was

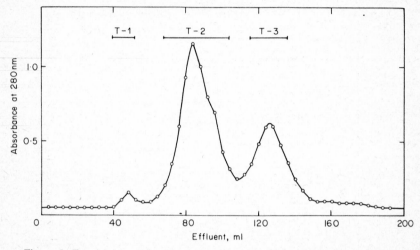

Figure 5. Fractionation of tryptic digest of peptide CB-23-AE by gel filtration. The digest was applied as a 1.5% solution in 0.3M CH_3COONH_4 at pH 6.0 to a column of Sephadex G-100 equilibrated with the same buffer. The ratio of sample volume to column volume was 1:40. A flow rate of 1.5 to 2.0 ml per cm^2 per h was maintained. Cuts were made as indicated by bars, desalted, and lyophilized.

therefore reduced with β-mercaptoethanol and the cysteines released were labeled with ethylene imine by the method of Raftery [24]. The aminoethylated peptide was digested with trypsin (4 h, 37°C, enzyme to substrate ratio 1:50) and the digest fractionated by gel filtration on Sephadex G-100. The elution pattern obtained is shown in Fig. 5. The fragment designated as T-3 was homogeneous and its N-terminal amino acid was isoleucine. Therefore, we assume that it represents the N-terminal moiety of the molecule of pepsin. The amino acid composition of this peptide is given in Table 2. The data obtained thus far, together with certain earlier findings [25], permit us to ascribe the 159-residue N-terminal moiety of the pepsin molecule the tentative partial structure shown in Fig. 6. Studies aimed at the determination of the amino acid sequence of fragment CB-23-AE-T3 are in progress.

    To complement the information obtained with the cyanogen bromide hydrolysate of S-sulfo-pepsin we are also studying at present the tryptic digest of

H-Ile-Gly-Asp-(Asp$_{14}$,Thr$_{10}$,Ser$_9$,Glu$_9$,Pro$_5$,Gly$_{11}$, Ala$_3$,Val$_7$,Ile$_7$,Leu$_{10}$,

Tyr$_7$,Phe$_7$,Trp$_2$)- Cys-Ser-Ser-Leu-Ala-Cys-(His,Asp$_6$,Thr$_2$,Ser$_{11}$,Glu$_5$,Pro$_3$,

Gly$_6$,Ala$_3$,Val,Ile$_2$, Leu$_4$,Tyr$_2$,Phe$_2$)-Met

Figure 6. Distribution of amino acids in the N-terminal region of hog pepsin. The scheme was derived from the results obtained in different studies [16, 23, 25].

reduced and aminoethylated pepsin. These studies will be reported elsewhere [26].

In conclusion, we would like to return to where we started and say that, unlike other proteolytic enzymes, the situation with respect to studies on the covalent structure of pepsin warrants very little optimism. The fractionation of large fragments which actually represent small denatured proteins is in our experience extremely troublesome and there is little doubt that novel techniques, such as, for instance, gel filtration in fluorinated solvents, will be needed. Furthermore, for the sequencing of these fragments the conventional techniques will most likely prove inadequate. The use of automatic sequenators might be a help. So far, however, the experimental problems involved in the application of these methods are formidable and their resolution will require considerable time and the joint effort of several laboratories.

REFERENCES

1. Hoch, H., *Nature, Lond.* **165** (1950) 278.
2. Ryle, A. P. and Porter, R. R., *Biochem. J.* **73** (1959) 75.
3. Stepanov, V. M. and Greil, T. I., *Biokhimiya* **28** (1963) 540.
4. Rajagopalan, T. G., Moore, S. and Stein, W. H., *J. biol. Chem.* **241** (1966) 4940.
5. Balls, A. K. and Jansen, E. F., *in* "Advances in Enzymology", Vol. XIII (edited by F. F. Nord), Interscience, New York, 1952, p. 321.
6. Crestfield, A. M., Stein, W. H. and Moore, S., *J. biol. Chem.* **238** (1963) 2413.
7. Pechère, J. F., Dixon, G. H., Maybury, R. H. and Neurath, H., *J. biol. Chem.* **233** (1958) 1964.
8. Ornstein, L. M., *Ann. N.Y. Acad. Sci.* **121** (1964) 310.
9. Stark, R. G., Stein, W. H. and Moore, S., *J. biol. Chem.* **235** (1960) 3177.
10. Sanger, F., *in* "Advances in Protein Chemistry", Vol. VII (edited by M. L. Anson, K. Bailey and J. Z. Edsall), Academic Press, New York, 1952, p. 1.
11. Niall, H. and Edman, P., *J. gen. Physiol.* **45** (1962) 185.
12. Spackmann, D. H., Stein, W. H. and Moore, S., *Analyt. Chem.* **30** (1958) 1190.
13. Gross, E. and Witkop, B., *J. biol. Chem.* **237** (1962) 1856.
14. Li, C. H., Liu, W. K. and Dixon, J. S., *J. Am. chem. Soc.* **88** (1966) 2050.

15. Neradová, V. and Kostka, V., to be published.
16. Kostka, V., Morávek, L. and Šorm, F., *Eur. J. Biochem.*, to be published.
17. Kostka, V., Morávek, L., Kluh, I. and Keil, B., *Biochim. biophys. Acta* **175** (1969) 459.
18. Edman, P., *Eur. J. Biochem.* **1** (1967) 80.
19. Dopheide, T. A. A., Moore, S. and Stein, W. H., *J. biol. Chem.* **242** (1967) 1833.
20. Perham, R. N. and Jones, G. M. T., *Eur. J. Biochem.* **2** (1967) 84.
21. Stepanov, V. M., "Abstracts of the Fifth FEBS Meeting, Prague, 1968", Academic Press, London and New York, 1968, p. 273.
22. Trufanov, V. A., Kostka, V., Keil, B. and Šorm, F., *Eur. J. Biochem.* **7** (1969) 549.
23. Kluh, I., unpublished results.
24. Raftery, M. A. and Cole, R. D., *J. biol. Chem.* **241** (1966) 3457.
25. Keil, B., Morávek, L. and Šorm, F., *Colln. Czech. chem. Commun.* **32** (1967) 1968.
26. Kostka, V., to be published.

FEBS Symposium, Volume 18, 1970, pp. 101-122

Some Topochemical Aspects of the Structure and Function of Aspartate Transaminase

A. E. BRAUNSTEIN

*Institute of Molecular Biology, Academy of
Sciences of the U.S.S.R. Moscow, U.S.S.R.*

Adequate description of the molecular mechanism of an enzyme reaction requires exact knowledge of the topochemistry of the active centre in the enzyme and in enzyme-substrate intermediates. In this context "topochemistry" connotes dynamic scanning of three-dimensional patterns, both geometrical and electronic, along the reaction coordinate, i.e. the time dimension.

The most straightforward information concerning chemical topography of protein molecules is provided by X-ray crystallographic investigations. This approach has hitherto been applied to no less than a dozen enzymes. In two cases at least, viz. lysozyme [1] and carboxypeptidase A [2], X-ray crystallography made possible the construction of three-dimensional models, resolved to the atomic level, for the complete molecular structure of the enzymes and of their complexes with pseudosubstrates. Such models, representing "frozen" approximative cross-sections of the catalytic system, do not constitute final proof for definite reaction dynamics. But they have provided a valuable basis for formulation of probable catalytic mechanisms. Unfortunately, the study of protein structure by X-ray diffraction is not only extremely laborious, but the scope of its application is limited, at the present, to enzymes of relatively low molecular weight.

For other enzymes, not likely to be studied by the X-ray method in the near future, plausible reaction mechanisms have to be deduced from indirect evidence obtained by the combined use of a variety of chemical, optical and kinetic methods.

In our Institute and elsewhere, such approaches have been applied extensively to the study of several highly purified enzymes dependent on pyridoxal phosphate (PLP); the one most thoroughly investigated to date is the cytoplasmic aspartate transaminase from pig heart.

Enzyme: Aspartate transaminase, L-aspartate:2-oxoglutarate aminotransferase (EC 2.6.1.1).

Detailed chemical and physical investigations of interactions between this enzyme and different types of pseudosubstrates and coenzyme analogues have contributed particularly valuable information, making it possible to derive an approximate picture of the configuration and functioning of the active site.

Several recent reviews adequately cover the chemistry of PLP-enzymes [3-8]. Recapitulation of some major features relating to aspartate transaminase will provide the background for a concise survey of current progress achieved in investigations of this enzyme by my associates, as well as in Dr. R. Khomutov's laboratory, and in some others.

The enzyme is a dimeric protein of approximately 90,000 molecular weight, made up of two identical, strongly associated subunits [9]. Each monomer consists of one peptide chain with an N-terminal alanine and a C-terminal glutamine (?) residue, and of a molecule of PLP attached by an aldimine bond to the ϵ-amino group of a lysine residue. Complete amino acid analyses of the enzyme have made by Turano, Polyanovsky, Wada, and others [3, 4, 8, 10, 13].

Studies of the primary structure, initiated by Fischer *et al.* [11] and by Polyanovsky and Keil [12], have thus far resulted in identification of the partially resolved hexadecapeptide sequence [10],

$$P\text{-Pyl}$$
$$|\epsilon$$
-Thr-(Gly,Thr)-(Val,Leu,Glx)-(Asx,Ser,Glx,Lys)-Lys-Ser-Asn-Phe-(Leu,Pro)-

including the coenzyme-linked lysine of the active site. Elucidation of the complete primary structure of the enzyme is in progress.

The estimated helical content of the aldimine form of the enzyme is about 25-30%, that of the apoenzyme is approximately the same. The compact conformation of the PLP-enzyme is significantly reduced on transition to the PMP-form, and especially in the apoenzyme, as revealed by alterations of several physical parameters and by the unmasking of a number of functional groups [9a, 13, 22].

Depending on the chemical state and mode of binding of the coenzyme, individual forms of aspartate transaminase and its substrate or inhibitor complexes differ in optical properties, including absorption and fluorescence spectra, rotatory dispersion, etc. The PLP-containing aldimine form behaves as a pH-indicator; in the acidic range its predominant absorption peak is situated at 430 mμ, in the alkaline, at 365 mμ; the spectral shift reflects dissociation from the coenzyme of one proton with a pK$_a$ of 6.2. The amino, or PMP-form, and other species having no double bond conjugated to the pyridine ring of the coenzyme, have an absorption maximum at 330-340 mμ. Some enzyme-substrate intermediates show a peak near 490 mμ.

Molecular species of the transaminase (and of other pyridoxal enzymes) with identical absorption spectra may differ in chemical structure. Variations in induced optical activity, or Cotton effects, associated with absorption peaks of the coenzyme, first studied by Torchinsky and Koreneva [14], provide valuable criteria for discrimination between spectrally similar forms or derivatives of the enzyme. This can best be achieved by measurements of circular dichroism [15-17]. Optical investigation of abortive complexes of aspartate transaminase with inhibitory substrate analogues that block the enzymatic reaction at different sequential stages has been of value for the assignment of spectral maxima and Cotton effects to transient reaction intermediates, to be considered later [16-18].

Much effort has been devoted to ascertaining the nature of the functional groups of the enzyme protein and of the coenzyme that are essential to the structure and activity of the transaminase molecule. In addition to the imine bond linking the carbonyl group of PLP to a lysine side chain, the coenzyme is fixed at the active centre by multiple weak forces (Fig. 1): an ionic bond

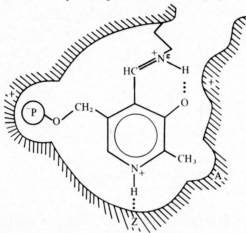

Figure 1. Groups involved in binding of PLP in the active centre of aspartate transaminase [18]. N^ϵ, ϵ-N atom of lysyl residue; +, cationic groups of the apoenzyme; Z, proton-accepting group; A, hydrophobic locus.

between the phosphate residue and a cationic site of the protein, and hydrogen bonds linking the pyridine nitrogen to a proton-donating group of the apoenzyme and the phenolic hydroxyl to the imino nitrogen atom. The phenolic group is essential for both enzymatic and model pyridoxal catalysis. The phosphate group is dispensable in model systems, but the enzyme protein has at least a thousand-fold higher affinity for B_6 phosphate esters than for the free vitamins [5, 7].

The methyl group in position 2 is of relative importance: 2-nor-PLP is an adequate substitute for the natural coenzyme in the case of aspartate trans-aminase, but displays much lower activity in other PLP-dependent enzymes

104 A. E. BRAUNSTEIN

[17-21]. Karpeisky and co-workers [17, 20, 21] have synthesized and tested a series of PLP-homologues substituted in positions 2 and 6 (Table 1). The physical and catalytic properties of their complexes with the apotransaminase indicate that the substituent (or H-atom) in position 2 contacts a hydrophobic area of the apoenzyme. When a larger, tight-fitting alkyl group is present in this position, catalytic activity is decreased [17, 18, 21].

Table 1. Physical and catalytic parameters of aspartate transaminase reconstituted from apoenzyme and PLP or its analogues [18, 20].

	PLP	2-ethyl	2-n-butyl	2-nor	N-oxide	5'-methyl
λ_{max}, mμ; pH 5.2	430	435	440	425	420	430
λ_{max}, mμ; pH 8.1	360	365	370	360	360	360
pK_a	6.25	6.5	6.4	5.8		
k_2, M^{-1} min^{-1}	1500	750	150	450		
V_m, relative	1.0	0.3	0.5	1.2	$\sim$0.4	0.6

The role of functional groups of the enzyme protein has been studied in our laboratory [10, 22, 23] and elsewhere [13, 24, 26] with the aid of conventional blocking reagents and specific bifunctional agents.

The following groups are of importance:

1. The coenzyme-linked lysine and, possibly, a second lysine residue located in the active site [12, 13, 17].

2. Polyanovsky et al. [10] and Martinez-Carrion et al. [26] have independently observed nearly complete inactivation of the enzyme on photooxidation of one or two histidine residues per monomer. Kinetic data indicate that *one* imidazole group is essential [26]; its role will be considered later.

3. Like many PLP-enzymes (but not all), aspartate transaminase is sensitive to thiol reagents. Alkylation or oxidation of 3 of its 5 sulphydryl groups does not impair the catalytic activity, but the blocking of one further, less reactive SH-group (in particular, by mercaptide-forming metal compounds) results in 70-90% inhibition. Modification of all thiol groups is associated with marked configurational alteration of the enzyme protein. The critical SH-group probably plays a conformational, rather than a catalytic role [14, 23].

4. Turano [24] has found that the binding of PLP is prevented by treatment of the apoenzyme with phenol reagents, and that one tyrosine residue is apparently situated in the active centre. Recent work by Karpeisky et al. [18,

21, 22] indicates that a tyrosyl residue participates in the binding of PLP. In the holoenzyme this residue displays a Cotton effect that responds to changes in coenzyme structure and functional state of the transaminase. On treatment of the holoenzyme with tetranitromethane under defined conditions, this group undergoes selective conversion to nitrotyrosine with shifting of the Cotton effect to the new absorption band and marked reduction in enzymatic activity, but without any release of coenzyme. In the apoenzyme many tyrosine residues undergo nitration [22, 24]; the protein thus modified fails to combine with coenzyme and shows no induced optical activity [18, 22].

The essential groups of aspartate transaminase are listed in Table 2. It is clear that further complementary groups of the protein must participate in the binding of ionizable and non-polar groups of the substrates and coenzyme, and in sequential transformations taking place in the active site.

Table 2. Essential groups of aspartate transaminase
(per monomer, mol. wt. ~ 45,000)

Coenzyme	Groups	Number total	essential
	Pyridine N (1) Carbonyl (4) Hydroxyl (3) Phosphate (5') Methyl (2) ±	1	1
Amino acid residues of apoenzyme			
	Thiol	5	1
	ϵ-NH$_2$	20	1
	Imidazolyl	7	1 (2?)
	p-Hydroxyphenyl	12	1
	Carboxyl	~30	?
	Amino and guanidino	~ 44	?

The major steps of the transamination reaction are well known. They include: (a) binding of the amino substrate through its ionized carboxyls to cationic groups of the active site; (b) conversion of the internal PLP-lysine imine to the PLP-substrate aldimine by way of a transaldimination reaction; (c) prototropic rearrangement of the PLP-substrate aldimine to the tautomeric PMP-ketimine, followed by (d) its hydrolysis to enzyme-bound pyridoxamine phosphate and the keto acid [3-7].

The same sequence of steps proceeds in the opposite direction in the second half-reaction. The reversible prototropic aldimine-ketimine rearrangement is the

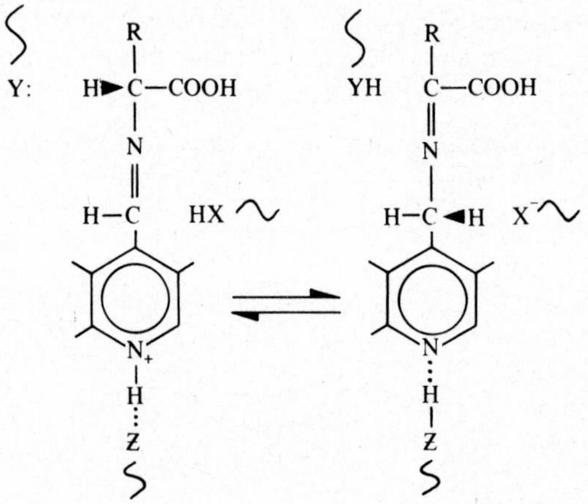

Figure 2. The aldimine ⇌ ketimine rearrangement.

rate-limiting step, both in enzymatic and in non-enzymatic transamination. The reaction rate in model systems is greatly enhanced, as demonstrated by Bruice and Topping [25], in the presence of imidazole buffer, which acts by a mechanism of concerted acid-base catalysis. In the active site of the enzyme this prototropic step may be similarly catalysed by suitably aligned acid-base groups of the protein. One likely candidate for such a function is the imidazole ring of the above-mentioned essential histidine [6, 10, 26].

In the enzyme-bound Schiff bases the proton shift is strictly stereospecific: the α-hydrogen atom of L-amino acids only, but not of the D-isomers, can react, and, as demonstrated experimentally by Dunathan et al. [27] and Snell (see ref. 27), only one of the two H-atoms in position 4 of PMP is subject to isotopic exchange in the enzymatic reaction (Fig. 2).

Investigation of interactions of the transaminase with a variety of substrate analogues has afforded some insight into topochemical details of events in the active site in the course of the enzymatic reaction. In the hands of Khomutov, Karpeisky, Severin and co-workers, the antibiotic cycloserine and related synthetic compounds proved of value for configurational inhibitor analysis of transaminases and other PLP-enzymes. The results have been presented in several experimental papers and reviews [28-32]; hence a brief survey of some essential points will suffice.

D- and L-cycloserine are sterically rigid cyclic analogues of the homonymous enantiomers of α-alanine [28, 29] (Fig. 3). Cycloserine derivatives substituted in position 5 of the isoxazolidone ring are analogous to the higher α-amino acids [30]. The unstable cyclic hydroxamate group imitates the properties of a

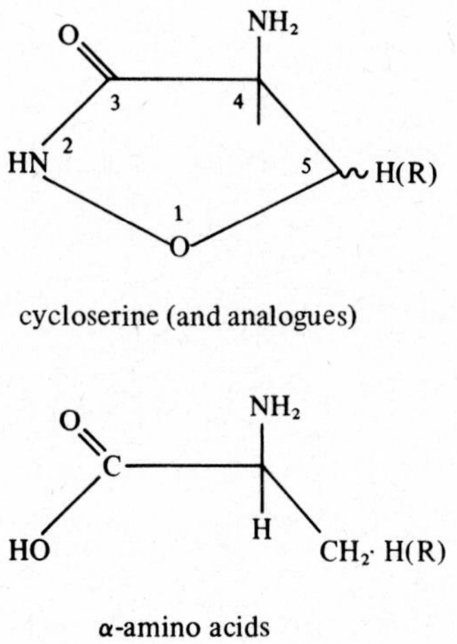

cycloserine (and analogues)

α-amino acids

Figure 3. Structural analogy between cycloserine (or its derivatives) and α-amino acids.

carboxyl group. Individual members of the series are irreversible inhibitors with markedly selective affinity for PLP-enzymes that act on the corresponding amino acids. These inhibitors were shown to be bound at the active site as pseudosubstrates and to form aldimines with the coenzyme.

As indicated in Fig. 4, this results in cleavage of the labile cyclic hydroxamate group. The resulting reactive acyl residue then forms a relatively stable bond with a nucleophilic group (X) of the active site. From aspartate transaminase inhibited with ^{14}C-labelled cycloserine, Karpeisky and Breusov [31] isolated a complex consisting of the degraded inhibitor and pyridoxamine phosphate. This fact, together with spectral data [28, 30], indicated an inhibition mechanism that includes abortive transamination of the inhibitor with the coenzyme up to the stage of PMP-ketimine; the reverse tautomeric shift is prevented by blocking of a required essential group (X), believed to be one of the acid-base catalysts.

Cycloserine analogues substituted in position 5 exist in stereoisomeric forms with rigid fixed configurations similar to definite rotational isomers of an amino acid. As a rule, a PLP-enzyme specific for such an amino acid has marked

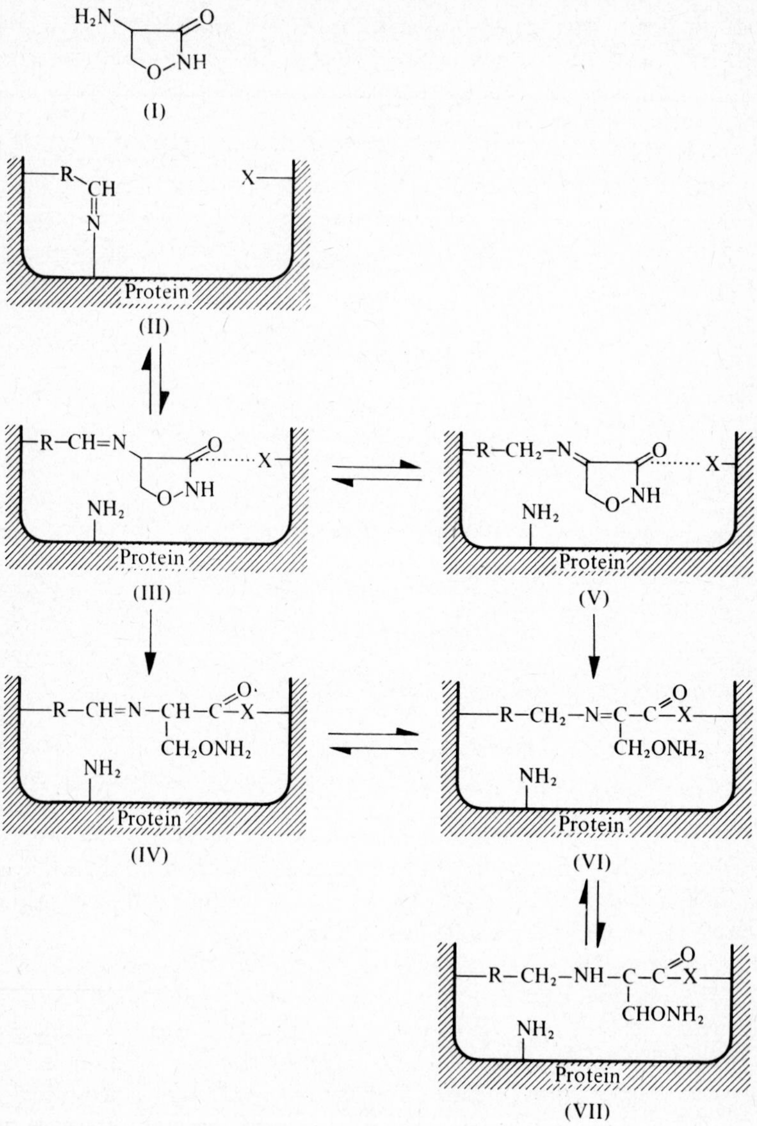

Figure 4. Scheme of the mechanism of inhibition of transaminase by cycloserine [5, 31].

preferential affinity to one of a stereoisomeric pair of the cyclic inhibitors. Khomutov and co-workers [30, 32] have shown that the active cycloserine analogues can be used as rigid templates, approximately replicating the complementary topography of the substrate-binding site and, *eo ipso,* defining the conformation of the amino acid rotamer actually fixed at that site. For example,

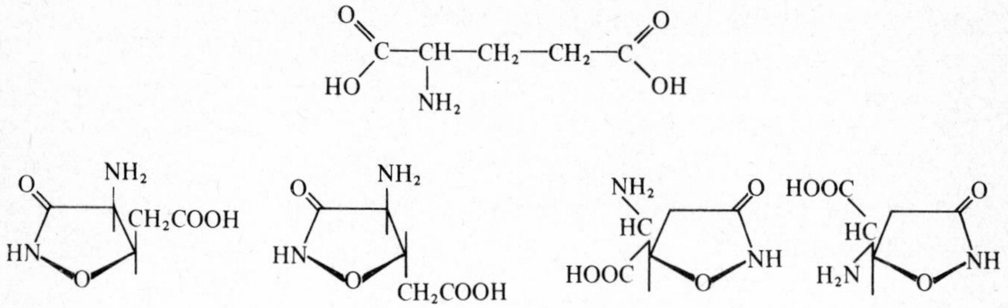

Figure 5. Projection formulas of stereoisomeric cycloglutamic acids [30, 32].

1 and 2, *cis-* and *trans-*α-cycloglutamates;
3 and 4, *threo-* and *erythro-*γ-cycloglutamates.

these authors synthesized the four possible isomeric isoxazolidones based on the skeleton of glutamic acid (shown in Fig. 5). In compounds 1 and 2, conventionally named *cis-* and *trans-*α-cycloglutamates, the α-carboxyl of glutamate is involved in the cyclic alkoxyamide group, whereas the cycle includes the γ-carboxyl in compounds 3 and 4, designated respectively as *threo-* and *erythro-*γ-cycloglutamates.

The substrate site of a glutamate-specific PLP-enzyme should presumably be complementary to at least one of these structures. Actually, aspartate-glutamate transaminase and alanine-glutamate transaminase are highly and selectively

Table 3. Molar concentrations of inhibitors reducing activity of the enzymes by 50% after a definite preincubation period (I_{50} values) [32].

| Enzyme | Cycloglutamate: | | | |
	1 *cis-*α	2 *trans-*α	3 *threo-*γ	4 *erythro-*γ
Alanine-glutamate transaminase	1×10^{-5}	1×10^{-3} (8%)	5×10^{-4}	1×10^{-3}
Aspartate-glutamate transaminase	2×10^{-5}	1.7×10^{-4}	7×10^{-5}	1×10^{-4}
γ-Aminobutyrate-glutamate transaminase	1×10^{-6}	1×10^{-3} (20%)	1×10^{-3} (20%)	1×10^{-3} (30%)
Phenylalanine-glutamate transaminase	1×10^{-3}	1×10^{-3}	1×10^{-3}	1×10^{-3}
Glutamate decarboxylase	1×10^{-3} (18%)	5×10^{-4}	1×10^{-4}	1×10^{-3} (28%)

sensitive to inhibition by compounds 1 and 3; the latter is somewhat less active
(Table 3), and the two enzyme-inhibitor complexes differ in properties and
structure, in accordance with the difference in orientation of the reactive
alkoxamide groups in compounds 1 and 3. Superposition of atomic models of
these compounds shows that their carbon chains have an identical steric
arrangement, which evidently approximates the conformation of enzyme-bound
glutamate.

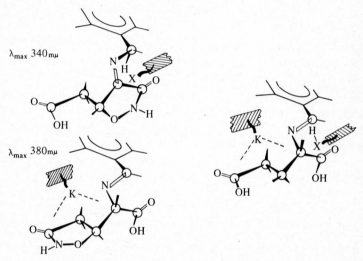

Figure 6. Configuration of compounds 1 and 3 are presumable conformation of
L-glutamic acid bound in the active centre of aspartate transaminase [31, 32].

The presumed topochemistry of the transaminase-glutamate aldimine, indicated by these results, is shown in Fig. 6. In devising this configurational model,
Khomutov [32] took into account the data on the stereospecificity of the proton
shift and the mutual orientation of substrate and coenzyme postulated by
Dunathan [33] on quantum-chemical grounds. For labilization of the α-C—H
bond, as in the case of transamination, conditions are optimal when this bond is
in a plane perpendicular to the pyridine ring of PLP (Fig. 7).

Results of considerable interest in relation to molecular properties of
PLP-enzymes, and to general aspects of enzyme catalysis, were recently obtained
by Severin et al. [39]. They studied the interactions of PLP-enzymes with a
series of open-chain O-alkyl derivatives of aminocarbhydroxamic acids, of the
type:

$$R^1-CH-CONH-OR^2$$
$$|$$
$$NHR^3$$

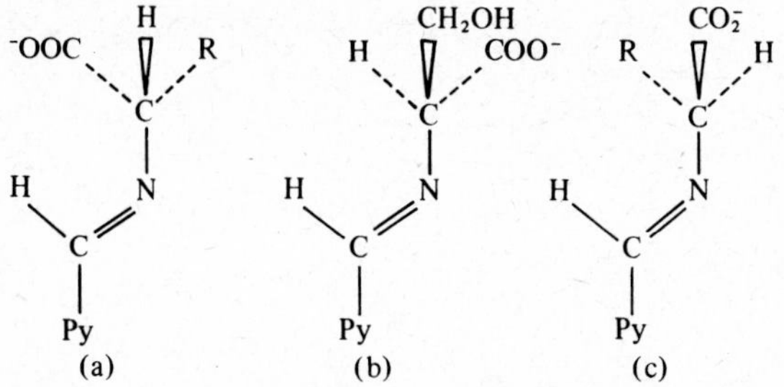

Figure 7. Optimum conformations of PLP-amino acid aldimines for cleavage of (a) the α-C—H bond (transamination, etc), (b) the α-C—β-C bond (serine hydroxymethyl transferase, threonine aldolase), and (c) the α-C—COOH bond (α-decarboxylation of amino acids) [33].

Some of these compounds formally correspond to molecules of cycloserine analogues dissected at the 4—5 bond, equivalent to the bond between α-C and β-C of an amino acid. Only those compounds of this series containing non-modified fragments of a substrate amino acid, for instance the O-alkyl glycinohydroxamic acids I-III, have the properties of strictly selective inhibitors of PLP-enzymes acting on that amino acid. Thus, glutamate decarboxylase and glutamate-specific transaminases are strongly inhibited by millimolar concentrations of compound II, the open-chain analogue of α-cycloglutamate; the inhibited enzymes have the properties of protein-bound oximes of PLP. In aqueous solutions the inhibitors of this type exist predominantly in the linear *all-trans* conformation; on interaction with pyridoxal or PLP they display no acylating properties, do not form oximes and are fairly inert substances in other respects.

$$\overset{\alpha}{CH_2}-CO-NH-O-\overset{\beta}{C}H_3 \tag{I}$$
$$NH_2$$

$$\overset{\alpha}{CH_2}-CO-NH-O-\overset{\beta}{C}H_2-CH_2-COOH \tag{II}$$
$$NH_2$$

$$\overset{\alpha}{CH_2}-CO-NH-O-\overset{\beta}{C}H_2-C_6H_5 \tag{III}$$
$$NH_2$$

It is evident that active carboxamate ethers, with aminoacyl and alkyl groups complementary to corresponding parts of the substrate site, are bound in the active centre of appropriate PLP-enzymes in a thermodynamically-unfavourable pseudo-cyclic *cis-cis* conformation with the potential α-C and β-C atoms in proximity, and form a PLP-aldimine. This structure and the transformations it undergoes are closely similar to those of a complex of a cycloserine analogue with the enzyme.

The mechanism of inhibition presented in Fig. 8 accounts for formation of a PLP oxime as the final product. Thus, we are dealing with a remarkable case of "compulsory fit" where an appropriate enzyme forms an inactive complex, resulting from interaction with an induced strained conformation of a relatively simple molecule whose structure and properties *per se* would not suggest any specific inhibitory capacity. The authors have proposed the generic term "conformational enzyme inhibitors" for agents of this type.

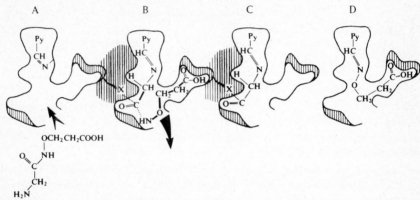

Figure 8. Mechanism of inhibition of a PLP enzyme (aspartate transaminase) by an O-alkyl aminocarbhydroxamic acid (compound III) [39].

Experiments with appropriate coenzyme analogues have likewise provided helpful clues concerning the topochemistry of aspartate transaminase and its reaction intermediates. As reported in 1967 by Khomutov [32a, 34], and observed independently by Turano and co-workers [35], the apoenzyme readily forms stable inactive complexes with phosphopyridoxyl derivatives of substrate amino acids, obtained by reduction of their PLP-aldimines (Fig. 9a). These

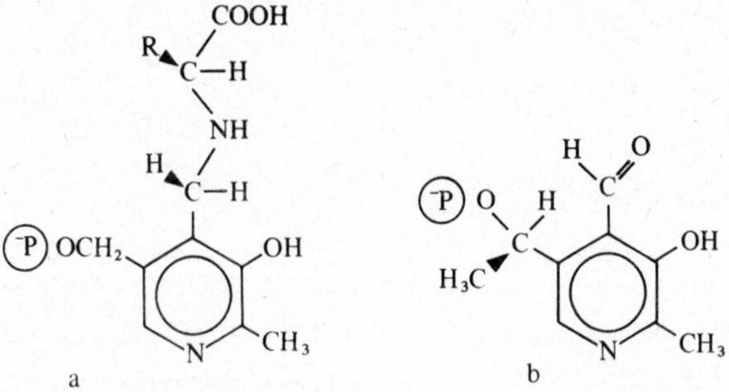

N-phosphopyridoxyl-
α-amino acids

5′-methyl-PLP

Figure 9. Formulas of N-phosphopyridoxyl-α-amino acids (a) [32a, 34] and of 5′-methyl-PLP (b) [18, 21].

compounds are similar stereochemically to the transient tetrahedral inter-mediates presumably formed in the transaldimination stage of the enzymatic reaction. As anticipated, phosphopyridoxyl-L-glutamate exhibits a particularly strong affinity for the apoenzyme.*

In my laboratory, Florentiev, Karpeisky and Ivanov [18, 21] have synthe-sized and studied, among other alkylated coenzymes, a PLP-homologue methy-lated in the $-CH_2-$ group of the 5-methoxyphosphate side chain (Fig. 9b). They found that only one of the enantiomers present in the racemic 5'-methyl-PLP combines with apotransaminase; the fully reconstituted artificial holoenzyme has fairly high catalytic activity ($\sim$ 60%). These facts, and structural consider-ations, suggest certain inferences as to the chemical topography of the phosphate group.

As stated in the introductory part of this paper, the catalytic function of an enzyme can be interpreted adequately only in terms including the topochemistry of sequential intermediates and their transitions. Enzyme-catalysed reactions usually proceed through a series of consecutive steps, each requiring a particular set of optimal conditions with respect to ionization of functional groups in the protein, coenzyme and substrates, to their orientation and interactions with neighbouring groups and solvents, etc.

In enzymatic transamination, as in other cases, the alternating requirements for sequential stages are different or even contradictory.

Ivanov, Karpeisky and others [17, 18, 36] have developed a tentative dynamic model for the molecular mechanism of enzymatic transamination. The model incorporates and details general features of reaction schemes proposed earlier, but amplifies them in essential points. It is based on the logical analysis, from physical-organic and stereochemical standpoints, of available information on aspartate transaminase and related enzymes, and new experimental evidence in support of this mechanism has been obtained by Karpeisky and his colleagues.

---

* At the 5th International Symposium on the Chemistry of Natural Products recently held in London (July 1968), Khomutov [32a] reported that the inactivation of aspartate apotransaminase, on interaction with phosphopyridoxyl-L-glutamic acid, is associated with transfer of the phosphate group to the enzyme protein; under definite conditions (storage at $pH \geqslant 7$) this is followed by spontaneous release of inorganic phosphate. Upon denaturation of the primary apoenzyme–inhibitor complex and digestion with pepsin and pronase, the authors detected free and peptide-bound O-phosphothreonine, in a yield approaching one equivalent per monomeric subunit of the enzyme. The threonine residue must be located in the neighbourhood of the coenzyme's phosphate group in the normal tetrahedral intermediate. A hypothetic reaction scheme accounting for this anomalous transphosphorylation was proposed, and it was suggested that the threonyl residue in the active centre might play an essential part in enzymatic transamination.

This conjecture has far-reaching implications; should it be verified by further experi-mental investigations, the catalytic mechanism to be discussed below may require partial revision. In any event, the facts observed so far by Khomutov's group afford an explanation for conflicting results previously obtained by several investigators who studied the fate of the phosphate group of the coenzyme upon reduction of PLP-enzymes and their substrate complexes with sodium borohydride (cf. [5]).

On close consideration it became evident that all the data relating to aspartate transaminase, some of them puzzling and apparently conflicting, can be given a consistent, unified interpretation on the assumption that structural and ionic transitions that take place during the course of the enzymatic reaction are associated with sequential changes in orientation of the coenzyme. The essential idea involved is that the active centre comprises two spatially and functionally distinct sites. In one of these sites the coenzyme, linked as a Schiff base of

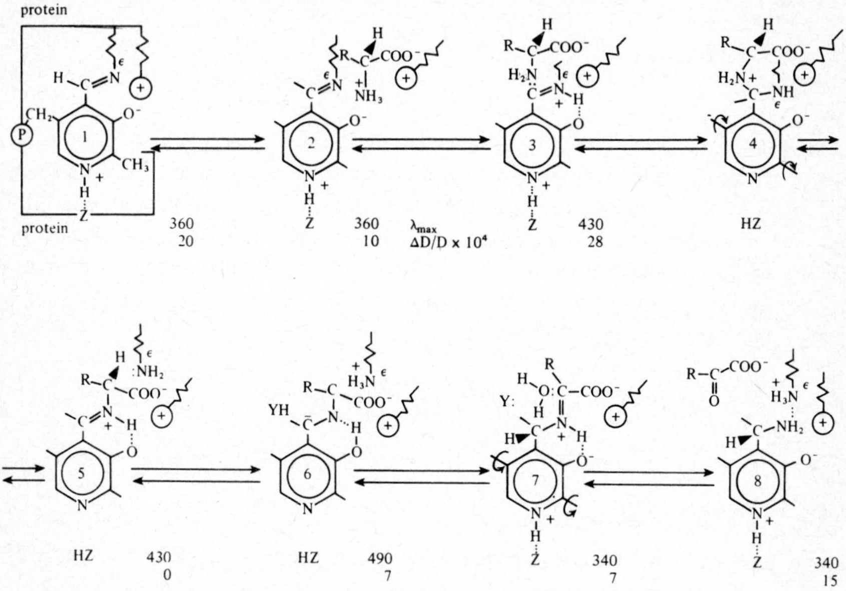

Figure 10. Dynamic model of the molecular mechanism of enzymatic transamination (Ivanov and Karpeisky [17, 18, 36]).

lysine, is "stored" in the absence of substrate; the other site includes the substrate-anchoring site and the catalytic groups participating in prototropic rearrangement of the coenzyme-substrate imines. On interaction of the enzyme with amino substrate, cooperative proton shifts involved in the trans-aldimination step, and in subsequent reaction stages, are linked with a reversible translocation, e.g. a pendulum-like or rotatory movement, of the coenzyme.

This feature, and other topochemical aspects of the reaction mechanism, will be explained in the following description, necessarily concise, of the dynamic scheme proposed by Ivanov and Karpeisky (Figs. 10-13).

The assignment of absorption maxima (λ_{max}) and factors of anisotropy ($\Delta D/D$) to definite structures, shown in the scheme, calls for some comment. It is based, (a) on the optical properties of stable forms of the enzyme, and (b) on investigations of the fast-reaction kinetics of the transamination system by spectrophotometric methods [6, 40]; assignments made for transient inter-mediates of the normal transamination reaction are supported (c) by the optical parameters of abortive complexes of similar structure between the enzyme and various inhibitory pseudosubstrates blocking the reaction at a definite stage. Examples include the stable PLP-aldimine of α-methylaspartate with its non-dichroic absorption peak at 430 mμ [14, 18]; the PMP-ketimine formed on interaction of transaminase and cycloserine or α-cycloglutamate, with the slightly dichroic band at 340 mμ [18, 31]; the red complex of the enzyme with erythro-β-hydroxyaspartate, displaying an absorption peak and negative dichroism at 490 mμ, attributable to a quinonoid structure [6] or a carbanion of type 6 (Fig. 10; cf. ref. 18), etc. We shall now consider the sequential steps of the dynamic model proposed by Ivanov and Karpeisky, as represented in Fig. 10:

(1) In the enzymatically-active aldimine form of aspartate transaminase (Fig. 10, Stage 1) the phenolic group of PLP is known to be ionized. Its pK_a is = 6.2, in sharp contrast to the pK_a of ~11.0 for the aldimines of free PLP. To explain this large difference in pK_a values, the presence of a hydrogen bond between the ring nitrogen atom and a proton-donating group (HZ) of the enzyme was postulated by Jenkins et al. [37]. Yet the properties of model aldimines methylated at the pyridine nitrogen atom show that a positive charge on this atom lowers the pK_a only to a value ~8.0. It is therefore suggested that further lowering of the pK_a to 6.2 is caused by the presence of a cationic group $\oplus$ of the protein in close proximity to the phenolic group.

(2) In the optimum range of pH for enzymatic transamination, all dissociable groups of the substrate (amino and carboxyl groups) are ionized. The enzyme's specific affinity for dicarboxylic substrates indicates that the substrate becomes attached through its two negatively charged carboxylate groups to suitably oriented cationic groups of the protein. The substrate (aspartate) is in a conformation with the carboxyl groups in proximity; this follows from the fact that the enzyme is competitively inhibited by maleic, but not by fumaric, acid.

It is assumed that the α-carboxylic group of the substrate binds to the same cationic group $\oplus$ of the protein (Stage 2), the interaction of which with the phenolic group of PLP was mentioned above as one of the two factors lowering the pK_a of the "internal" aldimine in aspartate transaminase. The effect of this factor is eliminated, and the pK_a of the "internal aldimine" rises to a value ~8.0, when the cationic group $\oplus$ is neutralized by the carboxyl group of the substrate (or of an inhibitory dicarboxylic acid). At the same time, neutrali-zation of the α-carboxyl of the amino substrate lowers the basicity of its amino group, shifting its pK from 9.6 to 7.4. This makes possible the transfer of a proton from the NH_3^+ group of the substrate to the coenzyme (Stage 3). The deprotonated amino group is now enabled to bind to the —C=N— double bond, while polarization of this bond, due to protonation of the imino nitrogen atom, markedly enhances the coenzyme's electrophilicity.

(3) The free NH_2-group attacks the protonated Schiff base. For this attack only one orientation is allowed by the stereochemistry of nucleophilic addition; viz. the orbital of the lone electron pair of the substrate's NH_2-group must be directed towards the $C_{4'}$ atom of the imino group perpendicular to the plane of the conjugated system. Formation of a tetrahedral transitional addition product (Stage 4) and, further, of the protonated substrate aldimine (Stage 5) requires rotation of the pyridine ring of the coenzyme by an angle of about 40° around an axis passing through positions 2 and 5 of this ring. One end of the axis is fixed at the 5'-phosphate group, its other end faces the hydrophobic site normally holding the 2-methyl group of the coenzyme. Freedom for rotation around the 2–5-axis requires that the hydrogen bond between the pyridine nitrogen and group Z be broken. The necessary conditions are provided in the preceding step (Stage 3), when the negatively-charged phenolic group accepts a proton from the amino group; this results, as shown by experiments with model systems, in lowering of the pK of the ring nitrogen to a value of ~8 (or to ~6.5 in the enzyme-substrate complex, Stage 3).

This enables the proton to shift to group Z, with rupture of the hydrogen bond. Now the ring can freely swing into the position required for covalent binding of the $C_{4'}$ atom to the substrate amino nitrogen.

The stereochemistry of the transaldimination step is illustrated by the photographs representing schematic (approximative) projection models of Stage 3 (Fig. 11) and Stage 5 (Fig. 12).

Evidence obtained by Karpeisky and Ivanov [17, 18] justifies tentative identification of the HZ group as the hydroxyphenyl group of the previously

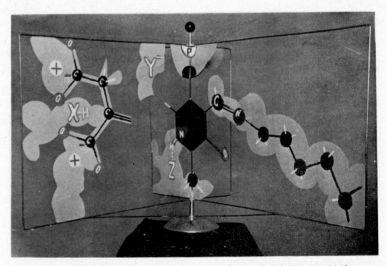

Figure 11. Schematic projection model of Stage 3 (cf. Fig. 10) [18].

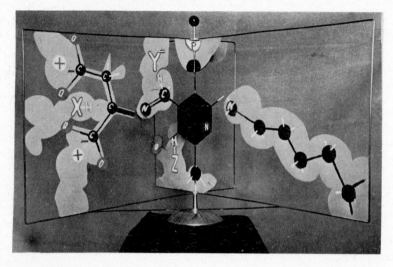

Figure 12. Schematic projection model of Stage 5 (cf. Fig. 10) [18].

mentioned tyrosine residue exhibiting induced optical rotation in the holo-enzyme. The aldimine form of the transaminase (Stage 1) has a negative dichroic peak at 295 mμ, which coincides with the absorption peak of a tyrosine anion. The negative Cotton effect is lowered and broadened in stable pseudosubstrate aldimines of the type of Stage 5, and is markedly increased in the artificial holoenzyme reconstituted with the 2-n-butyl analogue of PLP, evidently due to steric pressure of the bulky alkyl group upon the adjacent tyrosine side chain (Fig. 13). In the complex of apotransaminase with the N-oxide of PLP, an analogue incapable of hydrogen bonding with HZ, this negative Cotton effect is replaced by a peak of positive circular dichroism with λ_{max} 280 mμ–the absorption maximum of non-ionized tyrosine [18, 22].

(4) Let us now consider the subsequent step, namely, rearrangement of the PLP-substrate aldimine to the tautomeric PMP-ketimine (Fig. 10, steps 5 → 7). I mentioned earlier that two functional groups of the protein–an acidic and a basic one–are believed to participate in removal of the proton from the substrate α-C atom and in protonation of the coenzyme's $C_{4'}$ atom. In the present model (as in a scheme proposed by Snell [7]), the proton dissociating from the α-carbon is accepted by the ϵ-NH$_2$ group of lysine, and released at Stage 5. This is a suitable candidate for steric reasons. According to Dunathan's [33] postulate, the geometric reaction coordinate for dissociation of the α-hydrogen is in a plane perpendicular to that of the pyridine ring; removal of that protein results in formation of a carbanion (Stage 6) with an extensive resonance-stabilized system of conjugated bonds [38]. The necessary conforma-

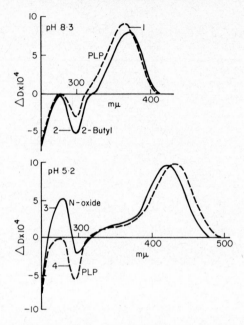

Figure 13. CD spectra of complexes of apotransaminase with PLP and some coenzyme analogues [18]. At pH 8.3:1, native holoenzyme; 2, complex with 2-*n*-butyl analogue of PLP. At pH 5.2: 3, complex with *N*-oxide of PLP; 4, native holoenzyme.

tion around the C_a–N bond has been ensured earlier, by fixation of the substrate's α-carboxyl group. As a result of reorientation of the coenzyme in the transaldimination step, its $C_{4'}$-atom is brought into proximity with the proton-donating group YH (possibly an imidazole group); proton transfer produces the ketimine (Stage 7). The tetrahedral configuration of atom $C_{4'}$ allows the ring of the coenzyme to revert into its initial plane by rotation around the 2–5 axis. In this position the N atom of the ring undergoes protonation by the HZ group, associated with an increase in acidity of the phenolic hydroxyl of the coenzyme.

(5) The proton of this group dissociates, and its binding to the imino nitrogen atom favours the attack on the α-carbon atom by a molecule of water (Stage 7). Moreover, the deprotonated group Y can attract a proton from this water molecule, thus producing a very active hydroxide ion. Hydration at α–C and cleavage of the transient hydrated ketimine yields the pyridoxamine form of the enzyme and the α-keto acid (Stage 8), thus completing the first half-reaction of enzymatic transamination.

(6) The amino group of PMP is now in close proximity to two cationic groups of the active centre—the ε-NH_2 group of the lysine residue to which the PLP was bound, and the cationic group ⊕ (Stage 8). The basicity of the NH_2 group of PMP, which normally has a pK value of 10.5, is thereby greatly decreased.

The amino group remains unionized and is, therefore, readily capable of interaction with the ketosubstrate in the reverse half-reaction of transamination.

In addition to the features considered above, the dynamic model proposed by Ivanov and Karpeisky makes it possible to explain certain well-known puzzling properties of aspartate transaminase [18, 36], for example:

(a) Why is the species with ionized hydroxyl in the coenzyme the enzymatically active form of the enzyme, whereas the protonated Schiff base has the highest activity in model systems? According to the mechanism discussed here, the necessary state of ionization is attained immediately before the stage when it is required.

(b) Why does no transamination occur between the coenzyme and the lysine residue forming the internal aldimine? Because tautomeric rearrangement of the substrate-aldimine is catalysed by two groups, one of which (the ϵ-NH$_2$ of lysine) is liberated only upon formation of the substrate imine and the other, YH, is spatially distant from the internal aldimine.

(c) Why are V_{max} of aspartate transaminase, and the relative concentrations of observable reaction intermediates, pH-independent within the pH range 5 to 10 [6, 37]? Because none of the reaction steps of this mechanism involve acceptance of protons from the medium or their release into the medium; all proton transfers occur between components of the intermediate complexes.

It is probable that certain features, at least, of this dynamic model for enzymatic transamination similarly apply to the molecular mechanisms of most reactions catalysed by other types of PLP-dependent enzymes, e.g. the existence of two spatially and functionally distinct loci in the active centre, implying the necessity for changes in orientation of the coenzyme in the course of each catalytic act.

Alternations in the mutual orientation of interacting components in enzyme-substrate complexes may have broader significance as a factor which can contribute, apart from eventual changes in conformation of substrate or enzyme protein, to the compulsory complementarity postulated by the "induced fit" hypothesis of Koshland.

One fundamental aspect of this dynamic scheme is presumably valid in a general sense for many types of enzymatic reactions involving sequential steps, namely, the implication that the optimum conditions for each stage of such reactions arise as a result of topochemical transitions, both conformational and electronic, occuring in the preceding step.

This very circumstance, namely, the fact that a polyfunctional system with relative conformational mobility, constituted from substrate, cofactors and a macromolecule of specific protein, has the potential capacity to provide different, or even incompatible, sets of conditions required for consecutive stages of a complicated chemical reaction, is probably one of the principal features that distinguish enzyme catalysis essentially from chemical transformations in solution or at the surface of a rigid heterogeneous catalyst [36a].

REFERENCES

1. Blake, C. C. F., Koenig, D. F., Mair, G. A., North, A. C. T., Phillips, D. C. and Sarma, V. R., *Nature, Lond.* **206** (1965) 757; Phillips, D. C., *Scient. Am.* **215** (1966) 78.

2. Reeke, Jr., G. N., Hartsuck, J. A., Ludwig, M. L., Quiocho, F. A., Stelitz, T A. and Lipscomb, W. N., *Proc. natn. Acad. Sci. U.S.A.* **58** (1967) 2220.

3. Snell, E. E., Fasella, P., Braunstein, A. E. and Rossi-Fanelli, A. (eds.), "Chemical and Biological Aspects of Pyridoxal Catalysis", Proceedings of 1st IUB Symposium, Rome, October 1962, Pergamon Press, Oxford, 1963.

4. Snell, E. E., Braunstein, A. E., Severin, E. S. and Torchinsky, Yu. M. (eds), "Pyridoxal Catalysis: Enzymes and Model Systems", Proceedings of 2nd IUB Symposium, Moscow, September 1966, Interscience, New York, 1968; Russian version, "Nauka" Publ. House, Moscow, 1968.

5. Braunstein, A. E., *Vitams. Horm.* **22** (1964) 451.

6. Fasella, P., *A. Rev. Biochem.* **36** (1967) 185; in "Pyridoxal Catalysis: Enzymes and Model Systems", Proceedings of 2nd IUB Symposium, Moscow, September 1966 (edited by E. E. Snell, A. E. Braunstein, E. S. Severin and Yu. M. Torchinsky), Interscience, New York, 1968, p. 1.

7. Snell, E. E., in Chemical and Biological Aspects of Pyridoxal Catalysis", Proceedings of 1st IUB Symposium, Rome, October 1962, (edited by E. E. Snell, P. Fasella, A. E. Braunstein and A. Rossi-Fanelli), Pergamon Press, Oxford, 1963, p. 1.

8. Morino, Y. and Wada, H., in "Chemical and Biological Aspects of Pyridoxal Catalysis", Proceedings of 1st IUB Symposium, Rome, October 1962 (edited by E. E. Snell, P. Fasella, A. E. Braunstein and A. Rossi-Fanelli), Pergamon Press, Oxford, 1963, p. 175; *Vitams Horm.* **22** (1964) 411.

9. Polyanovsky, O. L., in "Pyridoxal Catalysis: Enzymes and Model Systems", Proceedings of 2nd IUB Symposium, Moscow, September 1966 (edited by E. E. Snell, A. E. Braunstein, E. S. Severin and Yu. M. Torchinsky), Interscience, New York, 1968, p. 155; Polyanovsky, O. L. and Makarova, L. S., *Biokhimiya* **31** (1967) 372.

9a. Abaturov, L. V., Polyanovsky, O. L., Torchinsky, Yu. M. and Varshavsky, Ya. M., in "Pyridoxal Catalysis: Enzymes and Model Systems", Proceedings of 2nd IUB Symposium, Moscow, September 1966 (edited by E. E. Snell, A. E. Braunstein, E. S. Severin and Yu. M. Torchinsky), Interscience, New York, 1968, p. 171.

10. Vorotnitskaya, N. E., Lutovinova, G. F. and Polyanovsky, O. L., in "Pyridoxal Catalysis: Enzymes and Model Systems", Proceedings of 2nd IUB Symposium, Moscow, September 1966 (edited by E. E. Snell, A. E. Braunstein, E. S. Severin and Yu. M. Torchinsky), Interscience, New York, 1968, p. 131; Vorotnitskaya, N. E., Spyvack, V. A. and Polyanovsky, O. L., *Biokhimiya* **33** (1968) 375.

11. Fischer, E. H., Forrey, A. W., Hedrick, J. L., Hughes, R. C., Kent, A. B. and Krebs, E. G., in "Chemical and Biological Aspects of Pyridoxal Catalysis", Proceedings of 1st IUB Symposium, Rome, October 1962 (edited by E. E. Snell, P. Fasella, A. E. Braunstein and A. Rossi-Fanelli), Pergamon Press, Oxford, 1963, p. 543.

12. Polyanovsky, O. L. and Keil, B., *Biokhimiya* **28** (1963) 372.

13. Turano, C., Giartosio, A., Riva, F. and Vecchini, P., *in* "Chemical and Biological Aspects of Pyridoxal Catalysis", Proceedings of 1st IUB Symposium, Rome, October 1962 (edited by E. E. Snell, P. Fasella, A. E. Braunstein and A. Rossi-Fanelli), Pergamon Press, Oxford, 1963, p. 149; Turano, C., Giartosio, A., Riva, F., Bossa, F. and Baroncelli, V., *in* "Pyridoxal Catalysis: Enzymes and Model Systems", Proceedings of 2nd IUB Symposium, Moscow, September 1966 (edited by E. E. Snell, A. E. Braunstein, E. S. Severin and Yu. M. Torchinsky), Interscience, New York, 1968, p. 143.

14. Torchinsky, Yu. M. and Koreneva, L. G., *Biokhimiya* 28 (1963) 1087, and 29 (1964) 780; *Biochim. biophys. Acta* 79 (1964) 426.

15. Breusov, Yu. N., Ivanov, V. I., Karpeisky, M. Ya. and Morosov, Yu. V., *Biochim. biophys. Acta* 92 (1964) 388.

16. Braunstein, A. E., Torchinsky, Yu. M., Malakhova, E. M. and Sinitsyna, N. I., *Ukr. biokhem. Zh.* 37 (1965) 671; Torchinsky, Yu. M., Malakhova, E. A., Livanova, N. B. and Pikhelgas, V. Ya., *in* "Pyridoxal Catalysis: Enzymes and Model Systems", Proceedings of 2nd IUB Symposium, Moscow, September 1966 (edited by E. E. Snell, A. E. Braunstein, E. S. Severin and Yu. M. Torchinsky), Interscience, New York, 1968, p. 269.

17. Ivanov, V. I. and Karpeisky, M. Ya., *Molek, Biologiya* 1 (1967) 288, 588.

18. Ivanov, V. I., Dissertation, Institute of Molecular Biology, USSR Academy of Sciences, Moscow, 1968; Ivanov, V. I. and Karpeisky, M. Ya., *Adv. Enzymol.* 33 (1969).

19. Morino, Y. and Snell, E. E., *Proc. natn. Acad. Sci, U.S.A.* 57 (1967) 1692.

20. Bocharov, A. L., Ivanov, V. I., Karpeisky, M. Ya., Mamaeva, O. K. and Florentiev, V. L., *Biochem. biophys. Res. Commun.* 30 (1968) 459.

21. Florentiev, V. L., Karpeisky, M. Ya., in press.

22. Shliapnikov, S. V., Karpeisky, M. Ya., in press.

23. Torchinsky, Yu. M., *Biokhimiya* 29 (1964) 534; Polyanovsky, O. L. and Telegdi, M., *Biokhimiya* 30 (1965) 174.

24. Turano, C., paper presented at the Symposium on Vitamin B_6 and PLP-dependent Enzymes, Nagoya, 1967.

25. Bruice, T. C. and Topping, R. M., *in* "Chemical and Biological Aspects of Pyridoxal Catalysis", Proceedings of 1st IUB Symposium, Rome, October 1962 (edited by E. E. Snell, P. Fasella, A. E. Braunstein and A. Rossi-Fanelli), Pergamon Press Oxford, 1963, p. 29.

26. Martinez-Carrion, M., Turano, C., Riva, F. and Fasella, P., *J. biol. Chem.* 242 (1967) 1426.

27. Dunathan, H. C., Davis, L. and Kaplan, M., *in* "Pyridoxal Catalysis: Enzymes and Model Systems", Proceedings of 2nd IUB Symposium, Moscow, September 1966 (edited by E. E. Snell, A. E. Braunstein, E. S. Severin and Yu. M. Torchinsky), Interscience, New York, p. 325; Snell, E. E., ibid p. 335..

28. Khomutov, R. M., Karpeisky, M. Ya. and Severin, E. S., *Biokhimiya* 26 (1962) 772, and *in* "Chemical and Biological Aspects of Pyridoxal Catalysis", Proceedings of 1st IUB Symposium, Rome, October 1962 (edited by E. E. Snell, P. Fasella, A. E. Braunstein and A. Rossi-Fanelli), Pergamon Press. Oxford, 1963, p. 313; Karpeisky, M. Ya., Khomutov, R. M., Severin, E. S. and Breusov, Yu. N., *in* "Chemical and Biological Aspects of Pyridoxal Catalysis", Proceedings of 1st IUB

Symposium, Rome, October 1962 (edited by E. E. Snell, P. Fasella, A. E. Braunstein and A. Rossi-Fanelli), Pergamon Press, Oxford, 1963, p. 323.

29. Azarkh, R. M., Braunstein, A. E., Paskhina, T. S. and Hsu Ting-Seng, *Biokhimiya* 25 (1960) 954; cf. ibid 26 (1961) 882.

30. Khomutov, R. M., Severin, E. S., Kovaleva, H. K., Gulyaev, N. N., Gnuchev, N. V. and Sastchenko, L. N., *in* "Pyridoxal Catalysis: Enzymes and Model Systems", Proceedings of 2nd IUB Symposium, Moscow, September 1966 (edited by E. E. Snell, A. E. Braunstein, E. S. Severin and Yu. M. Torchinsky), Interscience, New York, 1968, p. 631.

31. Karpeisky, M. Ya. and Breusov, Yu. N., *Biokhimiya* 30 (1965) 153 (cf. ref. 18).

32. Khomutov, R. M., *Sixth Int. Congr. Biochem.*, New York, 1964, Abstr. Symposium IV.

32a. Khomutov, R. M., paper presented at the 5th International Symposium on the Chemsitry of Natural Products, London, July 1968.

33. Dunathan, H. C., *Proc. natn. Acad. Sci, U.S.A.* 55 (1966) 712.

34. Khomutov, R. M., paper presented at a Conference on Structure and Function of Peptides and Proteins, Riga, April 1967.

35. Turano, C., Borri, C., Orlacchio, A. and Bossa, F., this volume, p. 123.

36. Karpeisky, M. Ya. and Ivanov, V. I., *Nature, Lond.* 210 (1966) 493.

36a. Braunstein, A. E., Ivanov, V. I. and Karpeisky, M. Ya., *in* Pyridoxal Catalysis: Enzymes and Model Systems", Proceedings of 2nd IUB Symposium, Moscow, September 1966 (edited by E. E. Snell, A. E. Braunstein, E. S. Severin and Yu. M. Torchinsky), Interscience, New York, 1968, p. 291.

37. Jenkins, W. T. and Sizer, I. W., *J. biol. Chem.* 234 (1959) 1179, 235 (1962) 620, and *in* "Chemical and Biological Aspects of Pyridoxal Catalysis", Proceedings of 1st IUB Symposium, Rome, October 1962 (edited by E. E. Snell, P. Fasella, A. E. Braunstein and A. Rossi-Fanelli), Pergamon Press, Oxford, 1963, p. 123; Jenkins, W. T., *J. biol. Chem.* 239 (1964) 1742; Jenkins, W. T. and D'Aril, L., *in* "Pyridoxal Catalysis: Enzymes and Model Systems", Proceedings of 2nd IUB Symposium, Moscow, September 1966 (edited by E. E. Snell, A. E. Braunstein, E. S. Severin and Yu. M. Torchinsky), Interscience, New York, 1968, p. 317.

38. Perrault, A. M., Pullman, B. and Valdemoro, C., *Biochim. biophys. Acta* 46 (1961) 555; Pullman, B., *in* "Chemical and Biological Aspects of Pyridoxal Catalysis", Proceedings of 1st IUB Symposium, Rome, October 1962 (edited by E. E. Snell, P. Fasella, A. E. Braunstein and A. Rossi-Fanelli), Pergamon Press, Oxford, 1963, p. 103.

39. Severin, E. S., Gnuchev, N. V., Kovaleva, H. K., Gulyaev, N. N. and Khomutov, R. M., *in* "Pyridoxal Catalysis: Enzymes and Model Systems", Proceedings of 2nd IUB Symposium, Moscow, September 1966 (edited by E. E. Snell, A. E. Braunstein, E. S. Severin and Yu. M. Torchinsky), Interscience, New York, 1968, p. 651; Paper presented at a Conference on Structure and Function of Peptides and Proteins, Riga, April 1967.

40. Hammes, G. G. and Fasella, P., *in* "Chemical and Biological Aspects of Pyridoxal Catalysis", Proceedings of 1st IUB Symposium, Rome, October 1962 (edited by E. E. Snell, P. Fasella, A. E. Braunstein and A. Rossi-Fanelli), Pergamon Press, Oxford, 1963, p. 185, and *J. Am. chem. Soc.* 85 (1963) 3929; Fasella, P., Giartosio, A. and Hammes, G. G., *Biochemistry* 5 (1966) 197, and 6 (1967) 1793.

FEBS Symposium, Volume 18, 1970, pp. 123-131

Structural and Functional Aspects of the Active Site in Aspartate Aminotransferase

C. TURANO, C. BORRI, A. ORLACCHIO and F. BOSSA

Institute of Biological Chemistry, University of Rome, and
Institute of Biological Chemistry, Univeristy of Perugia,
Perugia, Italy

The mechanism of enzymatic transamination has been extensively investigated in several laboratories. In particular, aspartate aminotransferase (EC 2.6.1.1., "supernatant" enzyme from pig heart) has been the object of detailed studies (for a review see ref. 1). Although some general features of the process, and especially the role played by the coenzyme, are at present understood, and the nature of some intermediate complexes in the enzymatic reaction has been elucidated, the essential role of the protein in the highly efficient enzymatic catalytic process is still unknown.

In this respect a knowledge of the chemical composition and three-dimensional structure of the active site appears to be essential; this should be followed by a study of the interactions of the amino acid side chains present at the active site with both coenzyme and substrate or, preferably, with the substrate-coenzyme appearing during the course of the catalytic reaction.

We shall illustrate here two approaches to these problems: (a) chemical methods used for the elucidation of the structure of the active center, with particular reference to the spatial relationships between different amino acid side chains; and (b) the binding to the protein of compounds resembling a substrate-coenzyme complex studied in an attempt to obtain, from the thermo-dynamic parameters of binding, data on the nature of the interactions occurring at the active center.

CHEMICAL STUDIES ON THE ACTIVE SITE CONSTITUTION

In a systematic investigation of the importance of the different amino acid residues for the catalysis, we first studied, with the usual techniques, the effect of specific modifications of the functional groups of the protein on the activity. Similar experiments have been performed in other laboratories, particularly by Braunstein [12]. The results of this research indicate that thiol [2], indole [3] and carboxyl [4] groups are not directly involved in the catalytic process, even though reactions leading to their modification partially inactivate the enzyme as a result of conformational changes of the protein.

A lysine amino group, a histidine and a tyrosine appear instead to be essential for the enzymic function, either for the catalysis or for the binding of the coenzyme. The amino group, which is available to reagents only in the apoenzyme [5], participates in an aldimine bond with pyridoxal phosphate [6]. It has been proposed that this group plays some role as an acid-base catalyst for the rate-limiting step of the reaction [7].

Photo-oxidation experiments indicate that one histidine residue is essential for activity [8-10].

One tyrosine is apparently involved in the coenzyme-apoenzyme interaction, on the basis of nitration experiments with tetranitromethane [11]. Similar evidence has been obtained from circular dichroism measurements [12]. The problem in the interpretation of this kind of experiment, involving selective modifications with chemical reagents, is to differentiate between inactivation due to blockage of a group which is actually at the active site, and inactivation due to a protein conformational change. It is highly desirable, therefore, before drawing any conclusion as to the importance of a particular amino acid in the enzymic function, to establish, with a different method, the presence of that particular group at the catalytic site. With this approach in mind, we proceeded to test bifunctional reagents which may react first with one group, the presence of which at the active center has been established (for aspartate aminotransferase this group is the coenzyme-bound lysine), and then with a second closely adjacent group: such a crosslink should definitely prove that the particular side chain is located in the proximity of the active center and should therefore substantiate the conclusion reached with the previous experiments. Moreover, experiments of this sort would allow the building of a three-dimensional "map" of the active center, indicating the approximate (owing to possible side chains mobilities) distances between two groups at the active center. Compounds should be chosen with two reactive groups displaying different reactivities so as to allow the crosslinkage reaction to take place in two steps and to allow, if required, isolation of the product of the first step of the reaction; it is desirable, furthermore, to have a series of bifunctional compounds with different distances between the two reactive functions. Acid chlorides of ω-halogenated carboxylic acids fulfil these requirements; under particular conditions they can be used to acylate the lysine amino group at the active site of apo-aspartate amino-transferase selectively [5]; the second function is then able to alkylate a second amino acid side chain located nearby; this alkylated compound is likely to be stable to acid hydrolysis and can be identified on the amino acid analyzer. Difluorodinitrobenzene (FFDNB), a reagent extensively investigated by Zahn and Meierhofer [13] and Marfey et al. [14], is also useful, since upon reaction it leads to products with well-defined absorption spectra, stable to acid hydrolysis. Since this compound has an aromatic nature and a reactivity differing from that of the above-mentioned reagents, it is possible that the crosslinkage reaction will be directed towards a different amino acid, as in fact has been found with aspartate transaminase. The results of the use of such compounds with apoenzyme are summarized in Table 1: crosslinkages are formed between lysine and histidine with the reagent bromopropionyl chloride, and between lysine and

Table 1. Effect of bifunctional reagents on the apoenzyme.

Treatment of apoenzyme	Amino acid contents of protein hydrolysate			Amino acid derivative in protein hydrolysate
	Histidine	Tyrosine	Lysine	
3-Bromo-propionyl-chloride	6.7-7.1		‡	carboxyethyl histidine
3-Bromo-propionate*	8.1-8.5		‡	—
FFDNB		8.55-8.9	19.5-19-5	2,4-dinitrophenylen-1-N-lysine-5-O-tyrosine (and 1-fluoro-2,4-dinitrophenyl-5-O-tyrosine)
FFDNB†		10.2-10.3	20.7-20.3	1-fluoro-2,4-dinitrophenyl-5-O-tyrosine

The results of two experiments are shown. Experimental details are reported in refs. 11 and 34.

* No crosslinkage occurs because the reaction with the lysine at the active site cannot take place with this monofunctional reagent.

† No crosslinkage occurs because in these experiments the lysine at the active site was blocked by acetylation before the treatment with FFDNB.

‡ Acylated lysine is freed by the hydrolysis.

tyrosine with FFDNB. When the primary reaction with the lysine is precluded, by using a monofunctional alkylating reagent, or by blocking the lysine with an acetyl group, the reaction with that particular histidine or tyrosine does not occur. It is noteworthy that no crosslinkage reaction can be obtained by the use of reagents with shorter or longer chains (i.e. bromo acetyl chloride or bromo butyryl chloride). While these results set some limit to the distance separating the amino and the imidazole, or the phenol groups, it should be kept in mind before assuming a definite value for the distance between the groups that the apoenzyme may have a slightly different conformation from the holoenzyme and that the histidine side chain may have a certain degree of mobility; this is also shown by the formation of both 1- and 3-carboxyethyl histidine. These experiments, however, substantiate the view that both an imidazole and a phenol side chain take part in the formation of the active center. With respect to the role that these groups may play in the function of the enzyme, it can be assumed that histidine is required in the prototropic shift which constitutes the rate-limiting step in transamination, as shown by Bruice and Topping for model systems [15]. The role of the phenolic group is at present obscure, but is probably concerned with the interaction between the coenzyme and the protein.

The aromatic nature of this group might suggest that its interaction with the coenzyme leads to a stabilization of one of the intermediate complexes. Braunstein [12] has recently advanced the hypothesis that the phenolic hydroxyl forms a hydrogen bond with the pyridine nitrogen.

At the present time no reaction model has, to our knowledge, made use of phenolic groups to catalyze physidoxal-phosphate dependent reactions. It should also be mentioned that we found phenols (phenol and tyrosine derivatives) to be efficient quenchers of the fluorescence of pyridoxamine; this may be related to the strong quenching that occurs upon binding of the coenzymes to apotransaminase [16].

BINDING OF SUBSTRATE-COENZYME ANALOGS TO THE APOENZYME

While the identification of the amino acids at the active center is a necessary step in the understanding of the protein contribution to the catalysis, the way these amino acids work is not usually immediately apparent; the study of model systems may provide some suggestions, but definite progress would be made if the interactions of the various side chains with the different parts of the substrate and coenzyme were known. Thermodynamic parameters for substrate or coenzyme binding suggest the types of bonds established at the active center; moreover, this sort of data is essential for the quantitative description, not only of the binding, but also of the catalytic process itself.

Unfortunately, real substrates seldom provide useful thermodynamic data; the "binding constants" being measured are often composite constants, owing to the presence of several indistinguishable intermediate complexes.

This is probably also the case with some of the inhibitors, or quasi-substrates, which have been used with aspartate aminotransferase, such as α-methyl aspartate, [17, 18] or β-chloroglutamate [19].

Stable compounds structurally resembling the coenzyme-substrate complexes formed during the course of the enzymatic reaction should meet the requirements for binding studies. The pyridoxyl amino acids, reduction products of the imines formed between pyridoxal (or pyridoxal phosphate) and amino acids, are stable compounds, with structures similar to that generally postulated for the initial substrate-coenzyme complex formed during the enzymatic reaction; they actually resemble a hypothetical tetrahedral complex [7, 20] rather than the Schiff base between the amino acid and pyridoxal phosphate.

Phosphopyridoxyl amino acids have been isolated from transaminase reduced with sodium borohydride in the presence of substrates ([21], and Torchinsky and Melakova, quoted in ref. 20). Binding of phosphopyridoxyl amino acids to the apotransaminase has been demonstrated by Khomutov [22], and to apotryptophanase by Morino and Snell [23]. We have found that phospho-pyridoxyl-L-aspartate (PPyl-L-asp), PPyl-L-alanine and PPyl-D-alanine have a high affinity for the apotransaminase [24]. The rate of combination with the apoenzyme is similar in all these cases, although L-aspartate is the real substrate, L-alanine is a poor substrate and D-alanine is no substrate at all. At pH 7.5 in 0.07M triethanolamine-HCl buffer, the reaction is practically irreversible, since it prevents the reactivation of the apoenzyme even with a large excess of coenzyme. However, under different conditions (pH 5.3, 1M phosphate buffer), in which the equilibrium favors the dissociation, it has been possible to show that the PPyl-asp-apoenzyme complex has a markedly lower dissociation rate than the PPyl-ala-apoenzyme complexes, as expected.

Upon binding to the apoenzyme the fluorescence of the PPyl derivatives is strongly quenched, as is also that of the coenzymes [16]. All these data point to the conclusion that the PPyl derivatives bind at the active site of the enzyme in a very similar way to the substrate-coenzyme complex; therefore the interaction between these derivatives and the protein must be of the same kind as the interaction between the substrates and the coenzyme and the protein. Further-more, it is reasonable to assume that only a single complex is present: the equilibrium data and the thermodynamic parameters should therefore be of some interest in an understanding of the binding forces.

A further favorable characteristic of the PPyl derivatives is their high affinity for the enzyme, which allows the introduction at the active center of "poor" substrates, i.e. of amino acids lacking one or another of the functional groups of the real substrates; the contribution to the binding of the different functional groups should in this way be evaluated.

As a matter of fact, the dissociation constant of the complex apoenzyme-PPyl-asp is so low (less than 10^{-9}M) as to be hardly measurable. The non-phosphorylated form, however, (Pyl-asp), shows a measurable binding, with K_{diss} of the order of 10^{-5} M at pH 8.5 in triethanolamine-HCl buffer.

The binding has been observed in two ways, either by gel filtration experiments on Sephadex G-100 (which gives a rather low accuracy owing to the low concentration of ligand which may be used); or by kinetic measurements, by determining the inhibition caused by PPyl-asp on the following reactions catalyzed by the apoenzyme, described for the first time by Wada and Snell [25]:

$$\text{Pyridoxamine} + \alpha\text{-ketoglutarate} \; \rightleftarrows \; \text{Pyridoxal} + \text{glutamate.}$$

The reaction rates have been measured by the method of Wada and Snell [25]. Varying the pyridoxamine concentration (between 3 and 13 mM), straight lines are obtained in the double reciprocal plot ($1/v$ versus $1/[\text{Pyridoxamine}]$).

The initial low rate for this reaction is compatible with a mechanism implying the formation of a ternary complex [26]. Pyl-asp behaves as a competitive inhibitor towards pyridoxamine, according to the following rate expression:

$$\frac{E_0}{v} = \frac{1}{k_5} + \frac{1}{k_7} + \frac{1}{k_1[PM]} \left(1 + \frac{[Pyl\text{-}asp]}{K_I}\right) \left(1 + \frac{k_2}{k_3} \frac{\left(1 + \frac{k_4}{k_5}\right)}{[KG]}\right) + \frac{\left(1 + \frac{k_4}{k_5}\right)}{k_3[KG]}$$

where PM indicates pyridoxamine, KG ketoglutarate, K_I is the dissociation constant for Pyl-asp and the other notations are the ones used by Alberty [26].

The dissociation constant for the Pyl-asp-apoenzyme complex has been determined as a function of temperature. To estimate the contribution to the

Table 2. Interaction of pyridoxyl derivatives with apo-aspartate aminotransferase. Parameters for the formation of the complexes.

	$\Delta F°$ at 10°C kcal/mole	$\Delta H°$ kcal/mole	$\Delta S°$ e.u.
Pyridoxyl-L-aspartate	− 6.35 ± 0.05	+ 2.9 ± 0.6	+ 32.7 ± 2.2
Pyridoxyl-L-alanine	− 4.0 + 0.65		

binding of the β-carboxyl group of the substrate, the interaction between Pyl-L-alanine and apoenzyme has been tested; a measurable binding is observed also in this case although the low accuracy allows only a rough estimate of the equilibrium constant for the complex. The data obtained are reported in Table 2. A transaminase with a different specificity has also been tested, in order to see if these interactions present common features in different enzymes. Tyrosine apo-transaminase, the inducible enzyme from rat liver [27], permits a study of the interaction with both the phosphorylated and non-phosphorylated forms of the pyridoxyl derivative of its substrate (PPyl-tyrosine and Pyl-tyrosine). The dissociation constants have been determined as the competitive inhibition constants (towards pyridoxal phosphate) for the reaction:

Pyridoxal phosphate + apoenzyme $\rightleftarrows$ Holoenzyme.

The data obtained are reported in Table 3.

The data must, of course, be interpreted with caution since the effect of pH has not yet been studied, so that our values might include heats and entropies of ionization of some functional groups of the enzyme, or more probably of the pyridoxyl derivatives.

Effects of ionic strength should also be considered: in accordance with Jenkins and D'Aril's finding [28], an increase in ionic strength appears to

Table 3. Interaction of pyridoxyl derivatives with apo-tyrosine aminotransferase. Parameters for the formation of complexes.

	$\Delta F°$ at 10°C kcal/mole	$\Delta H°$ kcal/mole	$\Delta S°$ e.u.
Phosphopyridoxyl-L-tyrosine	− 10.24	+ 8.5 ± 1.1	+ 66 ± 4
Pyridoxyl-L-tyrosine	− 4.48	+ 2.3 ± 0.1	+ 24 ± 0.4

increase the dissociation constant for the Pyl-asp-apo-aspartate aminotransferase complex.

A common feature for all the complexes examined is that the driving force for their formation is the positive entropy factor, which overcomes the unfavorable enthalpy variation. The same is known to happen in the formation of complexes between the holo-aspartate transaminase and quasi-substrates, i.e. α-methyl aspartate [18] and β-chloroglutamate [19]. In this respect, interactions between substrates and enzymes display a highly variable behavior in different systems: thus the complex carbonic anhydrase-carbonic acid has very large negative $\Delta H°$ and $\Delta S°$ of formation [29]; acetylcholinesterase and its substrates have a negative $\Delta H°$ and a negative or slightly positive $\Delta S°$ [30]; pepsin and its synthetic substrates [31] and ribonuclease and cytidine-3′-phosphate [32] have, as in our case, positive $\Delta H°$ and $\Delta S°$. Again, great caution must be exercised in the interpretation of these data, since most of them, with the exception of the data on ribonuclease [32], have been obtained by assuming equilibrium constants equal to Km and using pH-dependent constants.

The origin of the increase in entropy in the formation of the complexes is still uncertain. In most cases [33], it has been interpreted as a loosening of the protein structure and this seemed plausible also in the case of formation of complexes between aspartate transaminase and quasi-substrates [18], since the formation of an aldimine bond between substrate and coenzyme, instead of that between protein and coenzyme, is in fact consistent with this view. It is, however, difficult to vizualize such loosening of the protein structure upon binding of the Pyl derivatives with the apoenzymes, which are notoriously much more labile than the corresponding holoenzymes. Probably local factors are here more important in determining the entropy increase.

It is interesting to note that the entropy factor is also solely responsible for the formation of the complex apotyrosine-transaminase-PPyl-tyrosine, which has a very low dissociation constant.

A similar, unusually high increase in entropy has been found for the formation of the complex between ribonuclease and cytidine-3′-phosphate [32], and it is perhaps not simply a matter of chance that in this case as well the substrate is a phosphoric ester.

More valuable for our purposes, perhaps, are the data on the contribution of a particular group to the binding. The comparison of binding of Pyl-asp and Pyl-ala to the apo-aspartate transaminase should give an estimate of the standard free energy of binding of the β-carboxyl of aspartate; this amounts to $-2.35 + 0.7$ kcal/mole at 10°C, at an ionic strength of 0.04M.

The comparison of binding of PPyl-tyr and Pyl-tyr to the apo-tyrosine transaminase should give an estimate of the standard free energy of binding of the phosphate group of the coenzyme; it amounts to about -5.8 kcal/mole at 10°C. If these data were also valid for aspartate transaminase (for which the binding of the phosphate forms cannot be measured, as mentioned above) the dissociation constant of the complex PPyl-asp-apo-aspartate transaminase should be of the order of magnitude of 10^{-10}M.

The interpretation of data on the binding of a particular group of the substrate or coenzyme is probably complicated by the fact that the binding of Pyl derivatives lacking one or another of their functional groups may lead to a different conformation of the substrate-coenzyme complex at the active center, or of the protein; so that the energetic contributions, for example, of the β-carboxyl of the Pyl-asp, as appears from our data, may in fact originate not only from the specific bond of the carboxylate group with its binding site, but also by a strain imposed on the substrate or on the complex (strain which is absent in the less "normal" Pyl-ala complex), or by a different protein conformation or similar additional factors.

It is clear that the data presented here are by no means complete or sufficient to present even a rough picture of the interactions at the active center. They are mainly intended to illustrate a possible approach, which seems to be particularly helpful in the case of the pyridoxal phosphate enzymes, and which we are pursuing at present not only on transaminase but also on enzymes with a different specificity of action.

REFERENCES

1. Fasella, P., A. Rev. Biochem. 36 (1967) 185.
2. Turano, C., Giartosio, A. and Fasella, P., Archs Biochem. Biophys. 104 (1964) 524.
3. Turano, C., Giartosio, A., Riva, F., Bossa, F. and Baroncelli, V., in "Pyridoxal Catalysis: Enzymes and Model Systems", Proceedings of 2nd IUB Symposium, Moscow, September 1966 (edited by E. E. Snell, A. E. Braunstein, E. S. Severin and Yu. M. Torchinsky), Interscience, New York, 1968, p. 143.
4. Bossa, F., Barra, D., Vecchini, P. and Turano, C., in preparation.
5. Turano, C., Giartosio, A., Riva, F. and Baroncelli, V., Biochem. J. 104 (1967) 970.
6. Jenkins, W. T. and Sizer, I. W., J. Am. chem. Soc. 79 (1957) 2655.
7. Snell, E. E., Brookhaven Symp. Biol. 15 (1962) 32.

8. Martinez-Carrion, M., Riva, F. and Fasella, P., Abstracts of the American Chemical Society Meeting, Chicago, 1966.
9. Martinez-Carrion, M., Turano, C., Riva, F. and Fasella, P., *J. biol. Chem.* **242** (1967) 1426.
10. Vorotniskaya, N. E., Lutovinova, G. F. and Polyanovsky, O. L., *in* "Pyridoxal Catalysis: Enzymes and Model Systems", Proceedings of 2nd IUB Symposium, Moscow, September 1966 (edited by E. E. Snell, A. E. Braunstein, E. S. Severin and Yu. M. Torchinsky), Interscience, New York, 1968, p. 131.
11. Turano, C., Giartosio, A., Riva, F., Barra, D. and Bossa, F., Abstracts of the International Symposium on Pyridoxal Enzymes, Nagoya, 1967, p. 2.
12. Braunstein, A. E., this volume, p. 101.
13. Zahn, H. and Meierhofer, J., *Makromolek. Chem.* **26** (1958) 126.
14. Marfey, P. S., Nowak, H., Uziel, M. and Yphantis, D. A., *J. biol. Chem.* **240** (1965) 3264.
15. Bruice, T. C. and Topping, R. M., *J. Am. chem. Soc.* **85** (1963) 1480.
16. Fasella, P., Turano, C., Giartosio, A. and Hammadi, I., *G. Biochim.* **10** (1961) 175.
17. Fasella, P., Giartosio, A. and Hammes, G. G., *Biochemistry* **5** (1966) 197.
18. Fasella, P., Khomutov, R. M., Brunori, M. and Antonini, E., in preparation.
19. Hammes, G. G., *Biochim. biophys. Acta* **146** (1967) 312.
20. Braunstein, A. E., *Vitams Horm.* **22** (1964) 451.
21. Riva, F., Vecchini, P., Turano, C. and Fasella, P., Abstract IV-140, Sixth International Congress of Biochemistry, New York, 1964, p. 329.
22. Khomutov, R. M., Symposium on Peptide and Protein Chemistry, Riga, 1967.
23. Morino, Y. and Snell, E. E., *J. biol. Chem.* **242** (1967) 5591.
24. Turano, C., Giartosio, A. and Orlacchio, A., in preparation.
25. Wada, H. and Snell, E. E., *J. biol. Chem.* **237** (1962) 127.
26. Alberty, R. A., *J. Am. chem. Soc.* **75** (1953) 1928.
27. Hayashi, S., Granner, D. K. and Tomkins, J. M., *J. biol. Chem.* **242** (1967) 3598.
28. Jenkins, W. T. and D'Aril, L., *J. biol. Chem.* **241** (1966) 5667.
29. Kiese, M., *Biochem. Z.* **307** (1941) 400.
30. Wilson, I. B. and Cabib, E., *J. Am. chem. Soc.* **78** (1956) 202.
31. Casey, E. S. and Laidle, K. J., *J. Am. chem Soc.* **72** (1950) 2159.
32. Cathou, R. E. and Hammes, G. G., *J. Am. chem. Soc.* **87** (1965) 4674.
33. Hammes, G. G., *Nature, Lond.* **204** (1964) 342.
34. Turano, C., Giartosio, A., Riva, F., Baroncelli, V. and Bossa, F., *Archs Biochem. Biophys.* **117** (1966) 678.

FEBS Symposium, Volume 18, 1970, pp. 133-150

Factors Affecting the Electrophoretic Pattern of Lactate Dehydrogenase Isoenzymes*

J. CLAUSEN

The Neurochemical Institute, Copenhagen,
Denmark

INTRODUCTION

Multiple molecular forms of enzymes are hybrids of parent subunits combinable to enzymic active polymers. The mechanism controlling the formation of these systems remains unknown. The lactate dehydrogenase (LDH)–(L-lactate; NAD oxidoreductase, E.C. 1.1.1.27)–consists of five electrophoretically-distinguishable fractions [1].

There are two primary parent subunits of LDH which are combined together in groups of four to give the active enzymic structure [2-4]. These two primary parent proteins have been designated H and M (or A and B) as they are individually prominent in heart muscle and skeletal muscle respectively.

Evidence has been gained in support of this theory, i.e. freezing experiments in which equal amounts of the H_4 and M_4 tetramers were frozen in phosphate buffer containing NaCl [5]. The isoenzyme pattern found after thawing of equal amounts of H and M subunits may be such that the isoenzyme activities were in the proportion 1:4:6:4:1, the binomial distribution. It has subsequently been found that guanidine-HCl and urea in high concentrations can also produce hybridization [6, 7]. Assuming dissociation, it should then be possible to measure the recombination of subunits by measuring the recovery of activity following freezing. Also, if it is assumed that the NADH molecule plays an intimate role in the conformational state of the enzyme, it might be possible to ascertain more explicitly certain aspects of this relationship [8-10, 35, 36].

As changes in LDH isoenzyme distribution occur under certain conditions, e.g. during foetal development [3, 11, 33], in malignancy [12-14, 37, 38], in muscle degeneration [15, 16], and under anaerobic conditions [17, 39], a study of the factors governing LDH recombination from the individual subunits may throw some light on these biologically-occurring changes in the LDH-pattern.

* These studies have been supported by a grant from the Danish Cancer Society, Copenhagen, Denmark.

The first part of the present communication is an investigation of the relationship between NADH and LDH subunit recombination as determined by using the procedure of hybridization (for details, cf. ref. 11).

MATERIALS AND METHODS

Materials

Macroscopically normal human tissue (heart and liver) was obtained at autopsy within twelve hours after death.*

All chemicals used were of the highest obtainable purity from British Drug Houses with the exception of ADP, NAD and NADH which were obtained from Boehringer, Germany.

Methods

Assay of total LDH activity and LDH-isoenzymes has previously been described [18].

Hybridization

To obtain an optimal hybrid effect with an approximately binomial distribution following hybridization, liver and heart samples, i.e. tissues with predominant and high M_4 and H_4 concentrations, were taken together to make the extract. 1 g samples of both tissues were homogenized in 10 ml of 0.1M Na-phosphate buffer, pH = 7.0, containing 0.01M NaI [10]. Homogenization was performed using a rounded glass stirring rod (40 rpm) to crush the tissue in a Hostalen centrifuge tube. The homogenate was then centrifuged at 4°C for 20 min at 18,000 g in a Beckman Centrifuge Model L-2. The supernatant was collected and diluted with phosphate-iodide buffer to the desired enzymic activity. 20 ml of this diluted extract was then dialyzed for 16 h at room temperature against 500 ml of the phosphate-iodide buffer containing 5 g of activated charcoal. This was found to remove essentially all of the "free" NADH as determined by absorption at 366 nm of a pure NADH sample. Freezing was performed in a deep-freeze at − 20°C for approximately 30 min to 2 h (see below).

Effect of NADH on the rate of activity recovery following freezing

It was found necessary to work with the enzymic activity of the extracts within certain limits, as too high an enzyme concentration resulted in immediate recovery, making kinetic studies impossible, whereas too low an activity resulted in complete inactivation with no subsequent recovery. An enzyme activity level of approximately 0.554 units/ml extract was found to be satisfactory.

* Gratitude is expressed to Prof. H. Poulsen, the Institute of Pathological Anatomy, Municipal Hospital, Copenhagen, for these samples.

Both undialyzed and dialyzed extracts were frozen for 2 h at $-20°C$. At time intervals (see below) during thawing, 0.5 ml samples were removed and added to 2 ml 0.9% NaCl at room temperature and total LDH activity measured. For the dialyzed extract, 250 μl were added to 1 ml 0.9% NaCl containing 2.82 μmoles NADH and total LDH activity measured as above. At the time of optimal recovery, the isoenzyme pattern of each extract was determined.

Effect of prior addition of NADH, nicotinamide and ADP on the extent of activity recovery following freezing

The enzyme activity was so diluted that the dialyzed extract exhibited complete loss of activity with no recovery following freezing. An original enzyme activity level of approximately 0.337 units/ml extract was found to lead to complete loss of activity with no recovery.

The binding of NADH was elucidated by studies of the influence of components possessing chemical groups identical with the end groups of NADH. Thus, the following extracts were examined: the undialyzed and dialyzed extracts to which (a) NADH was added prior to freezing to final concentrations ranging from 1-30 mM; (b) nicotinamide was added to final concentrations ranging from 1-20 mM; and (c) ADP was added to final concentrations ranging from 1-20 mM. These extracts were then divided into two fractions with the one fraction frozen as described above, and the other allowed to stand at room temperature. The frozen samples were thawed for 4 h at room temperature to allow for maximum activity recovery. Total activity and the isoenzyme patterns were determined on all samples.

Effect of coenzyme (NAD and NADH) on the mobility of isoenzymes

NAD and NADH were added to the dialyzed extracts to final concentrations ranging from 1-65 mM. They were then run in agar-gel microelectrophoresis and their relative mobilities determined, using albumin and dextran as reference markers.

RESULTS

Effect of NADH on the rate of activity recovery following freezing

As seen in Fig. 1, there is a 25% increase in the activity recovery in those dialyzed samples which were treated with NADH following freezing. Optimal activity was reached after 50 min.

The activity recovery in the undialyzed extract was equal to that found in the dialyzed extract to which NADH was added. In all instances, following freezing, the isoenzyme patterns showed complete hybridization.

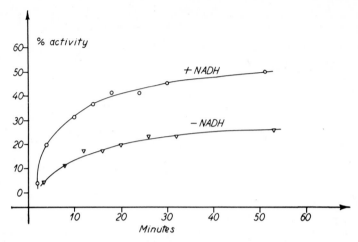

Figure 1. The relationship between recovery of lactate dehydrogenase activity after freezing of equal amounts of H and M subunits of lactate dehydrogenase from heart and liver tissue as a function of time after freezing.
Abscissa: time after freezing;
Ordinate: percentage of recovery of LDH activity.
The activity recovery was measured in an assay system in the absence or presence of the coenzyme NADH (for experimental conditions, see text).

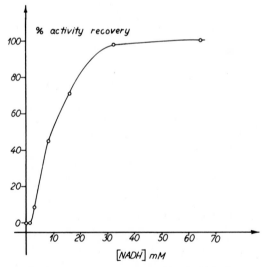

Figure 2. The relationship between percentage of lactate dehydrogenase activity recovery after freezing of equal amounts of H and M subunits from heart and liver tissue, respectively, as a function of concentration of NADH.
Abscissa: mM NADH (for experimental conditions, see text);
Ordinate: percentage of activity recovery.

Effect of prior addition of NADH, nicotanamide and ADP on the extent of activity recovery following freezing

When the dialyzed samples were frozen without prior addition of exogenous NADH, there was complete loss of activity with no recovery. As seen in Fig. 2, the addition of NADH to the dialyzed extract prior to freezing resulted in activity recovery which was a function of the amount of coenzyme added. The extent of recovery approached that of the undialyzed extract at high coenzyme levels (above 40 mM).

The addition of NADH to the undialyzed extract prior to freezing had no significant effect on activity recovery. In the undialyzed, unfrozen extract to which NADH was added there was no significant activity change.

When ADP and nicotinamide were added to the dialyzed extract prior to freezing there was no effect, i.e. there was still total loss of activity without recovery. When ADP was added to the undialyzed extract prior to freezing, there was similarly no effect on activity recovery. The addition of nicotinamide to the undialyzed extract produced, however, a progressive loss of activity, which was a function of concentration, as seen in Fig. 3. The inhibition was not selective; it depressed all five isoenzymes to the same extent, as determined by the H/M ratio.

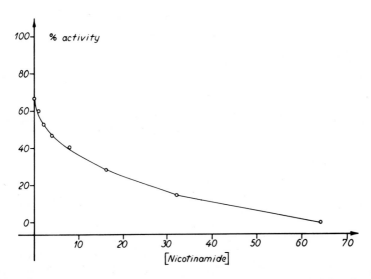

Figure 3. The percentage of lactate dehydrogenase activity recovered after freezing of equal amounts of H and M subunits from heart and liver tissue, as a function of concentration of nicotinamide.
Abscissa: mM nicotinamide;
Ordinate: percentage of activity recovery.

Effect of exogenous NAD and NADH on mobility of isoenzymes

As shown in Fig. 4, NADH has an enhancing effect on the mobility of all five isoenzymes. This effect is a function of the amount of coenzyme added and isoenzymes H_4, H_2M_2, HM_3 and M_4 appear to be affected to the same extent. Isoenzyme H_3M, however, experiences a much greater initial effect after which it too exhibits a mobility shift paralleling the other isoenzymes.

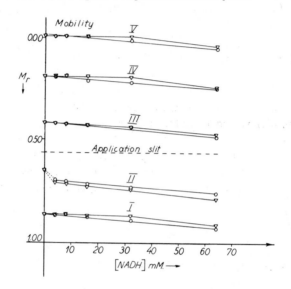

Figure 4. Relative electrophoretic mobilities of lactate dehydrogenase isoenzymes as a function of concentration of NADH in an extract.
Abscissa: concentration in mM NADH;
Ordinate: relative electrophoretic mobility using albumin and dextran as reference markers possessing electrophoretic mobilities M_r = 1.00 and 0.00 respectively.
I: H_4, II: H_3M, III: H_2M_2, IV: HM_3, V: M_4 isoenzymes.

DISCUSSION OF HYBRIDIZATION

The results of the present communication suggest that NADH plays a role in the enzymic reconstitution of LDH activity from LDH-subunits. The addition of NADH to an enzyme extract of heart and liver, either prior to, or following freezing, resulted in a more rapid and pronounced activity recovery. Indeed, it was possible by addition of NADH to recover activity in an extract which showed no activity without addition of NADH.

These results are supported by Chilson *et al.* [6, 19], who produced hybridization using guanidine-HCl or urea. They proposed that the NADH molecule served to stabilize the tertiary structure of the LDH molecule and that, in the absence of NADH, subunits remained free in solution and/or as inactive polymers in equilibrium with the enzymically-active tetramer.

The fact that the addition of either nicotinamide or ADP resulted in acitivity recovery suggests that the entire NADH molecule is necessary for the stabilizing effects.

The progressive loss of activity produced by increasing concentrations of nicotinamide in the undialyzed frozen extract suggests, furthermore, that it is the pyridine group which contains the site of attachment between the NADH and LDH molecules. This was also suggested by Shifrin et al. [20], using fluorescent studies.

The fact that the exogenous addition of NADH to an extract produces a progressive increase in electrophoretic mobility suggests that there are additional sites of attachment in the tertiary structure for NADH molecules which are not essential for enzymic activity. Evidently a number of these sites are most readily available in isoenzyme H_3M, possibly indicating a more unfolded active configuration.

In most tissues, with an optimal amount of NADH, the LDH isoenzyme pattern may be determined solely by the synthetic rate and concentration of the H and M subunits. However, in human testes and sperm cells a sixth LDH isoenzyme (LDH_x) has been described [21, 22, 34, 35]. As this isoenzyme also consists of subunits determined by a specific gene locus, and since it is present in tissue together with LDH containing H and M subunits [23], a special compartmentalization may explain why the LDH_x does not lead to the appearance of significant amounts of hybrids with the H and M subunits.

The second part of the present communication deals with this question in two experimental approaches, viz. studies of the possibility of a direct extraction of LDH from particulate fractions; and studies of LDH pattern influenced by the particulate changes induced by deficiency in essential fatty acids (EFA).

MATERIALS AND METHODS

Materials

Human sperm was obtained from the laboratory of the Copenhagen Health Insurance Society. For routine assays of LDH isoenzymes, the human sperm cells were isolated and extracted with 0.9% aqueous NaCl solution as previously described [18].

Adult Wistar rats (200 to 300 g) were used for studies of compartmentalization of LDH isoenzymes in particular fractions of testes. The testes were isolated by ligation and cutting off of the epididymis and associated vessels during mild ether narcosis. In the dietary studies mentioned below, adult male Wistar rats, 200 g in weight, were used. They were kept in plastic cages on a metal grating, allowing elimination of feces. During the experimental period, experimental and control animals were sacrificed by decapitation and the testes excised.

Diet

Normal female Wistar rats were bred with the same male. Fourteen days prior to delivery, one group (I) of pregnant females was fed a diet deficient in essential fatty acids (EFA) and a control group (II) was fed a normal diet (Table 1). The new-born rats of both sexes, respectively, were selected for continuous feeding from each of the two nutritional groups. During a three to eight month period the animals in both groups were reared *ad libitum* on the two diets.

Table 1. Composition of diets (g/100 g).

Normal diet*		Diet deficient in PUFA	
Barley	20.5	Casein low vitamin§	24.0
Oats	20.0	Sucrose	72.1
Ground wheat bran	20.0	Salt mixture†	3.9
Lucerne flour	20.0	Vitamin mixture	‡
Ground maise	10.0		
Ground sun-flower seed	5.0		
$CaCO_3$	1.0		
Fish flour	1.0		
Bone flour	1.0		
Peanut cakes	1.0		
Salt mixture†	0.5		
Vitamin mixture	‡		

* 26.1% of the total fatty acids is in series $C_{16:1\omega7}$ and $C_{18:1\omega9}$ unsaturated fatty acids 55.4% of total fatty acids is linoleic and linolenic acid derivatives
† No. 2 U.S. Pharmacopoeia XIII
‡ The Royal Danish Veterinary Food-control recommended diet containing per 100 g:

1,200 i.u.	A vitamin
400 i.u.	D vitamin
200 μg	thiamin
700 μg	riboflavin
4 mg	niacinic acid
1.5 mg	DL-panthothenic acid
40 mg	Choline-HCl
3.5 mg	DL-α-tocopheryl acetate
1 μg	vitamin B_{12}

§ Fat free (below 0.02% fat, extracted chloroform + methanol (2 + 1) polyenoic fatty acids below 0.001% of total diet).

Methods

Assay of LDH acitivity as above.

Assay of succinate dehydrogenase (SDH) was used as an arbitrary measure of mitochondria content of particulate fractions. The succinate dehydrogenase (succinate: (acceptor) oxidoreductase E.C. 1.3.99.1.) was assayed as described by Green *et al.* [24] using dichlorophenol-indophenol (DPIP) as electron acceptor. The reaction mixture (2.0 ml) for routine assay consisted of 40 μl particulate fraction, 20 mM Na-succinate, 35 μM 2,6-dichlorophenol-indophenol, 2.0 mM

KCN, 0.05M EDTA (ethylene-diamine-tetra-acetic acid (disodium salt)), and 0.5 g bovine albumin (Behringwerke, Germany) per litre 50 mM Na-phosphate buffer (pH 7.5). Dehydrogenase activity was assayed by reading the change in optical density at 30-s intervals for 5 min against a blank containing all components except succinate (temperature 25°C).

These assay conditions were fixed during systematic variation of pH and concentration of succinate and dichlorophenol-indophenol, as suggested by Singer and Kearney [25]. The concentrations used were those giving saturation with succinate and maximum activity with the indophenol. Higher concentrations of this component gave rise to inhibition of succinate dehydrogenase activity.

The optical density was read at 600 nm. On the basis of the extinction change at 25°C and the molar extinction coefficient of 2,6-dichlorophenol-indophenol at 600 nm of 1.91×10^7 cm^2/mole [26], the number of enzyme units present in 1 ml of the original sample was expressed as μmoles substrate transformed/min. The specific activity of the sample was expressed as μmoles substrate transformed/min/mg protein.

Isolation of head, middle piece and tail of human sperm cells
The sperm cells of pooled samples of fresh human sperm (6 to 10 ml) were collected by centrifugation at 2000 x **g** (10 min). The cells were washed three times with Ringer solution for spermatozoa [27]. The pellet of sperm cells was suspended in 10 vols. of 0.25M sucrose containing 1.0 mM EDTA and disintegrated in a Potter-Elvehjem disintegrator, equipped with a loosely fitting glass pestle, at 250 rpm for 2 min. The homogenate was fractionated by centrifugation as described by Mohri *et al.* [28]. The head fraction was obtained as the white bottom layer at 900 x **g** centrifugation (10 min), and the middle pieces were localized in the upper portion of the sediment. These two fractions were separated and isolated by centrifugation after resuspension in 0.25M sucrose containing EDTA. From the 900 x **g** supernatant of the crude homogenate, the tails were isolated by centrifuging in a Spinco Model L ultracentrifuge at 100,000 x **g** for 60 min. The fractions were controlled by electron microscopy.

Preparation of tissue suspension for ultracentrifugation and *preparation of subcellular fractions of testes* has been described previously [17].

The tissue suspensions were diluted with 30 ml of 0.32M sucrose. The nuclear fraction (N) was isolated by 10 min centrifugation at 900 **g** in a Sorvall RC-2 centrifuge. The collected supernatant was then centrifuged at 11,500 **g** for 20 min in a Spinco L-11 ultracentrifuge to obtain the crude mitochondrial sediment.

According to the density gradient centrifugation technique of DeRobertis *et al.* [29], the mitochondrial sediment (M) was resuspended in 10 ml of 0.32M sucrose and layered on a sucrose gradient 0.8–1.0–1.2–1.4M. Centrifugation at 50,000 **g** for 2 h produced in each gradient the corresponding mitochondrial subfractions, A-B-C-D and E respectively. The particulate mid-layers were removed separately for resuspension in 0.15M sucrose. After centrifugation at 10,000 **g** for 10 min, each subfraction was washed three times in 0.9% saline.

Enzyme extraction of particulate fractions and electron microscopy have been described previously [17, 40]. In some experiments the mitochondrial and microsomal pellets, after homogenization, were further extracted with 1.0 ml of 1.0% (W/V) Triton-X-100 in 0.9% (W/V) aqueous NaCl, rehomogenized, and centrifuged as above. The resulting supernatants were tested for total LDH activity, total protein, isoenzyme pattern and succinic dehydrogenase.

Gas chromatography

The fatty acid composition of the testicular lipids extracted with 20 vols. of chloroform + methanol (2 + 1) was evaluated by gas chromatography of the methyl esters. The fatty acids of the lipids were saponified with 2 ml 1N KOH in 96% ethanol for 30 min as described by Rathbone [30]. The fatty acids were liberated by addition of 1.5 ml 2N HCl + 5 ml H_2O. The fatty acids were dissolved in 10 ml ethyl ether and washed three times with 10 ml 0.1N HCl in a separatory funnel. The ether phase was dried above $CaCl_2$ for 16 h and concentrated to dryness below N_2. The fatty acids were afterwards methylated with 6 ml anhydrous methanol saturated with HCl gas. 250 μl dry benzene was added as catalyst. The methylation was performed for 30 min at 80°C in sealed bottles. The methyl esters were extracted and washed as described above. Finally, the esters were concentrated by evaporation of the ether under a stream of N_2. Gas chromatography was performed on samples containing 0.2 μl methyl esters in a Perkin-Elmer flame-ionization gas chromatograph F-11 (column dimensions 2 m x 1 mm, stationary phase 8% 1,4-butanediol succinate polyester on chromosorb W (80-100 mesh), carrier gas N_2, 190°C). The chromatograms were analyzed by triangulation and the proportions of the individual fatty acids expressed as percentages of total. The fatty acids were identified by logarithmic plotting and by means of standard markers of fatty acid methyl esters obtained from the Hormel Institute, Minnesota, U.S.A.

Statistical evaluation of results
The results obtained were evaluated on the basis of "Student's" t-distribution, using the following levels of significance: p = 0.001 as the level for highly significant data, p = 0.03 and 0.05 as levels for significant data.

RESULTS

Topographic studies of the occurrence of LDH in particulate fractions of sperm cells and rat testes homogenates

Topographic studies of sperm cells

Tables 2 and 3 demonstrate LDH activities of similar magnitude among the fractions obtained by ultracentrifugation. The supernatant from the differential centrifugation contains the highest LDH activity: six to fourteen times the activity of particulate fractions. The particulate fractions contain specific LDH activities of similar size. The specific SDH activity found was four times as high

Table 2. Total LDH and SDH activities of particular fractions of human sperm cells.

Fraction	LDH μmoles/min/mg prot.	SDH μmoles/min/mg prot.
Head: 0.9% NaCl ex.	1.12	0.31
Triton-X-100 ex.	1.16	
Middle piece:		
0.9% NaCl ex.	1.21	1.18
Triton-X-100	1.97	
Tail: 0.9% NaCl ex.	1.75	0.00
Triton-X-100	1.08	

in the middle piece fraction as in the head fraction. No succinate dehydrogenase activity was found in the tail fraction. The head fraction contains a higher ratio of total specific LDH activity to total specific SDH activity.

Agar-gel microelectrophoresis revealed the presence of both H- and M-containing LDH fractions in the supernatants from ultracentrifugation and in the head and tail fractions. However, the middle piece fraction contains only the LDH_x fraction, as seen in Table 4; therefore, the specific activity of the middle piece fraction refers exclusively to the LDH_x fraction. The LDH_x specific activity seems higher in the head fraction and in the supernatant (microsomal fraction) than in the middle piece fraction.

Electron microscopy of the particulate fractions obtained by ultracentrifugation revealed the presence of intact sperm cells beside pure head fragments in the head fraction. The middle piece fraction contained mainly small vesicles of variable size, sometimes equipped with internal cristae. The tail fraction contained long protracted fragments. The above-mentioned experiments seem to indicate that the middle piece fraction, containing the particulate SDH activity,

Table 3. The LDH and SDH activities of particulate fractions of rat testes.

Fraction	LDH μmoles/min/mg prot.	SDH μmoles/min/mg prot.
Nuclear fraction:		
0.9% NaCl ex.	0.75	0.00
Triton-X-100 ex.	0.06	
Mitochondrial fractions:		
A		
0.9% NaCl ex.	0.16	0.032
Triton-X-100	0.01	
B		
0.9% NaCl ex.	0.92	0.091
Triton-X-100	0.06	
C		
0.9% NaCl ex.	1.79	0.087
Triton-X-100	0.06	
D		
0.9% NaCl ex.	4.19	1.83
Triton-X-100	1.13	
E		
0.9% NaCl ex.	2.52	0.64
Triton-X-100	0.30	
Microsomes:		
0.9% NaCl ex.	0.27	0.034
Triton-X-100	0.46	

Table 4. Total LDH activity and distribution (%) of LDH isoenzymes in particular fractions of human sperm cells.

Fraction	Total LDH μmoles/min/mg prot.	H_4	H_3M	H_2M_2	HM_3	LDH_x	M_4
Head	0.73	0.0	0.0	21.0	5.3	73.7	0.0
Middle piece	0.29	0.0	0.0	0.0	0.0	100.0	0.0
Tail	0.31	0.0	0.0	26.1	6.5	65.2	2.2
Supernatant	4.26	0.0	0.0	19.4	3.1	77.5	0.0

also contains the LDH_x fraction. The presence of the LDH_x fraction in the head fraction may be explained by the admixture of the middle piece to that fraction.

After disintegration and extraction of the particulate fractions with 0.9% NaCl, the remnants isolated by centrifugation at 20,000 g for 30 min were extracted with 0.9% NaCl containing 1.0% Triton-X-100. In this way it was still possible to extract a high specific activity of the remaining parts of all particulate fractions (Table 3). As shown in Table 3, specific activities of the same magnitude as the NaCl extracts were obtained by re-extraction with Triton-X-100, but proportionally more LDH activity was extractable from the middle piece than from any other particulate fraction.

Topographic studies of the occurrence of LDH$_x$ *in particulate fractions of rat testes*

Table 3 demonstrates the highest specific SDH activity in the mitochondrial fraction D obtained by sucrose gradient ultracentrifugation of whole rat testes homogenate. These findings correspond well with the electron microscopical investigations revealing the mitochondrial fraction D to be rich in morphologic structures characteristic of mitochondria, e.g. of elipsoid structures internally equipped with cristae.

Table 5 demonstrates that the highest specific LDH activity is also localized in the mitochondrial subfraction D. However, the nuclear fractions also contain a high specific LDH activity. After extraction of the particulate fractions with 0.9% aqueous NaCl solution, the precipitate (washed three times) obtained by centrifugation at 20,000 g for 30 min was re-extracted with 1.0% Triton-X-100 dissolved in 0.9% aqueous NaCl. As shown in Table 5, the re-extraction with Triton-X-100 gives rise to a further liberation of LDH activity, especially pronounced in the mitochondrial subfraction D.

Agar-gel microelectrophoresis revealed an uneven distribution of LDH isoenzymes (Table 5) in the particulate fractions. Saline extraction of the particulate fractions resulted mainly in liberation of LDH isoenzymes containing H and M subunits. Thus the H$_4$ isoenzyme predominated in the nuclear and mitochondrial fractions A and B. However, the mitochondrial fractions C, D and E, and also the microsomal fraction, seem to exhibit a larger content of M subunit-containing isoenzymes. The LDH$_x$ was found only in saline extracts of the nuclear fraction and the mitochondrial fraction E.

Enzymic changes induced by EFA deficiency

Evaluation of EFA deficiency

During a period of three months in which group I rats were raised and fed a diet deficient in EFA, symptoms characteristic of EFA deficiency developed, including increased water intake, retarded growth, and skin eruptions. As further evidence of EFA deficiency, tissue analysis, carried out on these animals at the time of sacrifice, revealed decreased amounts of EFA (linoleic and linolenic acids) in testes tissue associated with a relative increase in oleic acid (Table 6). Thus the amount of oleic acid ($C_{18:1}$) was increased by 61%, while amounts of linoleic acid and linolenic acids decreased to 2.4 and 0.0%, respectively.

Table 5. LDH activity and LDH isoenzymes in particulate fractions of normal rat testes.

Fraction	0.9% NaCl ex. Total LDH-act.	Triton-X-100 ex. LDH-act.	0.9% NaCl ex. Distribution in per cent						
			H_4	H_3M	H_2M_2	HM_3	LDH_x	M_4
Nuclear	0.322	0.705	41.9	34.6	3.6	9.1	5.4	5.4
Mitochon-drial								
fraction A	0.112	0.414	100.0	0.00	0.00	0.00	0.00	0.00
fraction B	0.196	0.409	78.6	21.4	0.00	0.00	0.00	0.00
fraction C	0.596	0.244	46.6	53.4	0.00	0.00	0.00	0.00
fraction D	1.070	0.330	36.0	40.0	8.0	16.0	0.00	0.00
fraction E	0.535	0.406	16.7	10.4	6.3	29.2	18.7	18.7
Microsomes	0.430	0.437	41.6	58.4	0.00	0.00	0.00	0.00
Superna-tant of microsomes	0.393	0.830	34.6	46.2	11.5	7.7	0.00	0.00

Table 6. The fatty acid composition (%) of rat testes from a group of rats fed an optimal diet and from a group of animals fed a diet deficient in poly-unsaturated fatty acids (PUFA) during a period of twelve weeks.

Fatty acid	Normal diet (number 10)	Diet lacking PUFA (number = 10)
C_{12}	0.4 ± 0.3	0.5 ± 0.5
C_{13}	0.1 ± 0.07	1.1 ± 0.5
C_{14}	2.8 ± 0.8	3.4 ± 0.6
C_{15}	2.8 ± 1.9	0.3 ± 0.2
C_{16}	10.7 ± 3.1	43.9 ± 4.2
$C_{16:1}$	10.8 ± 2.1	11.3 ± 3.9
$C_yx)$	1.3 ± 0.9	0.0
C_{17}	0.9 ± 0.5	0.3 ± 0.3
C_{18}	2.5 ± 1.0	10.0 ± 3.2
$C_{18:1}$	16.3 ± 3.6	26.2 ± 3.2
$C_{18:2}$	17.9 ± 2.7	2.4 ± 0.4
$C_{18:3}$	8.5 ± 2.6	0.0
$C_xx)$	2.8 ± 1.8	0.0
C_{20}	7.3 ± 3.7	0.0
$C_{20:1}$	1.9 ± 1.0	0.0
$C_{20:2}x)$	2.6 ± 1.5	0.1 ± 0.09
$C_{20:3}x)$	4.7 ± 1.3	0.7 ± 0.3
$C_{20:4}$	0.8 ± 0.5	0.0
C_{22}	3.9 ± 3.9	0.0
$C_{22:4}$	0.3 ± 0.3	0.0
$C_{22:6}$	0.9 ± 0.6	0.0

Furthermore, the males showed a lack of sperm cells in the epididymic extracts, as evaluated by microscopy on samples obtained by puncture of the epididymis, followed by aspiration into a syringe.

Enzymic changes

Table 7 demonstrates the differences in total LDH and SDH activities in a group of rats raised on a normal diet and a group raised on a diet deficient in EFA. In

Table 7. The total LDH and SDH activities of normal and EFA-deficient rat testes (3 month EFA deficiency, see text). A statistically significant decrease in SDH occurs in the case of EFA deficiency.

	Total LDH μmoles/min/mg prot.	Total SDH μmoles/min/mg prot.
Normal diet N = 14	0.479 ± 0.026	0.104 ± 0.019
EFA deficient N = 16	0.541 ± 0.027	0.0697 ± 0.0058

the first group, 14 animals, and in the second group, 16 animals were studied. The means of the enzyme activities are indicated. From Table 7 it is obvious that a statistically significant decrease in SDH activity (p-level 0.1%, t-value 6.0) occurs in the group with EFA deficiency, correlated with that of SDH activity in the normal group. On the other hand, the total LDH activity seems slightly increased in the group with essential fatty acid deficiency, correlated with that of the normal diet. This increase is only statistically significant at the 5% level (t-value 2.3).

The changes in the LDH isoenzyme pattern induced by EFA deficiency are indicated in Table 8. During a three month period of deficiency the LDH_x

Table 8. The relative distribution (in per cent) of LDH isoenzymes in normal and in EFA-deficient rat testes (3 month EFA deficiency, see text). A statistically significant decrease in LDH_x occurs in the case of EFA deficiency.

	H_4	H_3M	H_2M_2	HM_3	LDH_x	M_4
Normal diet N = 16	35.6 ± 1.2	31.0 ± 1.9	11.8 ± 0.7	12.0 ± 1.4	7.4 ± 0.8	3.9 ± 0.4
EFA deficient N = 16	33.2 ± 2.7	39.9 ± 2.8	12.7 ± 2.4	8.7 ± 1.2	2.3 ± 0.8	3.3 ± 0.7

activity is decreased more than three times. This decrease is highly significant at the 0.1% level (t-value 7.3). The decrease in LDH_x activity seems associated with a slight relative increase in the H_3M isoenzyme fraction. This relative increase seems statistically significant at the 3% level (t-value 3.2).

Subcellular distribution of lactate dehydrogenase isoenzyme during essential fatty acid deficiency

Rat testes of essential fatty acid-deficient rats revealed, after differential ultracentrifugation, the presence of the LDH_x isoenzyme only in the nuclei and the mitochondrial subfraction E, similar to that found in normal rat testes.

DISCUSSION OF LDH_x-COMPARTMENTALIZATION

Studies on the topographic distribution of the cytochrome system in the sperm cell have revealed this enzyme system to be located in the middle piece (and tail), rather than in the sperm head [27, 31, 32]. Although the present data indicate the highest LDH activity in the supernatant of particulate fractions from sperm cells, the middle piece was found to contain LDH_x exclusively. The specific LDH_x activity of the middle piece exhibited 10% of the LDH-activity of the supernatant. The middle piece was also demonstrated to contain the highest

SDH activity. The LDH_x activity of the sperm head may be explained by the presence in this fraction of pieces of the middle piece or of broken sperm cells. Re-extraction of particulate fractions with Triton-X-100 revealed the possibility of liberating a specific activity as high as that already extracted with 0.9% aqueous NaCl from head and tail. Even higher specific activity could be liberated from the middle piece. This liberated activity was exclusively of the LDH_x type. These data thus support the idea that LDH_x is largely localized in the middle piece. The compartmentalization could conceivably be an artifact caused by an affinity of LDH_x for mitochondrial constituents during homogenization. This is hardly the case because, during fractionation of rat testes, the LDH_x was not preferentially localized in mitochondrial subfraction D of rat testes obtained by sucrose gradient centrifugation, although this fraction exhibited SDH activity higher than subfraction E containing LDH_x.

The studies of the isoenzyme distribution in particulate fractions of the rat testes are similar to those on calf kidney [17], indicating an affinity of kidney mitochondria for $LDH-H_4$. However, in contrast to the findings in kidney nuclei, rat testes nuclei fractions exhibited no preferential affinity for $LDH-M_4$.

As the two enzymes mentioned above thus seem more or less firmly attached to particulate fractions of the sperm cell and testes, their activity may be related to membrane structure, e.g. fatty acid composition. This relationship may be due either to loose binding of the enzyme (LDH_x) to the membrane, or probably through a direct (catalytic) involvement of the membrane in the enzymic function (SDH). This is supported by the present data indicating that EFA deficiency is associated with a highly significant decrease in both enzymic activities.

In conclusion, the present studies indicate that the LDH-isoenzyme pattern of a tissue is determined by at least three factors, i.e. the concentration of NADH, and LDH subunits, and finally the compartmentalization of the LDH subunits.

REFERENCES

1. Wieland, T. and Pfleiderer, G., *Biochem. Z.* **329** (1957) 112.
2. Apella, E. and Markert, C. L., *Biochem. biophys. Res. Commun.* **6** (1961) 171.
3. Cahn, R. D., Kaplan, N. O., Levine, L. and Zwilling, E., *Science, N.Y.* **136** (1962) 932.
4. Markert, C. L. and Møller, F., *Proc. natn. Acad. Sci. U.S.A.* **45** (1959) 753.
5. Markert, C. L. and Apella, E., *Ann. N.Y. Acad. Sci.* **103** (1963) 915.
6. Chilson, O. P., Costello, L. A. and Kaplan, N. O., *J. molec. Biol.* **10** (1964) 349.
7. Epstein, C. J., Carter, M. M. and Goldberger, R. F., *Biochim. biophys. Acta* **92** (1964) 391.
8. Sekuzu, I., Yamashita, J., Nozaki, M., Hagihara, B., Yonetani, T. and Okunuki, K., *J. Biochem., Tokyo* **44** (1957) 601.

9. Chilson, O. P., Kitto, G. B. and Kaplan, N. O., *Proc. natn. Acad. Sci. U.S.A.* **53** (1965) 1006.
10. Chilson, O. P., Costello, L. A. and Kaplan, N. O., *Biochemistry* **4** (1965) 271.
11. Clausen, J. and Hustrulid, R., *Biochim. biophys. Acta* **167** (1968) 221.
12. Goldman, R. D., Kaplan, N. O. and Hall, T. C., *Cancer Res.* **24** (1964) 389-399.
13. Gerhardt, W., Clausen, J., Christensen, E. and Riishede, J., *Acta neurol. Scand.* **39** (1963) 85-111.
14. Gerhardt, W., Clausen, J., Christensen, E. and Riishede, J., *J. natn. Cancer Inst.* **38** (1967) 343.
15. Brody, J. A., *Neurology, Minneap.* **14** (1964) 1091.
16. Garcia-Bunuel, L., Garcia-Bunuel, V., Green, L. and Sulvin, D. K., *Neurology, Minneap.* **16** (1966) 491.
17. Güttler, F. and Clausen, J., *Enzymol. biol. Clin.* **8** (1967) 456.
18. Clausen, J. and Øvlisen, B., *Biochem. J.* **97** (1965) 513.
19. Chilson, O. P., Kitto, G. B., Pudles, J. and Kaplan, N. O., *J. biol. Chem.* **241** (1965) 2431.
20. Shifrin, S., Kaplan, N. O. and Ciotti, M. M., *J. biol. Chem.* **234** (1959) 1555.
21. Zinkham, W. H., Blanco, A. and Kupchyk, L., *Science, N.Y.* **142** (1963) 1303.
22. Goldberg, E., *Science, N.Y.* **139** (1963) 602.
23. Blanco, A., Zinkham, W. H. and Kupchyk, L., *J. exp. Zool.* **156** (1964) 137.
24. Green, D. E., Mii, S. and Kohout, P. M., *J. biol. Chem.* **217** (1955) 551.
25. Singer, T. O. and Kearney, E. B., *Meth. biochem. Analysis* **4** (1965) 307.
26. Basford, R. E. and Huennekens, F. M., *J. Am. chem. Soc.* **77** (1955) 3873.
27. Mann, T., "Biochemistry of Semen and of the Male Reproductive Tract", Methuen, London, 1964, p. 347.
28. Mohri, H., Mohri, T. and Ernster, L., *J. exp. Cell. Res.* **38** (1965) 217.
29. DeRobertis, E., Pellegrino de Iraldi, A., Rodriguez de Lores Arnaiz, G. and Salganicoff, L., *J. Neurochem.* **9** (1962) 23.
30. Rathbone, I., *Biochem. J.* **97** (1965) 620.
31. Zittle, C. A. and Zitin, B., *J. biol. Chem.* **144** (1942) 105.
32. Hrudka, F., *Acta histochem.* **19** (1964) 346.
33. Clausen, J. and Hustrulid, R., *Biochem. J.* **111** (1969) 219.
34. Eliasson, R., Häggman, K. and Wiklund, B., *Scand. J. clin. Lab. Invest.* **20** (1967) 353.
35. Zinkham, W. H., Blanco, A. and Kupchyk, L., *J. exp. Zool.* **156** (1964) 137.
36. DiSabato, G. and Kaplan, N. O., *J. biol. Chem.* **239** (1964) 438.
37. Pfleiderer, G. and Wachsmuth, E. D., *Biochem. Z.* **334** (1961) 185.
38. Gerhardt-Hansen, W., "Lactate Dehydrogenase Isoenzymes in the Central Nervous System", Coster, Copenhagen, 1968.
39. Güttler, F. and Clausen, J., *Biochem. J.* **114** (1969) 839.
40. Clausen, J., *Biochem. J.* **111** (1969) 207.

FEBS Symposium, Volume 18, 1970, pp. 151-156

Comparison of Primary Structures of Pig Heart and Muscle Lactate Dehydrogenases

G. PFLEIDERER and K. MELLA

Abteilung für Chemie, Ruhr-Universität Bochum. *

Biochemical, immunological and biological methods have provided clear indications about differences in the structure of heart and skeletal muscle LDH isolated from a single organism. As it is established that the differences are due to differing chemical composition of the subunits of both types of LDH, the investigation of the primary structure is of interest: firstly, in connexion with the X-ray structural analyses being conducted by Rossmann *et al.* [1] in Lafayette; secondly, for providing indications on the relationship between structure and function; and thirdly, for the investigation· of the phylogenetic development of LDH. To emphasize the special nature of this enzyme, we should like to recall the comparison of protein structures, especially of enzymes with identical catalytical behavior, from various species, e.g. hemoglobins or cytochromes *c*; and that, in the cases of closely related species, frequently only individual amino acids have been exchanged in the primary structure. The development of the various hemoglobin chains can be understood by rough analogy with lactate dehydrogenase; these chains, according to present knowledge, have differentiated themselves from a common ancestor (probably myoglobin) by gene duplication and differing mutation ratios of the new genes. In the case of LDH we must deal with two chemically and physically different proteins, simultaneously present in a single cell of one single organism, which appear to be synthesized in larger or smaller quantities, depending on the physiological function of the organ. The hybrid from two heart and two muscle subunits, i.e. the isoenzyme 3 ($H_2 M_2$), can, for example, be compared with the hemoglobins which are also composed of two different subunits, for example, two α-chains and two β-chains ($\alpha_2 \beta_2$). As early as 1961, Wieland and Pfleiderer pointed out the differences in amino acid composition between heart and skeletal muscle LDH's [2, 3] and Kaplan *et al.* [4] have provided considerable detailed information about LDH from numerous sources.

* *Present address.* D 463 Bochum, Ruhr-Universität, Lehrstuhl für Biochemie.

Comparing our most recent amino acid analyses of pig heart and skeletal muscle LDH (Table 1), we especially note the nearly identical number of lysine and arginine residues, about 26 lysines and 9 arginines in each case related to the subunit of 35,000 molecular weight, and the equal number of 4 SH-groups, which are essential for the structure and function of this enzyme. The presence of the 4SH-groups has also been confirmed by optical methods. The number of

Table 1. Amino acid composition of different lactate dehydrogenases (number of residues per subunit, 35,000).

	Pig heart	Pig muscle	Chicken muscle*
Carboxymethyl-cysteine	4	4	4
Aspartic acid	40	30	31
Threonine	18	13	17
Serine	26	20	28
Glutamic acid	38	30	26
Proline	12	12	11
Glycine	26	25	26
Alanine	19	19	20
Valine	36	29	25
Methionine	8	7	8
Isoleucine	24	26	21
Leucine	37	34	30
Tyrosine	8	7	5
Phenylalanine	5	7	7
Lysine	26	27	28
Histidine	8	13	14
Arginine	9	10	11

* Kaplan, N. O., *J. biol. Chem.* **242** (1967) 2151.

cleavages recorded during tryptic digestion is 28 to 32 for both LDH's, within the limits of error. Fingerprints of both types of LDH, published mainly by Wieland and co-workers [5], and Kaplan *et al.* [6, 7], place in evidence marked differences in relation to peptide spots revealed by ninhydrin with both LDH-types. The peptide maps published by Kaplan show exactly 35 to 37 ninhydrin-positive spots expected with four identical subunits. However, during the course of sequence studies, we have found peptides with more than one lysine residue. Furthermore, we know that a number of larger peptides do not migrate in fingerprinting. It is generally known that similar, or identically migrating, peptide spots may have differing amino acid compositions or sequences.

The first homologies between heart and skeletal muscle LDH's of one single species were discovered in essential cysteine peptides. Kaplan [7] and our own group, particularly Holbrook [8], have isolated by different types of radioactive marking of essential SH-groups, a tryptic cysteine peptide showing almost the same amino acid composition in different organisms for heart and skeletal muscle LDH. However, we must correct Kaplan's results for chicken heart LDH; as we have shown recently together with Fölsche and Torff [9], the sequence of the essential cysteine peptide in pig heart LDH, published previously by us [10], is also valid for the cysteine peptides from other organs and species. We have established that the sequences of the essential cysteine peptides from pig heart, pig skeletal and chicken skeletal muscles are fully identical. The sequence is:

Val-Ile-Gly-Ser-Gly-Cys-Asn-Leu-Aso-Ser-Ala-Arg.

For a preliminary survey of further homologies in pig heart and skeletal muscle LDH's, we have tryptically digested trifluoroacetylated LDH. It is known that, after this treatment, trypsin attacks only the arginine carboxyl groups, since the ε-amino residues of lysine are blocked with the trifluoroacetyl group, thus obviating an attack by trypsin. The tryptic hydrolysate was further subjected to trichloroacetic acid precipitation. Fingerprints of the soluble fractions show only three out of nine possible heart muscle peptides with C-terminal arginine residues, viz. the essential cysteine peptide, one heptapeptide (Gln-Gln-Glu-Gly-Glx-Ser-Arg) and one hexapeptide (Leu-Asn-Leu-Val-Gln-Arg) (one Leu-Lys dipeptide might have escaped trifluoroacetylation by steric hindrance). These three peptides are also found, with identical sequences, in pig skeletal muscle LDH, as well as in the chicken skeletal muscle LDH (Fig. 1). In

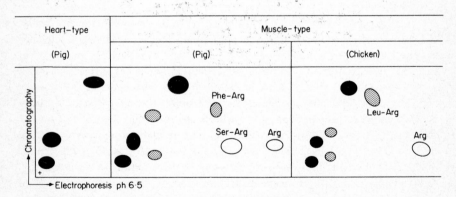

Figure 1. Peptide maps of special arginine-peptides of different lactate dehydrogenases.

Identical arginine peptides in heart and muscle LDH-isoenzymes.

Additional arginine peptides in muscle LDH-isoenzymes.

Additional arginine peptides in muscle LDH-isoenzymes, thought to be equivalent in LDH primary structures.

contrast to the heart enzyme, we find, in the relevant fingerprint of pig skeletal muscle LDH, seven low-molecular arginine peptides and free arginine, which means that almost all of the arginine peptides are arranged in close proximity in the primary structure of muscle-type LDH (Fig. 1) (if there were an acylated lysine ahead of one of the peptides mentioned, this would be revealed by amino acid analyses). There is consequently a clear difference between the primary structures of the heart and skeletal muscle enzymes of one single organism, together with a surprisingly close relationship between pig and chicken skeletal muscle LDH. Both contain, although far apart phylogenetically, free arginine and one arginine dipeptide, Leu-Arg or Phe-Arg. This small difference may be ascribed to a point mutation in the codons. Furthermore, we find two peptides with closely related amino acid compositions which migrate in a similar way in the fingerprint; however, tyrosine in pig muscle LDH is exchanged for histidine, again possible by a single point mutation. Our sequential analyses of chicken peptides, now under way, have revealed, however, that only the amino acid compositions, and not the sequences, are identical.

Another substantial difference appears to exist at the carboxyl end in the case of heart and skeletal muscle enzymes from one single animal species. We have now submitted for publication the sequence of the last 36 amino acid residues at the carboxyl end of heart muscle LDH and find that corresponding peptides are absent at the carboxyl end of the pig muscle enzyme [11]. The last four amino acids in the heart muscle are Leu-Lys-Asp-Leu and the carboxyl-terminal peptide in the LDH V has the sequence Glu-Leu-Gln-Phe. The exchange of the last amino acid, but not the others, can be ascribed to a point mutation.

In addition to finding clear differences between the arginine peptides of heart and skeletal muscle LDH, the divergence seems to be even greater in the case of the tryptic lysine peptides. Of about 250 amino acid residues whose sequence has been established by Mella and co-workers, and which will be published in the near future, only the three arginine peptide sequences mentioned above appear to coincide. Finally, there are indications of differences in the cysteine peptides other than the essential cysteine peptide. A second tryptic cysteine peptide with six residues has already been found in the pig skeletal muscle enzyme. This second cysteine peptide is not present in the corresponding fraction of heart muscle LDH. We suppose a different conformation in the vicinity of the essential SH-groups, whilst the essential cysteine residue of pig heart LDH reacts rapidly with various maleinimides; with loss of enzymic activity the thiol groups of pig muscle LDH were not attacked under the same conditions [8]. At the present stage of our investigations, it appears that an important difference in the amino acid sequences of heart and skeletal muscle enzymes may be expected. Jeckel has also found some difference in the amount of α-helix [12]. For other species, Kaplan has observed 30% α-helix formation in pig heart and 35% in pig muscle. The interesting question now is whether the homology is mainly limited to regions of the active center or not. We expect to be able to extend our knowledge about the "active center" of lactate dehydrogenases with an essential tyrosine peptide recently isolated by Jeckel and Pfleiderer [13] from heart muscle. Our aim will be to ascertain whether heart and muscle LDH types originate in a diverging or a converging row. From our results *now available,* a converging development of heart and muscle-type LDH's would appear to be more reasonable. On the basis of earlier comparative biochemical tests [3], we reported that the skeletal muscle enzymes, even of various species, are more closely related than the heart and skeletal muscle LDH from one single organism. These findings have now been confirmed. The homologies found may be partly responsible for the catalysis of one specific enzymatic reaction. Even here there exists a small variation in substrate specificity, illustrated by the different dehydrogenation rates of α-hydroxybutyrate by both types of LDH. Whilst several authors, in particular Bodansky, Kaplan, Markert, and our own group (in collaboration with Rajewsky and Grabar), have demonstrated absolute immunological differences between heart and skeletal muscle enzymes, Rajewsky [14] has been able to demonstrate an immunological relationship by using acetic anhydride-treated LDH for the immunization. Even the antibody against acetylated heart muscle LDH yields a cross reaction with the appropriately treated enzyme of skeletal muscle. From his results one may deduce that acetylation would reveal new determinants which are identical or closely related in both types of LDH.

REFERENCES

1. Wonacott, A. J., Mermall, H. L., Schevitz, R. W., Haas, D. J., McPherson, A. and Rossmann, M. G., *Fedn Proc. Fedn Am. Socs exp. Biol.* **27** (1968) 1716.
2. Wieland, T. and Pfleiderer, G., *Ann. N.Y. Acad. Sci.* **94** (1961) 691.
3. Wachsmuth, E. D., Pfleiderer, G. and Wieland, T., *Biochem. Z.* **340** (1964) 80.
4. Pesce, A., McKay, R. H., Stolzenbach, F. E., Cahn, R. D. and Kaplan, N. O., *J. biol. chem.* **239** (1964) 1753.
5. Wieland, T., Georgopulos, D., Kampe, H. and Wachsmuth, E. D., *Biochem. Z.* **340** (1964) 483.
6. Fondy, T. P., Pesce, A., Freedberg, I., Stolzenbach, F. E. and Kaplan, N. O., *Biochemistry* **3** (1964) 522.
7. Fondy, T. P., Everse, J., Driscoll, G. A., Castillo, F., Stolzenbach, F. E. and Kaplan, N. O., *J. biol. Chem.* **240** (1965) 4219.
8. Holbrook, J. J., Pfleiderer, G., Schnetger, J. and Diemair, S., *Biochem. Z.* **344** (1966) 1.

9. Mella, K., Fölsche, E., Torff, H. J. and Pfleiderer, G., *Hoppe-Seyler's Z. physiol. Chem.* in press.
10. Holbrook, J. J., Pfleiderer, G., Mella, K., Volz, M., Leskowac, W. and Jeckel, R., *Eur. J. Biochem.* 1 (1967) 476.
11. Mella, K. and Pfleiderer, G., in preparation.
12. Jeckel, D., unpublished.
13. Pfleiderer, G. and Jeckel, D., *Eur. J. Biochem.* 2 (1967) 171.
14. Rajewsky, K., *Biochim. biophys. Acta* 121 (1966) 51.

FEBS Symposium, Volume 18, 1970, pp.157-161

Structure and Action of Lactate Dehydrogenase

G. PFLEIDERER and D. JECKEL

*Abteilung für Chemie, Ruhr-Universität Bochum,
Frankfurt, West Germany**

Since we know the precise structure of the essential cysteine peptide of the lactate dehydrogenase [1], (see p. 151) we are interested in discovering further amino acid side chains located in the active center of LDH. In the case of dehydrogenases generally, the direct or indirect participation of SH-groups is known, as has also been demonstrated for alcohol dehydrogenase [2] and glyceraldehyde phosphate dehydrogenase [3]. Recently we reported on the possibility, after protecting all SH-groups of LDH by converting the heart-muscle enzyme to the mercury salt, of carrying out coupling reactions with diazotized ^{35}S-sulfanilic acid [4]. Following removal of the mercury ions, we found a decrease in enzymic activity without loss of SH-groups. Meanwhile we were able to demonstrate that about 1.5 moles of azo-LDH is formed per subunit extrapolated to 100% inhibition. The spectrum of the azo-enzyme, as well as that of an isolated labelled peptide which contains most of the radioactivity after trypsin digestion, resembles the spectrum of monoazotyrosine. We are exploring the amino acid sequence in the vicinity of this essential tyrosine peptide. It appears that a genuine side chain of the active group is involved, since, in contrast to Holbrook's [5] findings in the case of blocking SH-groups, no more coenzyme is bound in this case, nor is a DPN-sulfite-LDH complex formed. It must be established in these, and other tests involving masking of essential amino acids, that no alteration of conformation occurs under the conditions of pretreatment, i.e. of the mercury salt in this case. We have begun measurements of rotatory dispersion [6]. By comparison with α-helical poly-glutamic acid, a helix content of about 30% could be deduced for the native pig heart enzyme. In accordance with measurements of activity already reported by us, Jeckel has been able to show that no alteration of conformation occurs on blocking the essential SH-group, as, for example, with *p*-chloromercuribenzoate —which can subsequently be eliminated by adding an excess of thiol, leading to

* *Present address*. Institut für Biochemie im Institut für Organische Chemie, J. W. Goethe-Universität, Frankfurt, West Germany.

Table 1. Analysis of LDH acetylated with N-acetylimidazole.

Experiment no.	[LDH] mg/ml	[Acetyl-imidazole] mM	[Na$_2$SO$_3$] mM	Residual enzyme activity (per cent)	Residual SH-groups (moles/36,000 g protein)	Tyrosine acetylated (moles/36,000 g protein)	Residual NH$_2$-groups (moles/36,000 g protein)
1. Control	12	0	0	100	4.6	—	22-24
Control	16.8	0	27.4	100	4.6	—	22-24
2. a	11	32.8	0	50	3.2	0 ± 0.5	—
b	11	55.5	0	30	2.7	0 ± 0.5	—
c	11	33	0	1	1.6	0 ± 0.5	—
3. a	17.8	14.2	0	39	—	—	18.5
b	17.8	42.6	0	12.5	1.5	—	13.4
c	17.8	85.2	0	0	0.03	—	8.9
d	17.8	170.4	0	0	0	—	5.3
4. a	16.8	23.7	27.4	97	3.88	—	11.9
b	16.8	47.5	27.4	97	3.80	—	9.9
c	16.8	95.0	27.4	75	2.80	—	8.7
5. a	12	60.7	0	1.4	1.7	—	10.5
b	12	60.7	38.4	100	4.0	—	13.6

LDH, Na$_2$SO$_3$ and acetyl-imidazole, at the concentrations shown, were incubated in 67 mM phosphate buffer, pH 7.5, for 1 h and then analyzed for residual enzymic activity, SH-groups, amino groups, and, in one series, O-acetyltyrosine. Incubation mixtures containing sulfite were filtered through Sephadex prior to SH-analysis. Analytical procedures are described in the text. —,not measured.

Table 2. Effect of sulfite on the inhibition of LDH by diazotized sulfanilic acid.

Minutes after addition of [35S]-DS	Control LDH* Per cent residual enzyme activity	LDH + [35S]-DS* Per cent residual enzyme activity	Per cent DS remaining	LDH + [35S]-DS + Na_2SO_3* Per cent residual enzyme activity	Per cent DS remaining†
0	100	100	100	100	100
15	100	0	—	100	—
30	—	—	52	—	30
60	100	0	36	100	25
90	—	—	—	—	—
120	100	0	25	100	12.5
150	—	—	—	—	—

* 5.8×10^{-5} M LDH was incubated at 0°C in 0.1 M sodium pyrophosphate-HCl buffer, pH 8.5, either with no additions or with the additions of 1.7×10^{-2} M Na_2SO_3 and 5×10^{-5} M [35S]-DS as shown.

† The amount or DS remaining was measured by coupling an aliquot of the incubation mixture with resorcinol.

complete restoration of enzymic activity. If more than 1.5 moles p-chloro-mercuribenzoate is incorporated per mole of subunit, the helix content steadily decreases. Thus we can also confirm our previous finding that an irreversible unfolding of the molecule occurs when the non-essential SH-groups in the LDH have been blocked with aromatic mercurials. In contrast to this, the mercury salt itself shows, within the limits of error, a conformation identical to that of the native enzyme. This also explains why the full activity can be restored after the mercury has been removed. Unlike the findings of Theorell *et al.* [7] on the yeast ADH, which revealed an extrinsic Cotton effect with NAD and pyrazole, we find no alteration of the optical rotatory dispersion when lactate dehydro-genase forms a ternary complex with the very weakly dissociated NAD sulfite. Very interesting is the observation that optical rotatory dispersion of the ternary complex in 5M urea was the same as for the native enzyme, whilst the native enzyme, under identical conditions, was unfolded. The formation of this pseudo-ternary complex seems to hold together the entire LDH molecule like a seam and prevents the hydrogen bridges from being ruptured by urea. Finally, our more recent investigations allowed us to demonstrate the accumulation of polar groups on the enzyme surface [8] in analogy with the findings of Kendrew on myoglobin or Anfinsen on ribonuclease. We refer to an observation of Rajewsky [9] to the effect that LDH is not inactivated by introducing acetyl groups, probably by acetylation of ϵ-amino side chains of lysine, in the presence of a large excess of sulfite ions. We succeeded, with Rajewsky's conditions, in acetylating with acetyl imidazole 14 out of a total of 24 ϵ-amino groups of the heart muscle enzyme without any decrease in activity (Table 1). The acetylated enzyme migrates much more rapidly towards the anode and has a pH optimum of 5.3 instead of 7.15 for the native enzyme. The protecting effect of sulfite ions is also revealed in the case of the coupling reactions; no coupling occurs in the presence of sulfite, not even in the presence of a higher excess of diazotized sulfanilic acid (Table 2). The presence of sulfite ions may be assumed to protect the active center from chemical attack by an alteration of conformation. On the other hand, LDH also shows an accumulation of ϵ-amino groups, i.e. polar groups, on the enzyme surface.

REFERENCES

1. Holbrook, J. J., Pfleiderer, G., Mella, K., Volz, M., Leskowac, W. and Jeckel, R., *Eur. J. Biochem.* **1** (1967) 476.
2. Boyer, P. D., Lardy, H. H. and Myrbäck, K. (eds.), "The Enzymes", Vol. 7, Academic Press, New York, 1963, p. 73.
3. Velick, S. F., *J. biol. Chem.* **203** (1953) 563.
4. Pfleiderer, G. and Jeckel, D., *Eur. J. Biochem.* **2** (1967) 171.
5. Holbrook, J. J., *Biochem. Z.* **344** (1966) 141.
6. Jeckel, D. and Pfleiderer, G., XVIth Colloquim on Proteides of Biological Fluids, Brugge, 1968.

7. Rosenberg, A., Theorell, H. and Yonetani, T., *Archs Biochem. Biophys.* **110** (1965) 413.
8. Pfleiderer, G., Holbrook, J. J., Zaki, L. and Jeckel, D., *FEBS Letters* in press.
9. Rajewsky, K., *Biochim. biophys. Acta* **121** (1966) 51.

FEBS Symposium, Volume 18, 1970, pp. 163-167

Studies on pO$_2$-Dependent Changes in Lactate Dehydrogenase Isoenzyme Distribution

P. HELLUNG-LARSEN and V. ANDERSEN

*Chromosome Laboratory, Department of Obstetrics and
Gynaecology A-I, and Medical Department A, Rigshospitalet,
Copenhagen, Denmark*

INTRODUCTION

It is generally assumed that the overall intracellular lactate dehydrogenase (LDH) pattern of a given tissue is maintained through different rates of synthesis and/or catabolism of H- and M-subunits. In many *in vivo*, as well as *in vitro* systems, changes in LDH patterns may be induced by variations in the oxygen tension. Aeration with gas mixtures containing less than 20% oxygen results in a relative increase of the isoenzymes which are rich in M-subunits, whereas aeration with higher oxygen concentrations increases the H-containing isoenzymes. Until now, however, the intracellular turnovers of H- and M-subunits have not been determined.

In this paper the kinetics of changes in LDH patterns of cultured human lymphocytes, caused by variations in pO$_2$, are described.

MATERIALS AND METHODS

Suspensions of human leucocytes containing more than 99% lymphocytes were cultured with phytohaemagglutinin (PHA) [2, 3]. The different culture flasks were continuously aerated with gas mixtures containing 0, 3, 5, 9, 14, 20, 30, 50, or 70% O$_2$ + 5% CO$_2$ (balance N$_2$).

Experiments involving a change in oxygen tension or changes of medium were carried out as described previously [3].

The methods for determination of LDH isoenzyme distributions and total LDH activities have also been described earlier [1-3].

The isoenzyme patterns were quantitated as percentages of H-subunits [2].

163

RESULTS AND DISCUSSION

Figure 1 shows the percentage of H-subunits of LDH isoenzyme distributions in PHA-stimulated cultures maintained at different oxygen tensions. It is seen that the effect of changes in pO_2 is more pronounced at low oxygen tensions, and that the range in the proportion of H-subunits is 33% to 58%. The cells thrive poorly at oxygen concentrations higher than 50% [4].

The total LDH activity is constant between 5 and 20% O_2 (Table 1).

The relative decrease of H-subunits at low pO_2 might conceivably be due to accumulation of pyruvate or some other metabolite. However, changes of the

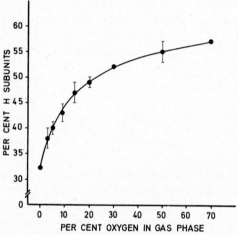

Figure 1. LDH isoenzyme pattern, expressed as per cent H-subunits, in human lymphocytes exposed to different oxygen tensions during culture for 72 h with phyto-haemagglutinin. Results, expressed as mean ± S.E., are based on 6 experiments; the others on 3 experiments.

Table 1. Effect of renewal of medium on the LDH isoenzyme pattern and the total LDH activity of human lymphocytes cultured with PHA for 72 h.

Per cent oxygen in gas phase	Isoenzyme distribution (per cent H-subunits)		Total LDH activity (arbitrary units)	
	undisturbed	change of medium	undisturbed	change of medium
5	40 ± 1.2	38 ± 1.4	100 ± 15	110 ± 12
9	43 ± 1.8	41 ; 46	112 ± 16	102 ; 166
20	49 ± 1.1	49 ± 1.5	100 ± 20	98 ± 24
50	55 ± 2.0	56 ± 1.8	69 ± 15	49 ± 10

Means ± S.E. are based on 6 experiments.

medium every twelve hours did not abolish the effect of pO_2 (Table 1). Addition of inhibitors of mitochondrial oxidation (antimycin A or oligomycin) to a final concentration of 20 μM likewise had no effect.

The results obtained indicate that changes in pO_2 influence the turnover of H-, as well as of M-subunits, since total LDH activities remain constant. The absence of any effect of frequent changes of medium indicates that oxygen

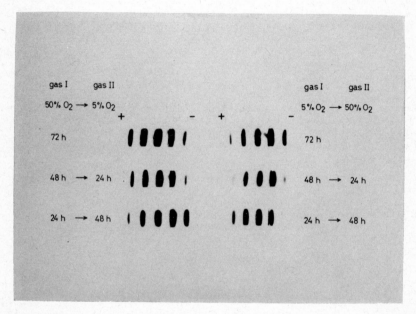

Figure 2. Effect of gas switch on LDH isoenzyme pattern of human lymphocytes cultured for 72 h with phytohaemagglutinin.

exerts its effect without a stable, diffusable mediator. Experiments on the rate of [^{14}C] thymidine incorporation demonstrate that the effect of pO_2 in our system is not due to interference with cellular growth [4].

Figure 2 illustrates patterns obtained when the cultures were switched from high to low oxygen concentration, or vice versa, during culture. The results are shown in Fig. 3. They indicate that a hypoxic pattern can be changed to a hyperoxic pattern and vice versa. If a correction is made for the isoenzyme changes due solely to the PHA-induced transformation of the cells (Fig. 4), the kinetics of the changes are alike.

A hypothesis based on the results obtained is presented in Fig. 4. According to this concept, the functional half-life of the isoenzymes is about 24 h and the lag period probably around 4 h.

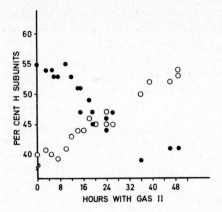

Figure 3. Kinetics of LDH isoenzyme changes illustrated in Fig. 2. All cultures harvested after 72 h. The gas switches took place between 24 and 72 h; the abscissa indicates the period at the new pO_2.

   • shift $50 \rightarrow 5\% \, O_2$
   ○ shift $5 \rightarrow 50\% \, O_2$

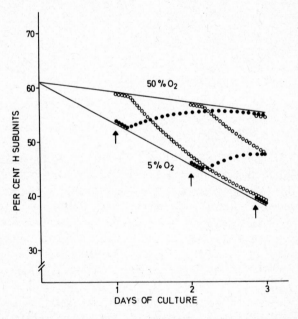

Figure 4. Hypothetical model based on the results of the gas switches (indicated by arrows).

ACKNOWLEDGEMENTS

We thank Mrs. Annelise Persson for excellent technical assistance.

This work was supported by the Danish State Research Foundation, Novo's Foundation and Christian X Foundation.

REFERENCES

1. Hellung-Larsen, P., *Acta chem. scand.* **22** (1968) 355.
2. Hellung-Larsen, P. and Andersen, V., *Exp Cell Res.* **50** (1968) 286.
3. Hellung-Larsen, P. and Andersen, V., *Exp Cell Res.* **54** (1969) 201.
4. Andersen, V., Hellung-Larsen, P. and Sørensen, S. F., *J. Cell Phys.* **72** (1968) 149.

FEBS Symposium, Volume 18, 1970, pp. 169-176

Temperature Dependence of the Mechanism of Heat Inactivation of Pig Lactate Dehydrogenase Isoenzyme H_2M_2

J. SÜDI

Institute of Biochemistry, Hungarian Academy of Science, Budapest, Hungary

Heat stability of the five isoenzymes of lactate dehydrogenase has been studied in a number of laboratories. The reported findings agree in two respects. It appears [1] that in most vertebrate species the heart-type homotetramer (isoenzyme H_4) is more stable at elevated temperature than the muscle-type homotetramer (isoenzyme M_4). Another, apparently general, conclusion indicates that the stability of the hybrid tetramers (isoenzymes H_3M, H_2M_2 and HM_3) is intermediate between those of the homotetramers H_4 and M_4 [2-4].

In recent experiments we have undertaken a detailed study of the temperature- and concentration-dependence of the rate of inactivation of pig lactate dehydrogenase isoenzymes. We find that isoenzyme H_2M_2 is less stable than both isoenzymes H_4 and M_4 [5]. We also find that, under certain conditions, isoenzyme M_4 is more stable than isoenzyme H_4 [5]. We have further observed that both temperature and protein concentration determine whether there is an apparent interaction between isoenzymes M_4 and H_4 when these are inactivated in a mixed solution [6]. It seems to us that these diverse observations can be fitted into a simple picture by considering the role which the tetrameric structure of lactate dehydrogenase might play in the process of inactivation. These mechanistic considerations are presented below, together with supporting evidence obtained in experiments with pig lactate dehydrogenase isoenzyme H_2M_2. We shall show that the kinetics of inactivation and the isoenzyme pattern of partially inactivated enzyme samples show the expected temperature- and concentration-dependence of the postulated mechanisms.

Non-standard abbreviations. H_4, H_3M, H_2M_2, HM_3 and M_4, the five isoenzymes of lactate dehydrogenase composed of four subunits of the heart type (H), and muscle type (M), respectively.
Enzyme. Lactate dehydrogenase or L-lactate: NAD oxidoreductase (EC 1.1.1.27).

RESULTS AND DISCUSSION

The scheme shown in Fig. 1 summarizes the basic considerations of our theoretical approach. The starting point of these mechanistic considerations is the reasonable assumption that inactivation is initiated by some practically irreversible change which affects only one of the four subunits. Accordingly, the product of this primary reaction is a tetramer composed of three active and one inactivated subunits. It is postulated that the subsequent fate of this unstable

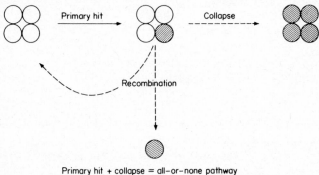

Primary hit + collapse = all–or–none pathway
Primary hit + recombination = one–by–one pathway

Figure 1. A general scheme of the alternative pathways which the inactivation of a tetrameric enzyme may follow. Open circles, active subunits; shaded circles, inactivated subunits.

intermediate might follow the alternative pathways shown by the broken arrows. One of these can be described as an *all-or-none pathway,* in which the primary step of inactivation leads to the collapse of the whole tetramer. In the alternative *one-by-one pathway* the primary event is followed by the separation of the inactivated subunit from the three intact subunits which, in turn, will recombine through a number of consecutive steps and lead to the formation of another stable tetrameric molecular species.

In Fig. 2 the general scheme is applied to a tetramer composed of two different subunits, as a working hypothesis for the present studies. It is seen that one can differentiate between the postulated alternative pathways if there is a difference between the primary stability of the two types of subunits. Evidently, unchanged tetramers will only be recovered from a partially inactivated sample if the overall process follows the *all-or-none pathway.* On the contrary, the *one-by-one pathway,* coupled with a differential primary stability of the two types of subunits, will result in new tetramers, enriched in the more stable subunit.

The predictions of Fig. 2 have been tested with pig lactate dehydrogenase isoenzyme H_2M_2 under our standard conditions [5, 6] over a wide range of temperature. The experimental procedures involved the preparation of iso-enzyme H_2M_2 by *in vitro* hybridization [7] from crystalline isoenzymes H_4 and M_4, and isolation of the hybrid by chromatography on DEAE cellulose. It should be noted that this procedure does not affect the heat stability of the homotetramers H_4 and M_4 [5]. The enzyme was exposed to elevated tempera-ture in 0.2M tris/HCl buffer, pH 7.5 (measured at 21°C).

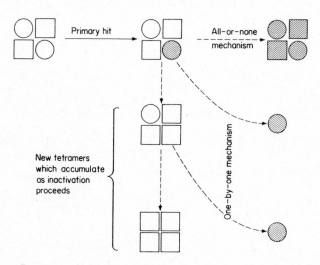

Figure 2. Scheme showing the postulated alternative pathways of inactivation of a tetramer composed of two types of subunits. The subunits with a lower primary stability are denoted by circles. Open symbols stand for active subunits, and shaded symbols for inactivated subunits, respectively.

A summary of some kinetic results obtained with isoenzyme H_2M_2 is shown in Fig. 3. In these experiments we have followed the inactivation process by determining residual enzyme activity with 1 mM pyruvate in the assay mixture. It is seen that the first-order rate constants for inactivation yield a straight-line Arrhenius function for the lower protein concentration (10 μg/ml), and an over-crossing curve with an inflexion point for a hundred times higher initial protein concentration (1 mg/ml). Figure 3 indicates that there is a concentra-tion-dependence in the mechanism of inactivation and, further, that at high protein concentration the overall process is also affected by temperature. One should note that on lowering the temperature we obtained a decrease of five orders of magnitude in the rate of inactivation.

Figure 3 also indicates that the situation is markedly different in the temperature ranges above and below 52°-54°C. Detailed representative results obtained in these two temperature ranges are shown in Figs. 4-6.

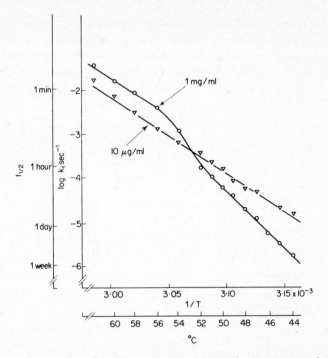

Figure 3. Arrhenius plot of the first-order rate constant for inactivation of pig lactate dehydrogenase isoenzyme H_2M_2. Rate constant is calculated from the semilogarithmic progress curve of the initial loss of enzyme activity. Solutions of the enzyme were incubated at the indicated temperatures in 0.2M tris/HCl buffer, pH 7.5. Residual enzyme activity was assayed with 1 mM pyruvate in the assay mixture. The two curves show results obtained with an initial protein concentration of 1 mg/ml (o), and 10 μg/ml (∇), respectively. A second abscissa gives temperature in °C, and a second ordinate gives the corresponding half-life of the enzyme.

High temperature range

Results obtained at 58°C, with an initial protein concentration of 1 mg/ml, are shown in Fig. 4. It is seen from the semilogarithmic progress curves in Fig. 4a that the residual activities determined with 0.33, 1 and 10 mM pyruvate yield parallel straight lines. This means that the ratio of enzyme activities determined with these pyruvate concentrations is constant and this, in turn, indicates [6, 10-12] that the ratio in which subunits H and M contribute to the measured total enzyme activities does not change during the course of inactivation. In full agreement with this conclusion, it is seen from the series of electrophorograms in Fig. 4b that isoenzyme H_2M_2 is the only protein present in the heat-treated solution, even after 2.5 min at 58°C, when about 75% of the original activity is already lost. These results indicate that the mechanism involves the *all-or-none inactivation* of the four subunits of the tetramer H_2M_2. This conclusion is

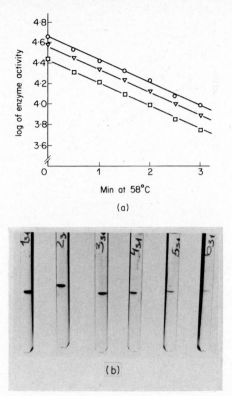

(a)

(b)

Figure 4. Loss of enzyme activity at 58°C with an initial protein concentration of 1 mg/ml. Heat treatment in 0.2M tris/HCl buffer, pH 7.5. The kinetics of inactivation are shown in Fig. 4a, where the logarithm of residual lactate dehydrogenase activity, as determined with 0.33 (∇), 1 (o) and 10 ($\square$) mM pyruvate in the assay mixture, is expressed in arbitrary units, and plotted against time. In Fig. 4b the corresponding series of electrophorograms is shown. Polyacrylamide gel electrophoresis was carried out according to Davis [8]. The gel cylinders were loaded with a 0.2 ml aliquot of the enzyme solution, treated for 0, 0.5, 1, 1.5, 2 and 2.5 min at 58°C, and stained for protein with amido black.

supported by the finding [5] that both isoenzymes H_4 and M_4 are more stable than isoenzyme H_2M_2 under the same conditions.

Similar experiments carried out in the temperature range 64°-54°C, and with initial protein concentrations ranging from 0.01 to 10 mg/ml, have yielded similar results with only the absolute value of the rate constant varying as indicated in Fig. 3.

Low temperature range

Figures 5 and 6 show markedly different results obtained at 51°C. It is seen from Fig. 5a that, with an initial protein concentration of 1 mg/ml, the residual activities determined with increasing pyruvate concentrations do not yield

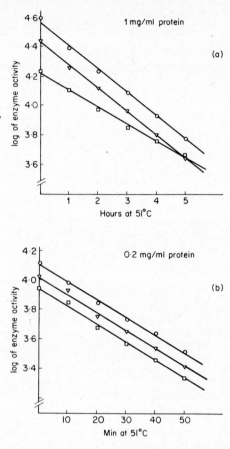

Figure 5. Kinetics of the loss of enzyme activity at 51°C with intial protein concentrations of 1 mg/ml (a) and 0.2 mg/ml (b). Further details as in Fig. 4a.

parallel semilogarithmic straight lines. In fact, the activity determined with 10 mM pyruvate in the assay mixture is lost at a slower rate than the activity determined with lower pyruvate concentrations. Since the activity determined with 10 mM pyruvate at zero time is mainly due to the presence of subunit M in isoenzyme H_2M_2 [5], this indicates that there must be an *M-shift* in the isoenzyme pattern, i.e. a shift towards isoenzymes enriched in isoenzyme M. In complete accordance with this, Fig. 6a shows that, after a 5-h incubation of a 1 mg/ml solution of isoenzyme H_2M_2, the residual lactate dehydrogenase activity is present in three main bands which correspond to isoenzymes H_2M_2, HM_3 and M_4. Accordingly, we may conclude that under these conditions the postulated *one-by-one pathway* is operating.

The above findings are characteristic of all results obtained between 52° and 44°C with an initial protein concentration in the range of 1-10 mg/ml. However, by lowering initial protein concentration to 0.2-0.01 mg/ml, results similar to those shown in Figs. 5b and 6b were obtained. It is seen from these figures that at low protein concentration an *M-shift* is no longer apparent, either from the activity ratios, or from the electrophoretic isoenzyme pattern. We should note

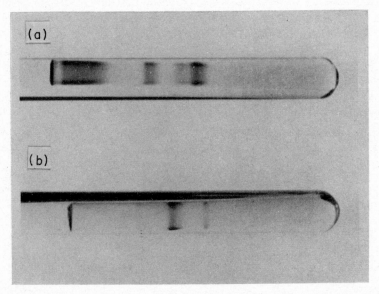

Figure 6. Electrophoretic isoenzyme pattern obtained after about 80% of the initial enzyme activity has been lost at 51°C, at different protein concentrations. Same experiment as Fig. 5. Staining for lactate dehydrogenase activity as described by Markert and Møller [9]. Load: 0.2 ml aliquot of the sample per gel. 5-h incubation of a 1 mg/ml solution of isoenzyme H_2M_2 (a), 50 min incubation of a 0.2 mg/ml solution of isoenzyme H_2M_2 (b).

that the faint bands which accompany that of isoenzyme H_2M_2 are identical with isoenzymes H_3M and HM_3 (Fig. 6b). The contaminations are also present in the original preparation, and in about the same negligible proportion.

Summarizing the observations in terms of our working hypothesis given in Fig. 2, we may conclude that the tetramer is apparently inactivated as such, by the *all-or-none mechanism,* under most conditions. These include low protein concentration in the entire range of temperatures studied, and even high protein concentration if the overall rate is rapid. However, if protein concentration is high and the overall rate of inactivation slow, the subunits of the tetramer appear to be inactivated in a *one-by-one manner.* This latter process involves the selective loss of H subunits coupled with a recombination, which is manifested in an *M-shift* of the isoenzyme pattern.

This proposal is supported by the observation that in our experiments a partial loss of enzyme activity appears to be an absolute requirement for an *M-shift*. Further support comes from the following considerations:

1. It is reasonable to assume that the temperature coefficient of the postulated reaction steps of recombination is lower than the temperature coefficient of inactivation. In this way we might account for the observation that the *one-by-one pathway* can successfully compete with the alternative *all-or-none pathway* only at relatively low temperature.

2. Induced recombination can only be observed at high protein concentration. This would, indeed, follow from the postulated reaction scheme of the *one-by-one mechanism*, since the overall rate of the complex consecutive steps of this pathway should be proportional to some higher exponent of protein concentration.

3. Induced recombination, and an increased stability at high protein concentration, are observed in the same temperature range (see Fig. 3). Indeed, this is to be expected on the basis of Figs. 1 and 2, since these schemes indicate that in a *simple all-or-none mechanism* one primary hit leads to the inactivation of four subunits, while in a *simple one-by-one mechanism* it will only result in the loss of one subunit.

REFERENCES

1. Kaplan, N. O., *Brookhaven Symp. Biol.* **17** (1964) 131.
2. Plagemann, P. G. W., Gregory, K. F. and Wroblewski, F., *Biochem. Z.* **334** (1961) 37.
3. Wachsmuth, E. D. and Pfleiderer, G., *Biochem. Z.* **336** (1963) 545.
4. Fondy, T. P., Pesce, A., Freedberg, I., Stolzenbach, F. E. and Kaplan, N. O., *Biochemistry* **3** (1964) 522.
5. Khan, M. G. and Südi, J., *Acta Biochim. Biophys. Acad. Sci. Hung.* **3** (1968) 409.
6. Südi, J., *Acta Biochim. Biophys. Acad. Sci. Hung.* (1969) in press.
7. Chilson, O. P., Costello, L. A. and Kaplan, N. O., *Biochemistry* **4** (1965) 271.
8. Davis, B. J., *Ann. N.Y. Acad. Sci.* **121** (1964) 404.
9. Markert, C. L. and Møller, F., *Proc. natn. Acad. Sci. U.S.A.* **45** (1959) 753.
10. Wieland, T. and Pfleiderer, G., *Adv. Enzymol.* **25** (1963) 329.
11. Markert, C. L., *in* "Cytodifferentiation and Macromolecular Synthesis" (edited by M. Locke), Academic Press, New York, 1963, p. 274.
12. Dawson, D. M., Goodfriend, T. L. and Kaplan, N. O., *Science, N.Y.* **143** (1964) 929.

FEBS Symposium, Volume 18, 1970, pp. 177-183

The Characterization of Lactate Dehydrogenase of Rat Brain After Electrophoretic Separation on Agar Gel

Z. MAŠEK, P. BREINEK and B. VEČEREK

1st Institute of Medical and Forensic Chemistry, Charles University, Prague, Czechoslovakia, and Department of Clinical Biochemistry, J. E. Purkyně University, Brno, Czechoslovakia

INTRODUCTION

Lactate dehydrogenase consists of at least five isoenzymes [1,2], which represent five possible tetramer combinations of two basic polypeptide chains [3]. They differ in electrophoretic mobility, K_m lactate, K_m pyruvate, K_m NAD$^+$, K_m NADH and other properties [4]. At present, investigation is concentrated on pure forms of isoenzymes, the anodically-migrating H_4 aerobic fraction and the cathodically-migrating M_4 anaerobic fraction. The multiple forms of LDH are studied chiefly after elution from the starch or agar gel [5, 6]. This paper presents the possibility of determining the main kinetic characteristics directly on the gel by means of nitrobluemonotetrazolium reduction.

MATERIALS AND METHODS

Adult rats of the Wistar strain from our inbred colony were decapitated. Tissue homogenates were prepared by repeated grinding of the whole brain in 3 to 10 volumes of 0.25M sucrose in the Potter-Elvehjem homogenizer. In all procedures the sample container was immersed in an ice bath. The homogenate was then centrifuged and the supernatant analyzed for enzyme content and separated by electrophoresis.

Total LDH activity was determined spectrally at 340 nm and expressed in mU/ml at 25°C (Biochemica Test, Boehringer). Agar gel electrophoresis was carried out according to Večerek *et al.* [7]. Barbital/HCl buffer 0.05M, pH 8.6,

Non-standard abbreviations. NBMT: 2-(*p*-nitrophenyl)-5-phenyl-3-(3,3′dimethoxy-4-diphenylyl) tetrazolium.
EC 1.1.1.27 1-lactate: NAD$^+$ oxidoreductase (Lactate dehydrogenase).

ionic strength 0.04, and 18 V/cm for 30 min separation, was used. For the visualization of enzyme activity on agar strips a new synthetized nitrobluemono-tetrazolium chloride (NBMT) was employed [8]. After electrophoresis, 5-mm wide strips were soaked in a 3.5 ml solution containing the following final concentrations—if not indicated otherwise: 10^{-2}M Na-L-lactate, 2×10^{-3}M NAD^+, 2×10^{-4}M PMS (phenazinmethosulphate) and 1.5×10^{-3}M NBMT. The incubation was interrupted after 30 min by adding 3% acetic acid in which the strips were washed for 3 h. Densitometric evaluations were carried out with a microphotometer MF 2 and a linear recorder. Densitometric areas are proportional to the amounts of nitrobluemonoformazane formed and are plotted as the sum total of extinctions on the graphs. Reproducibility of the results was about ± 4%.

RESULTS

The calibration curve for formazane formation is shown in Fig. 1. The total activities determined spectrally are plotted against amounts of nitrobluemono-formazane formed. After 30 min incubation only total activities higher than 50 mU/ml can be reliably detected densitometrically. The first part of the curve, up to an activity of about 1000 mU/ml, shows a linear relationship between both activity assays. At higher activities, at least two factors influence the relation between the spectral activity and the colour. Firstly, the developing formazane prevents further diffusion of the incubation medium into the gel. Secondly, if the colour is too intense, the Lambert-Beer law loses its validity and the change in activity does not correspond to the change in optical density. Hence, in our

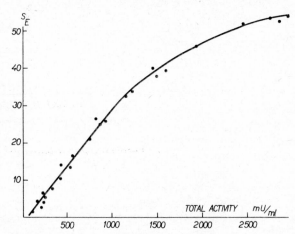

Figure 1. Calibration curve for formazane formation. The total activities determined spectrally are plotted against S_E, the amounts of nitrobluemonoformazane produced, expressed as the sum total of extinctions.

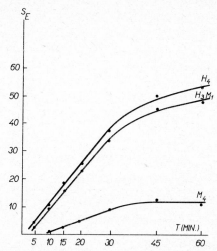

Figure 2. Time course of production of nitrobluemonoformazane. Reaction mixture as described in text, with the indicated production of nitrobluemonoformazane for H_4, H_3M_1 and M_4 forms.

experiments only activities of about 500 mU/ml were separated. The colour reaction was run from the lactate side, while the spectral test was performed from the pyruvate side. However, the curve enabled us to obtain rough information about the activity of the treated enzyme sample.

Using our procedure only during the first 30 min incubation, formazane production is linear with time, as shown in Fig. 2. As was previously described [8], a new tetrazolium salt NBMT was recommended for the determination of lactate dehydrogenase activity. Preparation with considerable yields and standard production of colour are advantages, as compared to INT [8, 9]. There are,

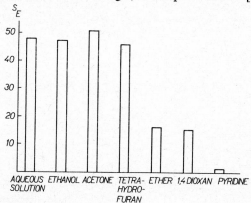

Figure 3. Effect of different organic solvents with the usual concentration of NBMT on formazane production. The different organic solvents were used in a minimal volume, 0.1 ml, added to incubation mixture.

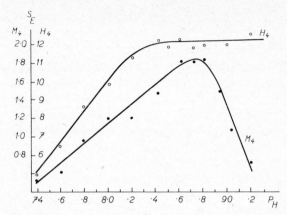

Figure 4. pH optima for the H_4 and M_4 forms. S_E, amounts of nitrobluemonoformazane formed, plotted against pH.

however, certain difficulties with dissolution in water, this being the only disadvantage, particularly for quantitative purposes. We tried adding the usual amount of NBMT in a minimal volume, 0.1 ml, of different organic solvents; this formed an opalescent solution with the incubation mixture. Figure 3 presents evidence that acetone and ethanol are most suitable and, with such minimal final concentrations, no changes of isoenzyme pattern were observed, as compared with aqueous solutions.

For the forward reaction the pH optimum was 8.5 to 8.8, which is in good agreement with other authors [10], as seen in Fig. 4. A higher spontaneous reduction of NBMT was noted only at a pH above 9.0 and even this pH does not essentially increase the background of the agar strips. Therefore we used pH = 8.6 for the incubation medium and the same kind of buffer as for the preparation of agar. Sensitive reagents (PMS, NBMT, NAD⁺) were added

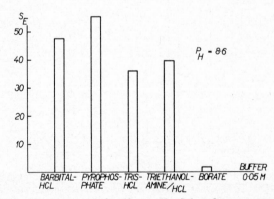

Figure 5. Effect of various types of buffer at pH = 8.6 on formazane production.

immediately before the start of the reaction. Effects of various types of buffer at pH = 8.6 are shown in Fig. 5. The best results were obtained with barbital/HCl and pyrophosphate buffers; borate buffer completely inhibited this reaction.

Under chosen conditions, i.e. with an acetone solution of NBMT/barbital buffer, pH = 8.6, we studied the H- and M-forms of LDH of whole brain. The average values for these isoenzymes corresponded to those found by Bonavita *et al.* [5]. Since the M_4 cathodic fraction is much weaker than the others in the rat brain, we prepared two samples, one for the H-, and the other for the M-form. The behaviour of these isoenzymes was examined with various concentrations of substrate and coenzyme. Two anodic and three cathodic isoenzymes can be detected. In both cases, only the H-form was evaluated. The whole activity of the M_4 sample was so high that we had to incubate only the

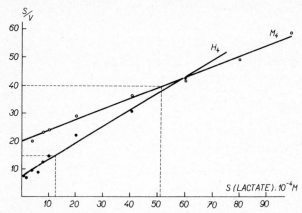

Figure 6. Determination of the K_m (lactate) for the M_4 and H_4 forms. S/V, ratio of substrate concentration to initial velocity, was plotted against substrate concentration.

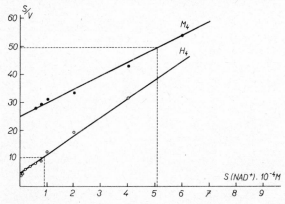

Figure 7. Determination of the K_m (NAD$^+$) for the M_4 and H_4 forms. S/V, ratio of NAD$^+$ concentration to initial velocity, was plotted against NAD$^+$ concentration.

fourth and fifth fractions to maintain approximately the same conditions—mainly the pyruvate concentration—as in the case of the H-form. From the densitometric data we attempted to calculate the K_m values for these two isoenzymes. Graphical evaluations are given in Figs. 6 and 7. The ratio of substrate concentration to the initial velocity was plotted against substrate concentration. K_m values for lactate concentration were 1.3×10^{-3}M for the H-form and 5.1×10^{-3}M for the M-form. The K_m values for coenzyme NAD$^+$ concentration amounted to 9×10^{-5}M for the H-form and 5.1×10^{-4}M for the M-form. The K_m values for the H_4 isoenzyme are approximately those obtained for the brain by Nisselbaum *et al.* [11]. Our values for M_4 differ slightly from these data, but they correspond with values valid for skeletal muscle tissue [11].

Figure 8 shows the relationship between the final concentration of PMS and the densitometric activity and Fig. 9 shows a similar dependence for various

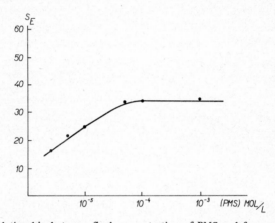

Figure 8. Relationship between final concentration of PMS and formazane production.

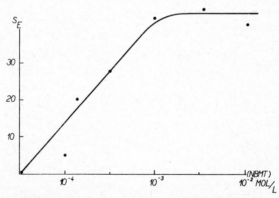

Figure 9. Relationship between final concentration of NBMT and formazane production.

NBMT concentrations. It may be concluded that sufficient amounts of both reagents were used, and that the reaction in the range of concentrations usually employed was relatively insensitive to changes in PMS and NBMT.

DISCUSSION

The coupled reduction of tetrazolium salts reported by many authors [12, 13] is suitable both for the determination of the total LDH activity in body fluids [9] and for histochemical [14] and electrophoretic purposes. The procedures, however, make use of a disadvantageous pH with regard to the equilibrium of the lactate-pyruvate reaction, due to the fact that higher pH values have a deleterious effect on the colour reagent stability. The NBMT seems to be more resistant to reduction at the pH optimum for the forward reaction and the background of the agar strips is not substantially coloured. This greater resistance enabled us to perform an investigation of isoenzyme properties under convenient conditions. A minimal amount of sample, 10 μl, is required for a simultaneous kinetic study of several isoenzymes. An electrophoretic apparatus can be connected to the microhomogenizer modified in our laboratory to use about 5 mg of tissue. It is, however, necessary to retain high accuracy during all the procedures to achieve good reproducibility of results. This report should be considered as an introduction to some quantitative aspects of agar gel electrophoresis of enzymes and the method described will be further tested and applied to other enzymes and tissues.

REFERENCES

1. Wieland, T. and Pfleiderer, G., *Biochem. Z.* **329** (1957) 112.
2. Markert, C. L. and Møller, F., *Proc. natn. Acad. Sci. U.S.A.* **45** (1958) 753.
3. Apella, E. and Markert, C. L., *Biochem. biophys. Res. Commun.* **6** (1961) 171.
4. Wilson, A. G., Kaplan, N. O., Levine, C., Pesce, A., Reichlin, M. and Allison, W. S., *Fedn Proc. Fedn Am. Socs exp. Biol.* **23** (1964) 1258.
5. Bonavita, V., Ponte, F. and Amore, G., *J. Neurochem.* **11** (1964) 39.
6. Krieg, A. F., Rosenblum, L. J. and Henry, J. B., *Clin. Chem.* **13** (1967) 196.
7. Večerek, B., Štěpán, J., Hynie, I. and Kácl, K., *Colln Czech. chem. Commun.* **33** (1968) 141.
8. Homolka, J., Křišťanová, D. and Večerek, B., *Clinica chim. Acta* **13** (1966) 125.
9. Babson, A. L. and Phillips, G. E., *Clinica chim. Acta* **12** (1965) 210.
10. Vesell, E. S., *Nature, Lond.* **210** (1966) 421.
11. Nisselbaum, J. S., Packer, D. E. and Bodansky, O., *J. biol. Chem.* **239** (1964) 9.
12. Nachlas, M. M., Margulies, S. I., Goldberg, J. D. and Seligman, A. M., *Analyt. Biochem.* **1** (1960) 317.
13. Raabo, E., *Scand. J. clin Lab. Invest.* **15** (1963) 405.
14. Lojda, Z. and Frič, P., "Abstracts of 5th FEBS Meeting", Academic Press, London and New York, p. 135.

FEBS Symposium, Volume 18, 1970, pp. 185-194

Detection of Basic Forms of Lactate Dehydrogenase *in situ*

Z. LOJDA and P. FRIČ

Laboratory of Angiology, 1st Department of Pathology,
2nd Research Unit of Gastroenterology, Medical Faculty,
Charles University, Prague, Czechoslovakia

Lactate dehydrogenase (LDH, E.C. 1.1.1.27) exists in multiple molecular forms (LDH isoenzymes) which are presumed to be tetramer molecules built up from two parent subunits. These subunits have been termed A and B [1] or M and H [2], the latter designation being widely used. After synthesis the monomeric subunits randomly associate to form five LDH isoenzymes: LDH 1 (= 4H), LDH 2 (= 3HM), LDH 3 (= 2H2M), LDH 4 (= H3M), and LDH 5 (= 4M). This accounts for five bands separated electrophoretically, the fastest anodal band being LDH 1 [3]. LDH isoenzymes differ in many properties [4-6], including their metabolic significance. The pattern of LDH isoenzymes is quite specific for each individual organ and shows species differences. These facts became apparent from comprehensive biochemical studies performed on homogenates of different organs. It must be realized, however, that such studies cannot reveal the contribution of individual cells of the sample examined. This fact is very important for a better understanding of many physiological and pathophysiological problems. As in other enzyme studies, the contribution of individual cells could be ascertained only by detection *in situ,* and this has already formed the subject of a number of papers [7-15].

There are two approaches for localizing the basic form *in situ.* The immunofluorescence method [7] demonstrates the enzymatic proteins as such and is based on the immunological differences between the basic subunits. The isolation of LDH 5 from tissues is not difficult, but problems arise in preparing it in an antigenically-pure form. Even then the localization studies are restricted to the species from which the antigen was prepared and must take into consideration the solubility of LDH, as well as other difficulties usually encountered with immunofluorescence methods (cf. ref. 16). Such studies are under way in our laboratory.

The second possibility is the localization of the enzymatic reaction catalysed by LDH. In this approach, different sensitivities towards analogues of substrate [8, 10, 13], analogues of coenzyme [12], as well as towards some inhibitory

185

agents such as urea [9-11, 14, 15], guanidine hydrochloride [11], 50% acetone [13], pyruvate [10, 15], lactate [9], or thermal inactivation [11], have been used. The reported results are not in complete agreement and the procedures used by most workers are subject to criticism. This prompted us to reinvestigate the problem.

MATERIALS AND METHODS

Our study was undertaken with the following materials:

1. Human material: (a) biopsies from jejunal and gastric mucosa, liver, pancreas; (b) material from autopsies (6-h post mortem) from liver, kidney, myocardium, aorta.

2. Animal material (removal immediately after killing the animals) from liver, myocardium, aorta, jejunum of rats and rabbits.

The material was processed in the following way. Parts of the specimens were dried with filter paper and weighed. 0.04M barbital buffer (pH 8.4) was added in the ratio 1:5 (aortae) or 1:10 (other organs) to the specimens and the samples homogenized in an all-glass homogenizer (Potter-Elvehjem type) at 0°C. The protein concentration of the supernatant was determined according to the method of Lowry et al. [17]. Electrophoresis was carried out according to Wieme [18] on microscope slides in 1% Difco Special Agar-Noble for 40 min at 140 V. The enzyme layer was then covered with an agar-gel medium (see below). The remaining parts of specimens were frozen on dry ice. A portion of the frozen samples was thawed and processed in the same way as described above. The remaining frozen blocks were cut in a cryostat at −15°C to sections 10-15 μ thick and transferred to warm slides. Sections from several blocks were transferred to the same side to ensure identical conditions for incubation. Incubation media were the same as in our previous studies [10, 13]. They were prepared from the universal stock solution [19, 20]:

nitroBT* (for sections) or INT† (for zymograms) 2 mg% . . . . 5 ml
0.1M Tris-HCl or phosphate buffer pH 7.4 . . . . 5 ml
0.1M potassium cyanide or sodium azide . . . . 4 ml
0.05M magnesium sulphate . . . . 4 ml
1M sodium lactate . . . . 4 ml
NAD‡ . . . . 14.5-110 mg
                                    pH adjusted to 7.4
PMS§ . . . . 0.3 mg

* 2,2′ di-p-nitrophenyl-5,5′-diphenyl-3,3′-(3,3′-dimethoxy-4,4′-biphenylene) ditetrazolium chloride; Loba Chemie, Wien.
† 2-p-nitrophenyl-3-p-nitrophenyl-5-phenyl tetrazolium chloride; Lachema, Brno.
‡ Nicotinamide adenine dinucleotide; Sigma, St. Louis or Boehringer, Mannheim.
§ N-methyl phenazonium methosulphate, Koch & Light, Colnbrook.

A part of this solution was carefully mixed with an equal part of the following solutions prepared in 0.1M Tris-HCl or phosphate buffer pH 7.4 (pH was checked): 2% agar (mixed at 70°C), 10% gelatine (mixed at 37°C), 10 or 15% polyvinylalcohol (PVA, mixed at room temperature). The gelatine and agar substrate mixtures were poured out on the cover glasses and allowed to gel. The cover glasses with gelled substrate films were then transferred onto the slides with sections and incubated in the dark at room temperature (gelatine) or at 37°C (agar). The PVA solutions were applied dropwise on the sections provided with paraffin frames and incubated at 37°C. Sections were inspected during the incubation and photographed if necessary. The zymograms were revealed with agar-gel media for 60 min at 37°C. After termination of the incubation, the slides were processed in the usual manner.

The following modifications of the incubation media were used:

(a) The concentration of nitroBT was increased x 3 (6 mg% instead of 2 mg% solution).

(b) The concentration of lactate was increased x 10-20.

(c) Sodium lactate was replaced by equimolar solutions of sodium α-hydroxy-butyrate or α-hydroxyvalerate (we are indebted to Dr. Heyrovsky for the synthesis of these substrates).

(d) NAD was replaced by 3-acetylpyridine-NAD or 3-acetylpyridine-deamino-NAD (Calbiochem).

(e) Inhibitory agents were added to the incubation mixtures (the final concentration is indicated): (i) Urea (3M, 2M, and 1M). Urea interfered with the gelatination of gelatine media and these media were therefore dropped onto the framed sections; (ii) Pyruvate (0.4M, 0.3M and 0.2M).

(f) Sections and slides with the enzyme layer were exposed after electrophoresis to 50% or 100% acetone, or transferred into the wet chambers at 80°C for 10-60 min.

RESULTS AND DISCUSSION

As pointed out in our previous paper [10], there are several prerequisites which must be fulfilled in studies of LDH isoenzymes *in situ:*

1. One must check if the spectra of LDH isoenzymes remain the same after freezing and thawing the specimens.

2. Owing to species, and even organ differences, the demonstration *in situ* must be run simultaneously with the demonstration after electrophoretic separation.

3. Owing to the fact that LDH isoenzymes are not firmly structurally bound, the same conditions as for soluble coenzyme-dependent dehydrogenases must be retained (cf. ref. 19), i.e. the use of gel media with PMS and with cytochrome

oxidase inhibitors. Even in this case only cellular (not intracellular) localization is possible.

These criteria have not been fulfilled in the majority of the papers quoted above.

Let us now consider some other features:

1. *The gel media.* PVA media [14, 21] are viscous and are not entirely free from diffusion and localization artifacts. The agar-gel media [10, 19] are excellent for the zymograms and for a rough orientation in sections. After a certain period, formazan is formed over the sections, but the diffusion artifacts are not great if a restricted period of incubation is used. Best results were obtained with gelatine media [22-24], the gelatine concentration being raised to 10% and the media supplemented with a cytochrome oxidase inhibitor and PMS. The sections can be readily inspected during incubation and, if necessary, a microphotograph can be made. The only drawback is that gelatination is impeded by urea.

2. *The variation in concentration of substances in the medium.* The media used in the histochemical demonstration of LDH are prescribed according to the authors' experience. They were not tested for ability to reflect different levels of LDH activities correctly, as for some media for hydrolytic enzymes [25]. The pattern obtained with our media compared favourably with the reported results. McMillan [15] stresses the use of a high concentration of nitroBT, and a lower concentration of substrate, to minimize diffusion artifacts. In our hands satisfactory results were obtained with the prescribed concentrations, and no apparent improvement in the tested material was observed by enhancing the nitroBT concentration to 6 mg/ml and lowering the substrate concentration. The point of a major importance, however, is the concentration of NAD. Pearse [26] stresses the higher concentration of NAD in the histochemical media, and higher concentrations of NAD are also recommended in other prescriptions [19, 27]. It must be realized, however, that higher concentrations of NAD interfere with some inhibitors so that the inhibitory effect is lower or entirely absent (see below).

3. *Substrate analogues.* From the work of Allen [8], it may be inferred that, with higher aliphatic α-hydroxy acids, cathodal bands (and thus M-subunits) only are revealed. On the other hand, in clinical practice [28], α-hydroxybutyric acid has been used as a substrate for the H-form (diagnosis of myocardial damage). In our experiments previously described [13] we demonstrated that, with α-hydroxybutyric acid, only bands intensely stained with lactate are revealed, irrespective of M- or H-subunits (cf. Fig. 1 a:d, b:e, c:f). As the α-hydroxyvalerate is oxidized even more slowly, the intensity of bands is weaker and only very few bands are thus revealed (see Fig. 1 g, h, i). In the myocardium one obtains the impression that the H-form is revealed, but in the liver the fifth band, i.e. the M-form, is shown. From this it is clear that

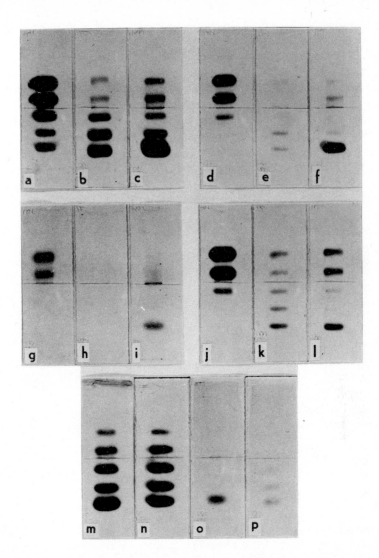

Figure 1. The pattern of LDH isoenzymes in human myocardium (a, d, g, j), aorta (b, e, h, k), liver (c, f, i, l), normal jejunal mucosa (m, o), and jejunal mucosa of a patient with celiac sprue (n, p) revealed with lactate (a, b, c, m, n), α-hydroxybutyrate (d, e, f), α-hydroxyvalerate (g, h, i), after pretreatment with 50% acetone for 30 min (j, k, l), and after inhibition with 0.3M pyruvate (o, p). Anode is at the top, point of application can be seen as a distinct line in the middle of each zymogram. With the substrate analogues only bands strongly stained with lactate are demonstrated (H-form in myocardium (d, g) M-form in liver (f, i)). Acetone inhibits M-form preferentially (j, k, l). Pyruvate inhibits H-form more than M-form (o, p).

α-hydroxybutyrate does not distinguish between M- and H-forms, and that its use instead of lactate in clinical studies has the same shortcomings as the use of lactate. Since in the aorta only the fourth and fifth bands are revealed, the use of this substrate with incubations can help in revealing the source of the M-form, which was demonstrated mainly in the muscle cells of the inner media and in cells of plaques (see ref. 13). As the distribution pattern of isoenzymes is similar if lactate and/or α-hydroxybutyrate are used (only intensities differ), we do not imagine that a special enzyme is responsible for the oxidation of α-hydroxybutyrate, as has been claimed [29].

4. *Coenzyme analogues.* It was concluded [12] that, with acetylpyridine-NAD, only M-LDH type was demonstrated. We could not obtain these results in our material, however (see Fig. 2 e-h). As with substrate analogues, only the most

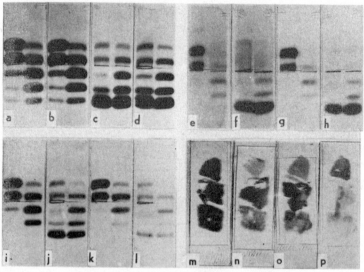

Figure 2. The pattern of LDH isoenzymes in human myocardium (a-left, b-left, e-left, g-left, i-left, k-left), aorta (a-right, b-right, e-right, g-right, i-right, k-right), liver (c-left, d-left, f-left, h-left, j-left, l-left), and jejunal mucosa (c-right, d-right, f-right, h-right, j-right, l-right) revealed with NAD concentrations of 2.5 mg/ml (b, d), 0.3 mg/ml (a, c), with acetylpyridine-NAD (e, f), with acetylpyridine-deamino-NAD (g, h), and in the presence of 3M urea with NAD concentrations of 2.5 mg/ml (i, j) and 0.3 mg/ml (k, l). No apparent changes in the pattern with various concentrations of NAD (a, b, c, d). After application of coenzyme analogues only strongly stained bands are revealed (e, f, g, h). The inhibitory effect of urea depends on the concentration of NAD—greater inhibition of M-form in media with a lower concentration of NAD (k, l). The staining pattern in sections of human liver (top sections), aorta (2nd sections), myocardium (3rd sections), and kidney (bottom sections) revealed with the standard gelatine medium for LDH (m), after 60 min pretreatment with 50% acetone (n), inhibited with 3M urea (o), and with 0.3M pyruvate (p). In liver sections, decrease of staining intensity after acetone (n) and urea (o), slight decrease after pyruvate (p). In myocardium the staining is affected with pyruvate only (p); note complicated diminution of the staining in kidney.

intensely stained LDH isoenzymes are revealed. It is not clear whether a cytochrome oxidase inhibitor was used in the studies of Benitez-Bribiesca *et al.* [12] but this could explain the discrepancies.

5. *Inhibitory agents.* (a) *Urea.* As described [9-11, 14, 15], urea inhibited preferentially cathodic bands, i.e. the M-form, in our present experiments. Optimal concentration was found to be 3M. The results of the inhibition experiments are inversely proportional to the concentration of NAD in the incubation medium. This is shown in Fig. 2 i:k, j:l, in comparison with Fig. 2 a:b and c:d, the latter showing no substantial differences in the pattern if the reduced concentration of NAD is used without the inhibitor. Urea interferes with the gelatination of gelatine media, however. (b) *In this connection* the effect of acetone [13] is worth mentioning. 50% acetone (in contradistinction to 100% acetone) suppresses preferentially cathodic bands (Fig. 1 j,k,l); 60 min was found to be the best time in our studies. (c) *Pyruvate.* The optimal concentration was found to be 0.3M. This concentration inhibits mainly the anodal fractions (H-form), the cathodic bands being inhibited to a lesser degree (Fig. 1 m:o, n:p). Because pyruvate also inhibits the M-form to some extent, it is less suitable for the selective demonstration of the basic forms of LDH than acetone or urea. Increasing the concentration of lactate did not significantly change the LDH isoenzyme pattern, in contradistinction to the statement of Brody and Engel [9] who used aqueous media. Such a procedure cannot be recommended owing to the solubility of LDH.

The results of the analysis of zymograms can be applied to the sections. Some examples are shown in Fig. 2m (control), 2n (acetone), 2o (urea), and 2p (pyruvate), demonstrating the effect in the human liver, aorta, myocardium and kidney. Acetone and urea inhibit predominantly the LDH in liver, pyruvate in myocardium and kidney. In Fig. 3 a-d human liver, and in Fig. 3 e-h human myocardium, are shown at a somewhat higher magnification. Acetone (Fig. 3 b,f) and urea (Fig. 3 c,g) inhibit enzyme activity in liver cells and in muscle cells of a branch of the coronary artery, pyruvate (Fig. 3 d,h) in the myocardial cells. Thus, in myocardial cells mainly H-form, and in liver and muscle cells of the coronary artery mainly M-form, is demonstrated.

Our previous paper [13] demonstrated the differences in distribution of basic LDH forms in human aorta: M-form derives mainly from the muscle cells of the inner part of the media and from the cells of plaques; while the muscle cells of the outer media contain a somewhat higher amount of H-form than the muscle cells in the inner media. This was demonstrated on the basis of acetone pretreatment and with the use of α-hydroxybutyrate as the substrate.

In the small intestinal mucosa, a shift of LDH isoenzymes towards the faster-moving fractions was observed in celiac sprue patients (both children and adults) [30]. Cellular infiltration of the lamina propria in these conditions makes a simultaneous histochemical analysis a prerequisite for a correct

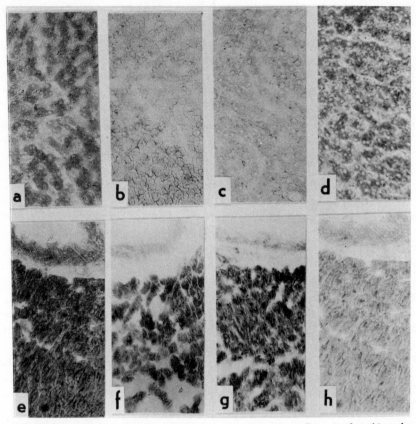

Figure 3. Sections of human liver (a, b, c, d) and myocardium (e, f, g, h) under a higher magnification (x 100). Standard gelatine medium (a, e), pretreatment with 50% acetone for 60 min (b, f), inhibition with 3M urea (c, g), and with 0.3M pyruvate (d, h). Acetone and urea decrease the staining in liver cells (b, c) and in muscle cells of the coronary artery (top of f, g). Pyruvate affects mainly the myocardial cells (h).

interpretation of biochemical findings. On the basis of our histochemical studies, the mature enterocytes covering the villi in healthy subjects contain predominantly the M-form of LDH, whereas the enterocytes of the crypts reveal a higher participation of the H-form [31, 32]. This was concluded from results of inhibition studies with acetone, urea and pyruvate. In celiac sprue enterocytes the effect of these agents is somewhat different [32]: greater inhibition with pyruvate than with urea and acetone. This shows that the poorly differentiated enterocytes contain a·higher amount of H-subunits. The shift in zymograms is thus due mainly to the enterocytes.

Our histochemical studies on gastric mucosa (in collaboration with Procházka) revealed a prevalence of the H-form in parietal cells.

These few examples illustrate the usefulness of the *in situ* studies of the distribution of basic LDH forms for a correct interpretation of biochemical findings. They also show the necessity and usefulness of a close correlation between analyses *in vitro* and *in situ.* The histochemical approach demonstrates the H- or M-form in cells without the possibility of distinguishing their proportions in cells as such, or as the hybrid isoenzymes. This would be possible if selective inhibitors of hybridic fractions could be devised.

REFERENCES

1. Markert, C. L. and Ursprung, H., *Devl. Biol.* **5** (1962) 363.
2. Cahn, R. D., Kaplan, N. O., Levine, L. and Zwilling, E., *Science, N.Y.* **136** (1962) 962.
3. Webb, E. C., *Nature, Lond.* **203** (1964) 821.
4. Kaplan, N. O. and Ciotti, M. M., *Ann. N.Y. Acad. Sci.* **94** (1961) 701.
5. Appella, E. and Markert, C. L., *Biochem. biophys. Res. Commun.* **6** (1961) 171.
6. Dawson, D. M., Goodfriend, T. L. and Kaplan, N. O., *Science, N.Y.* **143** (1964) 929.
7. Nace, G. W., *Ann. N.Y. Acad. Sci.* **103** (1963) 980.
8. Allen, J. M., *Ann. N.Y. Acad. Sci.* **94** (1961) 937.
9. Brody, I. A. and Engel, W. K., *J. Histochem. Cytochem.* **12** (1964) 687.
10. Lojda, Z., Histochemistry of enzymes, Thesis, Charles University, Faculty of Medicine, 1965.
11. Ikawa, Y., *Gann* **56** (1965) 201.
12. Benitez-Bribiesca, L., Horwitz, C. H. and Fahimi, D. H., *J. Histochem. Cytochem.* **14** (1966) 763.
13. Lojda, Z. and Frič, P., *Čsl. Patol.* **3** (1967) 161.
14. Kunze, K. D., *Histochemie* **11** (1967) 350.
15. McMillan, P. J., *J. Histochem. Cytochem.* **15** (1967) 21.
16. von Mayersbach, H., *in* "Handbuch der Histochemie" (edited by W. Graumann and K. Neumann), Gustav Fischer Verlag, Stuttgart, 1966, Bd. I, T. 2, p. 188.
17. Lowry, O. H., Roseborough, N. J., Farr, A. L. and Randall, R. J., *J. biol. Chem.* **193** (1951) 265.
18. Wieme, R. J., "Studies on Agar Gel Electrophoresis", Arscia N.V., Brussels, 1959.
19. Lojda, Z., *Folia morph.* (Prague), **13** (1965) 84.
20. Lojda, Z., "Histochemical Detection of Enzymes", 2nd part, Czechoslovac Society for Histochemistry and Cytochemistry, Brno, 1968.
21. Altmann, F. P. and Chayen, J., *Nature, Lond.* **207** (1965) 1205.
22. Fahimi, H. D. and Amarasingham, C. R., *J. Cell Biol.* **22** (1964) 29.
23. Benitez, L. and Fischer, R., *J. Histochem. Cytochem.* **12** (1964) 858.
24. Pette, D. and Brandau, H., *Enzymol. biol. clin.* **6** (1966) 79.
25. Lojda, Z., van der Ploeg, M. and van Duijn, P., *Histochemie* **11** (1967) 13.
26. Pearse, A. G. E., "Histochemistry, Theoretical and Applied", 2nd edn., Churchill, London, 1960.

27. Barka, T. and Anderson, P. J., "Histochemistry", Harper & Row, New York, Evanston, and London, 1963.
28. Wilkinson, J. H., "An Introduction to Diagnostic Enzymology", E. Arnold, London, 1962.
29. Sanwald, R. and Kirk, J. A., *Nature, Lond.* **209** (1966) 912.
30. Frič, P. and Lojda, Z., *Clinica chim. Acta* **12** (1965) 111.
31. Lojda, Z. and Frič, P., Abstracta Congressus Gastroenterologiae Internationalis A.S.N.E.M.G.E., StZdN, Prague, 1968, p. 211.
32. Lojda, Z. and Frič, P., Abstracta Congressus Gastroenterologiae Internationalis A.S.N.E.M.G.E., StZdN, Prague, 1968, p. 212.

FEBS Symposium, Volume 18, 1970, pp. 195-202

Properties of Horse Liver Alcohol Dehydrogenase Isoenzymes

J. P. von WARTBURG, P. M. KOPP and
U. M. LUTSTORF

*Medizinisch-chemisches Institut, University of Berne,
Berne, Switzerland*

Heterogeneity of horse liver alcohol dehydrogenase was first reported by Dalziel [1], who separated crystalline enzyme preparations into two distinct fractions by electrophoresis and chromatography. Three isoenzymes are found in rhesus monkey [2]. Two of them occur in the liver [3]; the third is localized mainly in the gastro-intestinal tract. The two isoenzyme preparations isolated from liver show marked differences in their catalytic properties [3]. Five fractions of ADH could be detected in horse liver by means of electrophoresis on agar gel [3, 4]. Similar electrophoretic patterns were obtained by McKinley-McKee and Moss on starch gel [5]. An improved chromatographic procedure permits the separation of up to seven different fractions [6, 7].

Previous studies have revealed two identical active sites in horse [8] and human liver alcohol dehydrogenase [9], suggesting a dimeric structure of these enzymes. If two non-identical subunits exist, their random aggregation would result in the formation of three active isoenzymes. This is the number of isoenzymes actually found in rhesus monkey and man [6, 10]. The existence of more than three fractions of ADH in horse liver might be expected on the basis of a tetrameric structure of the enzyme in this particular species. Four subunits in horse liver ADH have been suggested by Vallee and his collaborators [11]. Evidence for only two subunits arises from studies on the primary structure of the main fraction of horse liver ADH carried out by Jörnvall and Harris [12]. The results presented in this paper are consistent with a dimeric structure of ADH isoenzymes in horse liver. The separation and purification of the iso-enzymes of horse liver ADH was achieved by chromatography on carbo-

Non-Standard Abbreviations: Alcohol dehydrogenase, ADH; gram atoms, gm. at.; molecular weight, mol. wt.; isoelectric point, IEP; 5-dihydrotestosterone, DHT; phenyleth-anediol, PED.
    Enzymes. Alcohol dehydrogenase or alcohol: NAD oxidoreductase, EC 1.1.1.1; Peroxidase or donor: hydrogenperoxide oxidoreductase, EC 1.11.1.7.

xymethyl- and diethylaminoethyl-cellulose [7] . Three fractions (IIc, III and IV) have been crystallized from methanol-containing buffer. The electrophoretic mobility on agar gel of the purified isoenzymes is shown in Fig. 1. The three fractions IIa, IIb and IIc move close to each other, so that no complete electrophoretic separation can be obtained on crude horse liver extracts. Larger differences exist, however, between the electrophoretic mobilities of isoenzymes III, IV and V. The model of a tetramer built by random combination of two polypeptide chains with unequal charge would predict five fractions with corresponding differences in electrophoretic mobility. The existence of seven distinct chromatographic fractions with unequal distances between the electrophoretic bands can only be explained on the basis of a more complex model.

One possibility is the existence of active protomers with molecular weights of 40,000 or 160,000. The molecular weight of the main fraction (isoenzyme III) is known to be approximately 80,000 [11, 13]. The molecular weights of the purified ADH fractions were estimated by gel filtration on a Sephadex G-200 column (Table 1). The values for the elution volumes of all fractions are nearly identical, and distinct from those obtained from horse radish peroxidase (EC 1.11.1.7, mol. wt. 44,000 [14]) and yeast alcohol dehydrogenase (EC 1.1.1.1, mol. wt. 150,000 [15]). The liver and yeast enzymes are known to be

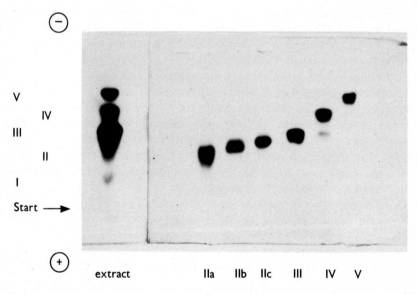

Figure 1. Agar gel electrophoresis of horse liver ADH isoenzymes. Electrophoresis at pH 9.0, Tris-HCl buffer [3]. Staining for ADH-activity by incubation with nitrotetrazolium blue 0.42 mg/ml, phenazine metasulfate 0.034 mg/ml, ethanol 2.5×10^{-2}M, NAD 1.25×10^{-3}M at $37°$C..
Left: crude liver extract; right: fractions separated and purified by chromotography.

Table 1. Gel filtration of horse liver ADH isoenzymes.

Enzyme	Horse liver alcohol dehydrogenase						Peroxidase	Yeast ADH
	IIa	IIb	IIc	III	IV	V		
v_e/v_O	1.82	1.94	1.89	1.82	1.90	1.90	2.20	1.61

Mean value of two determinations of v_e/v_O on a Sephadex G-200 column (1.9 x 27.5 cm); flow rate 12 ml/h; fraction volume 2.0 ml; Na-phosphate, pH 7.0, 0.05M; sample, 2 mg protein in 2 ml buffer; determination of v_O with blue dextran; 4°C.

metallo-enzymes. It is conceivable that a variable zinc content could lead to different electrophoretic mobilities. In their purest enzyme preparations, Vallee and his collaborators [11] found 3.4-3.6 gm. at. of zinc per mol. wt. 80,000. Values of 3.8-4.2 gm. at. per mol. wt. 84,000 were found in Theorell's laboratory [16] and by Oppenheimer et al. [17]. The zinc content of the four fractions IIc, III, IV and V of our preparations ranged from 3.3-3.6 gm. at. per mol. wt. 80,000 (Table 2) when the enzyme is crystallized at pH 7.0-7.5. However, crystallization at pH 8.8 leads to the incorporation of 4 gm. at. of zinc. An identical zinc content of fractions IIc (the minor fraction in Dalziel's preparation [1]) and III is in accord with the determinations reported by Åkeson [16]. It seems that the zinc content depends on the conditions of crystallization but that a differential zinc content is not relevant to the occurrence of isoenzymes.

Horse liver ADH has previously been shown to be active against 3β-hydroxy-5β-cholanic acid [18]. This activity was attributed by Theorell et al. [19] to a

Table 2. Zinc content of horse liver ADH isoenzymes.

Enzyme	Conditions of crystallization	Zinc gm. at./mol. wt. 80,000
IIc	Na-phosphate, pH 7.0	3.4
III	Na-phosphate, pH 7.0	3.3-3.6
IV	Na-phosphate, pH 7.5	3.4-3.6
V	Na-phosphate, pH 7.5	3.5
III	Tris-HCl, pH 8.8	4.1-4.3
III	Åkeson [16]	4.0-4.2
IIc	Åkeson [16]	4.2
III	Oppenheimer et al. [17]	3.8-4.0
III	Drum et al. [11]	3.4-3.6

Zinc determinations by atomic absorption (Unicam No. SP 90) of enzyme solutions dialyzed against metal free buffer [9].

particular enzyme, the steroid-active liver alcohol dehydrogenase. This sub-fraction is more basic than the main fraction and corresponds to isoenzyme IV in our preparations. The amino acid composition of both fractions was found to be nearly identical. Therefore it was considered by these authors that the difference in electrophoretic mobility might be due to different amide nitrogen contents. According to our determinations, isoenzymes III, IV and V all contain 28 ± 1 acid amide residues (Table 3).

Table 3. Acid amide content* of horse liver ADH isoenzymes.

Enzyme	III	IV	V
Mean value	28.1	28.5	27.1
1st determination	27.0	29.3	27.8
2nd determination	29.2	27.6	26.5

*Acid amide groups per mol. wt. 80,000, determined according to Knight and Blomstrom [27].

Various buffer ions are known to influence the electrophoretic mobility of proteins. The IEP of pH 6.8 for the main fraction of horse liver ADH (isoenzyme III) was determined by Dalziel [1] by free boundary electrophoresis in phosphate buffer. When the pronounced endosmosis in agar gel is taken into account, isoenzyme III shows only a small anodic mobility at pH 9.0 (Fig. 1), indicating that the IEP might be higher than pH 6.8 in Tris buffer. Isoenzyme IV migrates slowly to the cathode, suggesting an IEP above pH 9.0. Therefore, the isoelectric points of both isoenzymes were determined by the method of isoelectric focusing (Electrofocusing column LKB 8101 and ampholytes from LKB-Produkter AB, Stockholm-Bromma 1, Sweden) according to Vesterberg and Svensson [20]. In this system, which is free of buffer ions, an IEP of 8.7 was found for isoenzyme III and 9.3 for isoenzyme IV. This result demonstrates that the binding of buffer ions can alter the electrophoretic mobility of ADH. Nonetheless, the possibility of a differential binding of buffer ions can be excluded as a cause for the occurrence of ADH fractions with distinct electrophoretic mobilities. The separation of isoenzymes can be achieved by zone electrophoresis on several carriers, such as polyacrylamide, starch and cellulose acetate in the presence of a variety of buffer ions. Furthermore, the electrophoretic mobility of isoenzymes is unaffected during purification.

Ursprung and Sofer [21] have shown that new electrophoretic bands arise after the preincubation of purified ADH from Drosophila melanogaster in the presence of NAD. They are due to the formation of non-specific enzyme-coenzyme complexes. The extent of this conversion depends on time and

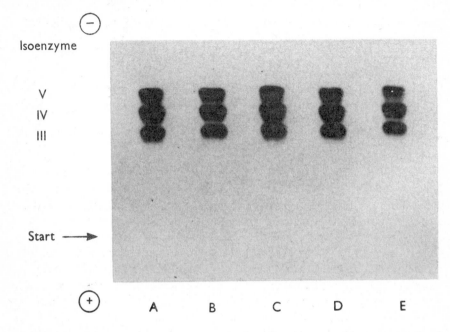

Figure 2. Electrophoretic pattern of horse liver ADH isoenzymes after preincubation with coenzymes. For the electrophoretic procedure see Fig. 1. Conditions of incubation: control without coenzyme (A), 1.0×10^{-2}M NAD (B), 1.0×10^{-3}M NAD (C), 1.0×10^{-2}M NADH (D), 1.0×10^{-3}M NADH (E), in Na-pyrophosphate pH 8.8, 2×10^{-2}M at 4°C for 2 h. Identical patterns were obtained after 24 and 48 h incubation.

coenzyme concentration. McKinley-McKee and Moss [5] previously proposed a similar mechanism in order to explain the heterogeneity of horse liver ADH. We therefore incubated a mixture of pure isoenzymes III, IV and V with coenzyme in the oxidized and reduced form. No shift in the position of the activity of the single isoenzymes, and no formation of new bands at the localization of isoenzymes IIa, b and c, could be observed (Fig. 2).

Structural interrelations between isoenzymes may also be reflected in the catalytic properties, provided that the enzymatic function of the single polypeptide chains is not grossly changed when combined to a protomer. Therefore, some catalytic parameters of the purified ADH fractions were determined. Pyrazole has been shown by Theorell and Yonetani [22] to be an inhibitor of ADH, which is competitive with the substrate ethanol. According to Vallee *et al.* [23], *o*-phenanthroline inhibits ADH by chelating the functional zinc, thus competing with the coenzyme. No marked differences are found between fractions I, IIa, b and c in respect to the inhibitor concentrations needed for a 50% inhibition of ADH-activity (Table 4). Increasing concentrations are needed for both inhibitors to achieve the same effect with isoenzymes III, IV and V.

Table 4. Inhibition of horse liver ADH isoenzymes by pyrazole and o-phenanthroline.

Isoenzyme	I	IIa	IIb	IIc	III	IV	V
Pyrazole	0.4	0.6	0.4	0.7	35	73	125
o-Phenanthroline	170	250	320	190	310	820	9000

Inhibitor concentrations (μM) required for 50% inhibition (I_{50}). Conditions: Na-pyrophosphate-buffer, pH 8.8, 3.3 x 10^{-2}M; NAD, 1.6 x 10^{-3}M; ethanol, 1.6 x 10^{-2}M; initial velocity measurements at 25°C.

According to Theorell *et al.* [19] isoenzyme IV is, at least in part, responsible for the steroid activity of horse liver ADH. By partial separation of the five electrophoretic fractions, Pietruszko *et al.* [24] demonstrated that the fraction with the most basic character exerts even a higher steroid activity, whereas the two more acidic fractions and the main fraction show only little activity with 5β-androstan-17β-ol-3-one (5-dihydrotestosterone, DHT). The results of the determination of the activity ratios of DHT to acetaldehyde reduction for pure fractions IIa, IIb, IIc, III, IV and V are listed in Table 5. No activity with DHT was detectable for isoenzyme III, thus confirming the observation made by Theorell *et al.* [19]. Isoenzyme IV reduces DHT at a rate which is 50 times lower than the one for acetaldehyde. For isoenzyme V the rate with the steroid is almost half that with acetaldehyde. Analogous results are also obtained with phenylethanediol as a substrate. It seems that the same steric requirements as for 3β-hydroxy of 3-keto-steroids must be fulfilled for this compound. In this context, it is noteworthy that Winer [25] observed 2-phenylethanol to be a good substrate for the main fraction of horse liver ADH (isoenzyme III).

Table 5. Enzymatic activities of horse liver ADH isoenzymes in the presence of 5-dihydrotestosterone and phenylethanediol.

Isoenzyme	$\dfrac{V_{AcAld}}{V_{DHT}}$	$\dfrac{V_{EtOH}}{V_{PED}}$
IIa	n.d.	n.d.
IIb	n.d.	n.d.
IIc	n.d.	n.d.
III	n.d.	n.d.
IV	50	30
V	2.3	2.1

Conditions: Na-phosphate buffer, pH 7.0, 3.3 x 10^{-2}M; NADH, 3.3 x 10^{-4}M; acetaldehyde (AcAld), 1.0 x 10^{-3}M; 5-dihydrotestosterone (DHT), 3.3 x 10^{-4}M; Na-pyrophosphate-buffer, pH 8.8, 3.3 x 10^{-2}M; NAD, 1.6 x 10^{-3}M; ethanol (EtOH), 1.0 x 10^{-3}M; phenylethanediol (PED), 1.0 x 10^{-3}M.

n.d. = no detectable activity with DHT or PED.

These results indicate that isoenzyme V, as would be expected from the structural model, is much more active as a steroid dehydrogenase than isoenzyme IV. They also suggest, together with the inhibition studies, that the group of fractions IIa, b and c is distinct from isoenzymes III, IV and V. It is probable that isoenzymes III, IV and V correspond to the structural models AA, AB and BB. Monomerization and reactivation of isoenzyme III under the conditions described by Drum et al. [11] yield only one enzyme with an electrophoretic mobility identical with that of the original material. An analogous result is obtained with isoenzyme V. If isoenzyme IV is treated in the same way, a mixture of isoenzymes III, IV and V is formed.

All three isoenzymes are also found after monomerization and reactivation of a mixture of pure isoenzymes III and V, representing an *in vitro* hybridization. No enzymatic fractions corresponding to fractions IIa, b or c were formed in either of the experiments. These hybridization studies [26] are consistent with the proposed model for isoenzymes III, IV and V. Together with the investigation on the catalytic parameters, they provide evidence in support of a dimeric structure for the isoenzymes of horse liver alcohol dehydrogenase. The possibility that the occurrence of the fractions I, IIa, IIb and IIc is due to factors other than genetic determination cannot be ruled out.

ACKNOWLEDGEMENTS

The authors wish to express their gratitude to Prof. H. Aebi for his continued interest in this work, and to Mrs. E. Chuit and J. Ellenberger for skilful technical assistance. This investigation was supported by Schweizerischer Nationalfonds zur Förderung der wissenschaftlichen Forschung, Project No. 5021.3.

REFERENCES

1. Dalziel, K., *Acta chem. scand.* **12** (1958) 459.
2. von Wartburg, J. P. and Papenberg, J., *Psychosom. Med.* **28** (1966) 405.
3. Papenberg, J., von Wartburg, J. P. and Aebi, H., *Biochem. Z.* **342** (1965) 95.
4. Aebi, H., Richterich, R. and von Wartburg, J. P., *Helgoländer wiss. Meeresunters.* **14** (1966) 343.
5. McKinley-McKee, J. S. and Moss, D. W., *Biochem. J.* **96** (1965) 583.
6. von Wartburg, J. P., Abstracts, Third FEBS Meeting, Warsaw, April 1966, Abstract F. 180.
7. von Wartburg, J. P., Papenberg, J., Kopp, P. M., Schürch, P. M. and Lutstorf, U. M., manuscripts in preparation.
8. Sund, H. and Theorell, H., in "The Enzymes", Vol 7 (edited by P. D. Boyer, H. Lardy and K. Myrbäck), Academic Press, New York, 1963, p. 25.
9. von Wartburg, J. P., Bethune, J. L. and Vallee, B. L., *Biochemistry* **3** (1964) 1775.

10. Blair, A. H. and Vallee, B. L., *Biochemistry* **5** (1966) 2026.
11. Drum, D. E., Harrison, J. H., Li, T. K., Bethune, J. L. and Vallee, B. L., *Proc. natn. Acad. Sci. U.S.A.* **57** (1967) 1434.
12. Jörnvall, H. and Harris, J. I., Abstracts, Fifth FEBS Meeting, Prague, July 1968, Abstract 759.
13. Ehrenberg, A. and Dalziel, K., *Acta chem. scand.* **12** (1958) 465.
14. Theorell, H., *Adv. Enzymol.* **7** (1947) 265.
15. Hayes, J. E. and Velick, S. F., *J. biol. Chem.* **207** (1954) 225.
16. Åkeson, A., *Biochem. biophys. Res. Commun.* **17** (1964) 211.
17. Oppenheimer, H. L., Green, R. W. and McKay, R. H., *Archs Biochem. Biophys.* **119** (1967) 552.
18. Waller, G. and Theorell, H., *Archs Biochem. Biophys.* **111** (1965) 671.
19. Theorell, H., Taniguchi, S., Åkeson, A. and Skursky, L., *Biochem. biophys. Res. Commun.* **24** (1966) 603.
20. Vesterberg, O. and Svensson, H., *Acta chem. scand.* **20** (1966) 820.
21. Sofer, W. H. and Ursprung, H., *J. biol. Chem.* **243** (1968) 3110; Ursprung, H., personal communications.
22. Theorell, H. and Yonetani, T., *Biochem. Z.* **338** (1963) 537.
23. Vallee, B. L., Williams, R. J. P. and Hoch, F. L., *J. biol. chem.* **234** (1959) 2621.
24. Pietruszko, R., Clark, A., Graves, J. M. H. and Ringold, H. J., *Biochem. biophys. Res. Commun.* **23** (1966) 526.
25. Winer, A. D., *Acta chem. scand.* **12** (1958) 1695.
26. Lutstorf, U. M., unpublished observation.
27. Knight, E. Jr. and Blomstrom, D. E., *Biochim. biophys. Acta* **89** (1964) 553.

FEBS Symposium, Volume 18, 1970, pp. 203-208

Malate Dehydrogenase Isoenzyme Distribution in Rat Tissues

C. J. R. THORNE and N. J. DENT

*Department of Biochemistry, University of Cambridge,
Cambridge, England*

Malate dehydrogenase (L-malate: NAD oxidoreductase E.C. 1.1.1.37) is found in both the cytoplasm and mitochondria of many animal tissues [1-5]. The enzyme of the cytoplasm (S-MDH) differs from that of the mitochondria (M-MDH) in catalytic [2-5] and physico-chemical [4-8] properties, in chromatographic [3, 5] and electrophoretic [5, 9] behaviour. Gel electrophoresis has revealed that M-MDH is often a family of catalytically-active components [9]. It has recently been demonstrated that S-MDH and M-MDH are under differential genetic control [10, 11], confirming what had been suspected from their significantly different properties.

Changes in the proportions of M-MDH and S-MDH during mammalian development have been reported by Wiggert and Villee [12, 13], who used starch-block electrophoresis to separate the two MDH isoenzymes. Employing subcellular fractionation, Moorjani and Lemonde [14] have observed changes in the MDH isoenzyme proportions of a developing beetle, while by using essentially the same method, Berkěs-Tomašević and Holzer [15] have reported MDH isoenzyme changes in rat liver on starvation. The present paper describes a rapid and accurate chromatographic method for the quantitative assessment of the MDH isoenzymes in a tissue, and reports some observations on the isoenzyme proportions in rat organs.

MATERIALS AND METHODS

L-malic acid was from Calbiochem; NAD from C. F. Boehringer & Soehne GmbH; Sephadex G-25 medium from Pharmacia; Amberlite CG-50 (XE 64) resin, 200 mesh, was a product of the Rohm and Haas Co. distributed by British Drug Houses; Triton X-100 was from British Drug Houses. Other chemicals were of analytical grade.

MDH was assayed in a Beckman DK 2 recording spectrophotometer at 30°C in a medium containing 90 mM glycine buffer, pH 10, 33 mM sodium L-malate and 0.38 mM NAD. The increase in absorption at 340 mμ was measured. One

MDH unit is the amount of enzyme catalysing the oxidation of one μmole of malate under these conditions.

Starch gel electrophoresis, and subsequent staining for MDH activity made use of methods elsewhere described [9].

RESULTS

(a) Quantitative separation of M-MDH from S-MDH

Extracts of rat tissues, containing assayed amounts of MDH activity, were rapidly passed through a column of Sephadex G-25, equilibrated in and eluted with 0.025M sodium phosphate, pH 7.0. The eluate front was collected and found to contain all the MDH applied to the Sephadex. Material obtained from the gel filtration was applied to a small (6 cm x 2.2 cm diameter) column of Amberlite CG-50 equilibrated with 0.025M sodium phosphate, pH 7.0. The column was eluted with 30 ml of the same 0.025M sodium phosphate, and then with 60 ml of 0.2M sodium phosphate, pH 7.0. Examination of the effluent from the column by starch gel electrophoresis showed that S-MDH was unretarded by the Amberlite column, while M-MDH was eluted only by the stronger phosphate buffer. The M-MDH recovered from the column, when examined electrophoretically, was seen to be a family of three cathodal components; the S-MDH from the Amberlite appeared as a single anodal component.

(b) Subcellular distribution of MDH isoenzymes in rat liver

Livers from adult rats were homogenized at 4°C, in 5 volumes of a medium of 0.25M sucrose containing 0.0025M Tris-chloride, pH 7.4, employing six full strokes of a Teflon-in-glass Dounce homogenizer. The homogenate was divided into two equal portions, one of which was subjected to further homogenization for six min at 4°C, in a tight fitting glass-in-glass homogenizer. The two portions were subjected to differential centrifugation, and the various fractions obtained were, after resuspension in 0.025M sodium phosphate, pH 7.0, when necessary, treated with Triton X-100 to a final concentration of 0.5% (w/v), passed through Sephadex G-15, and their M-MDH and S-MDH determined by quantitative chromatography on Amberlite CG-50. Table 1 shows the MDH isoenzyme distribution observed, recoveries being recorded as percentages. It can be seen that, on gentle homogenization, most of the mitochondrial MDH is M-MDH, while most of the cytoplasmic MDH is S-MDH, but that a large amount (over 40% of the total) remains in the debris. More vigorous homogenization releases additional MDH from the debris, but the cytoplasmic fraction now contains a significant amount of M-MDH, probably released from broken mitochondria.

Table 1. MDH isoenzymes in rat liver subcellular fractions.*

		Dounce only		Dounce, then 6 min glass-in-glass	
		M-MDH	S-MDH	M-MDH	S-MDH
1,800 rpm ppt	"Debris"	36.5	5.7	18.7	2.3
10,000 rpm ppt	"Mitochondria"	24.3	1.6	39.6	2.5
30,000 rpm ppt	"Somes"	(1.6)		(2.2)	
30,000 rpm sup	"Cytoplasm"	1.2	29.0	6.4	28.3
		62.0	36.3	64.7	33.1

* Recoveries in percentages

(c) MDH isoenzyme distribution in whole rat organs

Rats were taken at various ages, or after various dietary regimes; organs were removed, weighed, and homogenized at 4°C for 2 min in a tight fitting glass-in-glass homogenizer in a medium containing 0.025M sodium phosphate, pH 7.0, and 0.5% Triton X-100. After centrifugation at 30,000 rpm for 30 min, the supernatant was retained and the precipitate resuspended and homogenized as before. After centrifugation, the second supernatant was pooled with the first. The pooled supernatant was passed through Sephadex G-25 and subjected to Amberlite chromatography, as previously described, to separate the M-MDH and S-MDH.

Tables 2-4 show the proportion of the two MDH isoenzymes in rat organs under various conditions. Table 2 shows the isoenzyme ratios in four organs of adult female rats; Table 3 the ratios in liver and kidney of developing postnatal rats. Allowing for the changes in total MDH per unit weight of liver, it can be seen that liver S-MDH remains fairly constant for the first two weeks of life, and then increases towards adulthood, while the M-MDH rises to a peak at two weeks

Table 2. MDH isoenzyme distribution in nine week-old female rats.

	MDH units per g wet weight	%M	%S
Liver	185	69	31
Kidney	211	63	37
Heart	273	76	24
Brain	90	59	41

Table 3. Variation of MDH isoenzyme distribution with age.

	Rat liver			Rat kidney		
Age	MDH units per g wet weight	%M	%S	MDH units per g wet weight	%M	%S
1 day	116	69	31	68	64	36
1 week	163	78	22	74	67	33
2 weeks	190	83	17	115	61	39
3 weeks	182	77	23	136	62	38
4 weeks	170	75	26	205	61	39
9 weeks (♀)	185	69	31	211	63	37

Table 4. Effect of diet on MDH isoenzyme balance in rat liver.

	MDH units per g wet weight	%M	%S
9 week female rats			
Controls	185	70	30
2 days starvation	216	75	25
3 days starvation	185	77	23
2 days starvation, followed by 2 days carbohydrate refeeding	187	74	26
12 week male rats			
Controls	189	77	23
6 days fatty diet	206	80	20

and then falls. But because liver weight as a proportion of body weight falls to a trough at two weeks and then rises (see also ref. 16), it may also be said that liver M-MDH, expressed per unit of body weight, rises progressively during growth, to remain steady after three weeks, while S-MDH falls and then rises. In kidney there is a progressive and parallel increase in both isoenzymes when expressed either in terms of body weight or of organ weight. Table 4 shows that starvation and fatty diet both result in a small increase in the M-MDH:S-MDH ratio. Carbohydrate feeding after starvation allows the M-MDH:S-MDH ratio to fall a little.

DISCUSSION

Subcellular fractionation, if carried out carefully, enables most of the S-MDH to be separated from the M-MDH. But there is always some cross-contamination;

this, together with the problems of mitochondrial breakage and of residual MDH in the low-speed precipitate (debris), renders the method inadequate for accurate quantitative assessment of the MDH isoenzymes (Table 1). Rapid homogenization of tissues in Triton X-100, followed by a chromatographic separation of the isoenzymes seems to be a better method. Triton, at the concentrations used, does not affect either MDH stability or MDH assay (unpublished experiments), and isoenzyme recoveries are quantitative.

Using this method, the M-MDH:S-MDH ratio in some rat organs has been studied. Conditions leading to increased rates of gluconeogenesis (early post-natal life [17], starvation [18]) or decreased lipogenesis (starvation, high fat diet [19]), result in slight increases in the M-MDH : S-MDH ratio of liver. Conditions leading to increased lipogenesis (later post-natal life [20], carbohydrate refeeding after starvation [21]), result in a decrease in the M-MDH : S-MDH ratio. It may be, therefore, that M-MDH plays a part in gluconeogenesis [15, 22], while perhaps S-MDH, in conjunction with the malic enzyme, has a role in lipogenesis. But it should be stressed that the changes in isoenzyme ratio are small, and that both MDH isoenzymes are present in such ample amounts that any regulatory role would be rather improbable.

Most of the isoenzyme ratios reported (Tables 2-4) show values for M-MDH : S-MDH of approximately 2-3. These values, it must be noted, are very dependent on the MDH assay conditions used. By the assays of Wiggert and Villee [13], and of Berkĕs-Tomašević and Holzer [15], respectively, the M-MDH activity would be 130% and 450% and the S-MDH activity would be 115% and 770% of that reported here. These differences are partly due to the fact that M-MDH is subject to high substrate inhibition when assayed from the oxaloacetate side, while S-MDH is similarly affected by high substrate when assayed from the malate side. Although these differences exist, under any fixed assay conditions the MDH isoenzyme ratios are real, and observed changes in these ratios may provide information on the metabolic state of a tissue.

REFERENCES

1. Christie, G. S. and Judah, J. D., *Proc. R. Soc.* Ser. B, **141** (1953) 420.
2. Delbrück, A., Schimassek, H., Bartsch, K. and Bücher, Th., *Biochem. Z.* **331** (1959) 297.
3. Thorne, C. J. R., *Biochim. biophys. Acta* **42** (1960) 175.
4. Siegel, L. and England, S., *Biochem. Biophys. Res. Commun.* **3** (1960) 253.
5. Grimm, F. C. and Doherty, D. G., *J. biol. Chem.* **236** (1961) 1980.
6. Siegel, L. and England, S., *Biochim. biophys. Acta* **64** (1962) 101.
7. Thorne, C. J. R. and Kaplan, N. O., *J. biol. Chem.* **238** (1963) 1861.
8. Thorne, C. J. R. and Cooper, P. M., *Biochim. biophys. Acta* **81** (1964) 397.
9. Thorne, C. J. R., Grossman, L. I. and Kaplan, N. O., *Biochim. biophys. Acta* **73** (1963) 193.
10. Davidson, R. G. and Cortner, J. A., *Science, N.Y.* **157** (1967) 1569.

11. Davidson, R. G. and Cortner, J. A., *Nature, Lond.* **215** (1967) 761.
12. Wiggert, B. and Villee, C. A., *Science, N.Y.* **138** (1962) 509.
13. Wiggert, B. and Villee, C. A., *J. biol. Chem.* **239** (1964) 444.
14. Moorjani, S. and Lemonde, A., *Can. J. Biochem.* **45** (1967) 1393.
15. Berkĕs-Tomašević, P. and Holzer, H., *Eur. J. Biochem.* **2** (1967) 98.
16. Vernon, R. G. and Walker, D. G., *Biochem. J.* **106** (1968) 321.
17. Ballard, F. J. and Hanson, R. W., *Biochem. J.* **102** (1967) 952.
18. Weber, G., *Adv. Enzyme Regulation* **1** (1963) 1.
19. Lowenstein, J. M., *in* "Biochemical Society Symposium 27" (edited by T. W. Goodwin), 1968, p. 61.
20. Taylor, C. B., Bailey, E. and Bartley, W., *Biochem. J.* **105** (1967) 717.
21. Foster, D. W. and Srere, P. A., *J. biol. Chem.* **243** (1968) 1926.
22. Lardy, H. A., Paetkau, V. and Walter, P., *Proc. natn. Acad. Sci. U.S.A.* **53** (1965) 1410.

FEBS Symposium, Volume 18, 1970, pp. 209-214

Changes in Activity and Isozyme Patterns of Glycolytic Enzymes in the Developing Rat Liver *

T. TEPPER and F. A. HOMMES

Department of Paediatrics, University of Groningen, School of Medicine, Groningen

Non-Standard Abbreviations and Enzymes.
Hexokinase or ATP: glucose-6-phosphotransferase (EC 2.7.1.2), HK;
Glucose-6-phosphate dehydrogenase or D-Glucose-6-phosphate:NADP oxidoreductase (EC 1.1.1.49), G-6-PDH;
6-Phosphogluconate dehydrogenase or 6-phosphogluconate:NADP oxidoreductase (EC 1.1.1.44), 6-PGDH:
Phosphoglucomutase or α-D-glucose-1,6-diphosphate:α-D-glucose-1-phosphate phosphotransferase (EC 2.7.5.1), PGLuM;
Phosphoglucoisomerase or D-glucose-6-P ketolisomerase (EC 5.3.1.9), PGI;
Phosphofructokinase or ATP:D-fructose-6-P-1-phosphotransferase (EC 2.7.1.11), PFK;
Fructokinase or ATP:fructose-1-P-phosphotransferase (EC 2.7.1.4), FK;
Aldolase or D-fructose-1,6-di P:G-glyceraldehyde-3-lyase (EC 4.1.2.13), Ald;
α-Glycerophosphatedehydrogenase or L-glycerol-phosphate:NAD oxidoreductase (EC 1.1.99.5), α-GPDH;
Phosphoglycerate mutase or 2,3-diphosphoroglycerate-D-glycerate: 2-phospho-D-glycerate phosphotransferase (EC 2.7.5.3), PGM;
Phosphoglycerate kinase or ATP:3-phospho-D-glycerate-1-phosphotransferase (EC 2.7.2.3), PKG;
Glyceral-dehydephosphatedehydrogenase or D-glyceraldehyde-3-P:NAD oxidoreductase (EC 1.2.1.13), GAPDH;
Enolase or phosphopyruvate hydratase (EC 4.2.1.11), EN;
Pyruvate kinase or ATP:pyruvate phosphotransferase (EC 2.7.1.40), PK;
Lactate dehydrogenase or L-lactate:NAD oxidoreductase (EC 1.1.1.27), LDH.

According to Weber *et al.* [1], the synthesis of glycolytic enzymes of liver is regulated by at least two operons, one for the bifunctional enzymes, catalysing reactions of glycolysis and glyconeogenesis, and a second for the key glycolytic

* This investigation was supported by grants from the Netherlands Organization for the Advancement of pure research (Z.W.O.)

enzymes, HK, PFK and PK. Such a control would be very efficient: a single molecule may activate or inhibit the coordinated expression of functionally-related genic units. With this so-called "Functional Genome Unit", Weber *et al.* [1] were able to explain the synchronous behaviour patterns of PK, HK and PFK after administration of insulin and other external factors to the rat. The object of this communication is to report evidence for coordinated enzyme synthesis by a study of the influence of an internal variable; i.e. age.

MATERIALS AND METHODS

Homogenates of rat liver in 0.25M sucrose were prepared in a loosely-fitting Potter-Elvehjem homogenizer. The crude homogenate was centrifuged at 0°C for 1 h at 40,000 g. The clear supernatant was used for enzyme assay. Activity determinations were carried out fluorimetrically, using a recording fluorometer as described by Estabrook and Maitra [2], by coupling to an NAD- or NADP-dependent system in such a way that the enzyme to be tested was made rate-limiting [3].

Isozymes were separated by electrophoresis on agar or cellulose acetate strips as described by Latner [4], with minor modifications.

PK isozymes were made visible by layering the cellulose acetate strip on agar containing all the substrates and enzymes necessary for the PK reaction coupled to the LDH reaction. Disappearance of NADH was followed under UV light.

RESULTS AND DISCUSSION

Our results were compared with the above-mentioned hypothesis of Weber *et al.* [1] on the supposition that synchronous activity changes indicate coordinated synthesis, controlled by one operon.

Changes in activities of the glycolytic enzymes, and of some other enzymes closely associated with this pathway, are summarized in Fig. 1. Parallel changes in activity were observed for seven enzymes, namely HK, PGK, TPI, PFK, EN, PGM and PK. The behaviour of HK, PFK and PK is in agreement with the hypothesis of Weber *et al.* [1]. By varying both external and internal factors, we observe synchronous activity changes. As regards the other four enzymes, PGI, TPI, EN and PGM, there are no data available as to the influence of insulin on their synthesis. It is therefore not possible to decide whether the synthesis of these enzymes is also controlled by the operons responsible for HK, PFK and PK synthesis.

The activity changes of Ald, GAPDH, PGK and LDH (Fig. 1) are not synchronized with any other enzyme investigated. This would seem to suggest that the synthesis of these enzymes is regulated by four independent operons. The behaviour of Ald, TPI, GAPDH, PGK, PGM and EN is in disagreement with the theory of Mier and Cotton [5], who, following Bücher *et al.* [6, 7], arranged

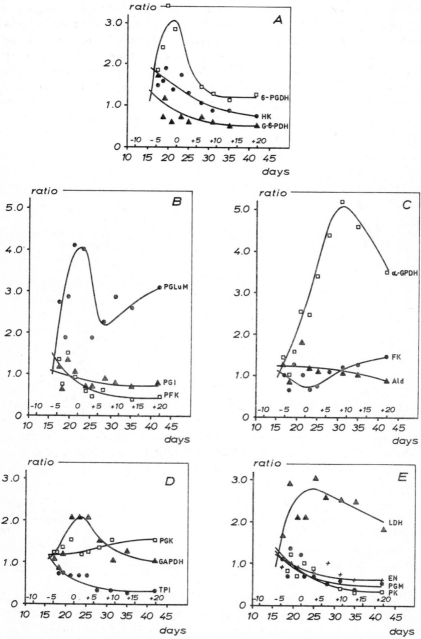

Figure 1. Changes in activities of glycolytic and other enzymes in the developing rat liver, from five days before birth to adult age. For abbreviations see text.

these six enzymes in a so-called "constant proportion group". The ratio of activities of enzymes of such a group should be constant, whatever organ or animal is chosen. According to the calculations of Mier and Cotton [5], these six enzymes are synthesized in equimolar quantities. Therefore they concluded that one operon is responsible for their synthesis. Hence one might expect to find synchronous activity changes in our experiments. This was not observed for Ald, GAPDH and PGK. The synthesis of these constant proportion group enzymes is, in our opinion, probably controlled, at least during development, by more than one operon.

Table 1.

Group I	Group II	Group III
Hexokinase	Phosphoglucoisomerase	Aldolase
Phosphofructokinase	Triosephosphate isomerase	Glyceraldehyde phosphate dehydrogenase
Pyruvate kinase	Phosphoglycerate mutase	Phosphoglycerate kinase
	Enolase	Lactate dehydrogenase
	Glucose-6-phosphate dehydrogenase	α-Glycerophosphate dehydrogenase
		Fructokinase
		6-Phosphogluconate dehydrogenase
		Phosphoglucomutase

Lastly, four enzymes related to glycolysis, α-GPDH, FK, 6-PGDH and PGLuM, do not show synchronous activity changes. The results are summarized in Table 1. Group 1 includes those enzymes whose synthesis is influenced by insulin and probably regulated by one operon. Group II enzymes shows parallel changes in the course of development. However, it cannot be decided if these enzymes are indeed controlled by one operon or if more than one is involved. Group III consists of those enzymes whose synthesis is apparently regulated by independent operons, because their activity changes do not parallel any of the other enzymes.

Since the overall results refer to total enzyme activities, it seemed of interest to study the developmental changes of isozyme patterns of some glycolytic enzymes.

Figure 2 shows the large variation in the LDH_1/LDH_2-ratio during the period just before birth to 25 days after. As is illustrated in Fig. 3, a considerable shift in isozyme composition of hexokinase was observed. Of the two isozymes of PK, the L-type disappears almost completely during the period 5-25 days after birth and reappears at adult age, whereas the M-type is present throughout develop-

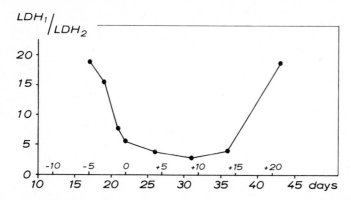

Figure 2. Ratio of LDH isozymes in the developing rat liver.

ment. If the synthesis of PK and HK is exclusively regulated at the transcription level by one operon, then their isozyme pattern should remain constant during development. Because of the changes in enzyme activity observed in this study, we must, however, conclude that the synthesis of PK and HK is regulated by a much more complicated mechanism. Recent work of Tanaka *et al.* [8] on PK points in this direction. They found that the level of L-type PK varied greatly under various physiological conditions, whereas that of M-type varied only slightly. Weber *et al.* [1] left open the possibility for regulation of enzyme synthesis on a level different from transcription. Our results seem to substantiate this possibility because an approach on the basis of Weber's hypothesis uncovers at least as many as nine operons for the synthesis of the glycolytic enzymes: two for PK, three for HK, one for PGI, TPI, PGM and EN, one for LDH, one for PGK and one for Ald.

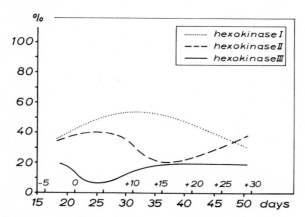

Figure 3. Isozyme composition of the glucose phosphorylating system of the developing rat liver. Isozymes are numbered according to Holmes *et al.* [9].

REFERENCES

1. Weber, G., Singhal, R. L., Stamm, N. B., Lea, M. A. and Fisher, E. A., *Adv. Enzyme Regulation* 4 (1965) 59.
2. Estabrook, R. W. and Maitra, P. K., *Analyt. Biochem.* 3 (1962) 369.
3. Bergmeyer, H. U. (ed.), "Methods of Enzymatic Analysis", Academic Press, New York, 1963.
4. Latner, A. L., *Adv. clin. Chem.* 9 (1967) 69.
5. Mier, P. D. and Cotton, W. K., *Nature, Lond.* 209 (1966) 1022.
6. Bücher, Th. and Pette, D., Proc. Vth Intern. Congr. Biochem., Moscow, 1961, I.U.B. Symp. Series, Vol. 23, 271.
7. Pette, D., Luh, W. and Bücher, Th., *Biochem. biophys. Res. Commun.* 7 (1962) 419.
8. Tanaka, T., Harano, Y., Sue, F. and Morimura, H., *J. Biochem., Tokyo* 62 (1967) 71.
9. Holmes, E. W., Malone, J. I., Winegrad, A. I. and Oski, F. A., *Science, N.Y.* 156 (1967) 646.

FEBS Symposium, Volume 18, 1970, pp. 215-226

Multiple Forms of Hormone-Sensitive Lipase in Different Rat Tissues

M. CHMELAŘ and M. CHMELAŘOVÁ

First and Second Departments of Medical Chemistry,
Charles University, Prague, Czechoslovakia

Hormone-sensitive lipase, first described by Rizack [1] and Hollenberg *et al.* [2] in 1961, represents the key enzyme in the regulation of adipose tissue metabolism. Although the experiments with intact adipose tissue [3], as well as with isolated adipocytes [4], revealed the physiological importance of the effects of individual hormones on the release of free fatty acids or glycerol from endogenous substrate under different experimental conditions, it was impossible to characterize the hormone-sensitive enzyme itself. Further progress was made in experiments with adipose tissue homogenates. Rizack [5], Vaughan *et al.* [6-8], Gorin and Shafrir [9], Björntorp [10] and many others concluded that adipose tissue contains, in addition to lipoprotein lipase, at least two other lipases, the monoglycerides and triglycerides. It seems that only triglyceride lipase is hormone-sensitive, and it is presumed to be the rate-limiting factor in lipolysis [6, 7]. Among inhibitors used for its characterization, iodoacetic acid (10^{-4}M), 2-propanol [6] and amytal [11] were found to be typical for hormone-sensitive lipase. The mode of action of the individual lipolytic hormones was believed to be mediated through cyclic $3',5'$-AMP formed by cyclase from ATP [12, 13]. From indirect evidence it was also suggested that hormone-sensitive lipase had two forms, active and inactive [8, 10].

However, hormone-sensitive lipase has not yet been isolated and purified. The detailed mode of action of individual hormones in lipolysis, as well as the participation of cyclic $3',5'$-AMP in the activating system of hormone-sensitive lipase, could therefore not be unambiguously determined. Moreover, on the basis of previous experiments, it could not be decided whether hormone-sensitive lipase was represented by the only enzyme sensitive to all regulatory agents, or whether the hormone-sensitive lipolytic system involves more enzymes with different properties, including differing sensitivities to individual hormones or to cyclic $3',5'$-AMP. Since hormone-sensitive lipase was originally described, and for years studied in adipose tissue and, recently, also in adrenals [14], it was of importance to investigate in parallel whether hormone-sensitive lipase was present also in other organs with intensive lipid metabolism, and, if so, whether

8   215

it was different from that in adipose tissue and whether it was heterogenous or not. We earlier described hormone-sensitive lipase in aortic tissue [15] and in skeletal and heart muscle [16]. In this paper we have also studied liver tissue. The analysis of lipolysis and its regulation in samples of liver tissue is difficult when we take into consideration the high activity of glycerokinase, which leads to rapid reutilization of glycerol released. The turnover of fatty acids in liver tissue is also extremely high. Therefore we applied to the study of hormone-sensitive tissue lipases our original method, based on the use of synthetic substrates and extremely diluted homogenate. This method has been discussed elsewhere [17].

MATERIALS AND METHODS

Albino Wistar rats weighing 160-220 g were used for the experiments. After decapitation, samples of epididymal fat pads, liver tissue, skeletal muscle (femoral muscles), or aorta (free of adventitia), were promptly weighed and homogenized in a ground-glass grinder (or sea sand was used) in distilled water. Homogenates were centrifuged at 2,000 x g for 5 min; 0.2 ml of the supernatant fluid (from adipose tissue, whole supernatant fluid, including the fat cake, was shaken for several minutes in a water bath at 37°C so that the fat was finely dispersed), corresponding to 4 mg of adipose tissue, 5 mg of skeletal muscle, 0.3 mg of liver tissue or 4 mg of aortic tissue, was then preincubated in 2 ml buffered medium (Tris-HCl buffer, pH 7.5) containing 40 mM KCl, for 10 min at 37°C, and then 5 ml of 3.5×10^{-4} M substrate (2-naphthol laurate or a mixture of triglycerides C^{14}-C^{18}) was added and incubated for 30 min at 37°C. The released 2-naphthol was estimated by the method of Nachlas and Seligman [18] and the released glycerol by the method of Lambert and Neish [19]. Lipolytic activity was expressed in units of μmoles of released 2-naphthol or glycerol per min.

For the separation of tissue lipases, zone electrophoresis in starch medium was used [20]. For this purpose the tissue samples were homogenized in 0.05M veronal:veronal sodium, pH 8.6, which was also used as the medium for electrophoresis. The concentration of homogenate was 40%. After separation (60 min, 14 V/cm) the starch block (13.6 x 2.8 x 0.4 cm) was cut into strips 0.5 cm wide and each strip was eluated with 0.15M KCl.

Gel filtration made use of a Sephadex G-200 column (78 x 1.2 cm) [21]. It was equilibrated with 0.1M Tris-HCl buffer, pH 7.5. The samples (eluates from starch medium electrophoresis), either not centrifuged or centrifuged at 105,000 x g for 120 min at 2°C in a volume of 1 ml, were applied to the top of the column and column effluents were collected in 2 ml fractions with a collector. Both zone electrophoresis and gel filtration were carried out at 0°C.

The results were statistically evaluated by means of t-tests.

RESULTS

We first attempted to establish whether there is one or more hormone-sensitive lipase in rat adipose tissue. In contrast to other authors, who differentiated lipases by means of inhibitors, centrifugation or differential elution of crude homogenates [6, 7, 22], we utilized zone electrophoresis on starch. The whole homogenate of fresh rat adipose tissue was applied to the starch block and enzymes were separated under the described conditions. Figure 1a demonstrates that lipase activity, measured in eluates of individual strips, was separated into two fractions, cathodic and anodic. Recovery of lipase activity from the starch block was 97%. From Table 1 it is apparent that both the lipase fractions were sensitive to lipolytic hormones *in vitro*. Cyclic 3′,5′-AMP increased the activity only of the cathodic fraction, although both fractions, especially the anodic one, were stimulated by norepinephrine. Insulin slightly inhibited the cathodic fraction and stimulated the anodic one. The antilipolytic effect of insulin in the presence of ACTH or norepinephrine, however, was not demonstrable. From the results we concluded that both fractions are hormone-sensitive.

In the next series of experiments, gel filtration was used for further purification of isolated enzymes. Both fractions were separately applied to the column of Sephadex G-200. As shown in Fig. 2, the cathodic fraction (full line) was separated into seven subfractions with different molecular weights, whereas the anodic one (dotted line) was homogenous. Since such a large number of fractions, confirmed in a number of experiments, was unexpected and highly surprising, the respective eluates from starch medium, both cathodic and anodic, were centrifuged for 2 h at 105,000 x g before application to the Sephadex G-200 column. Ultracentrifugation gave analogous results but the activities were considerably lower.

It was of interest to ascertain the sensitivity of these fractions to hormones. From the results summarized in Fig. 3, it is apparent that the fractions with the lower molecular weights are activated only with ACTH and are not stimulated by norepinephrine and cyclic 3′,5′-AMP. The fractions with high molecular weights, on the other hand, are stimulated with norepinephrine and the first and second fractions also by cyclic 3′,5′-AMP. The third fraction, the only one with anodic mobility, was highly activated by norepinephrine, but cyclic 3′,5′-AMP was without effect. The effects of insulin were as follows: whereas in whole homogenates insulin slightly inhibited lipase activity, its inhibitory effect on fractions was demonstrated especially in the first, and to a lesser extent in the second, fifth and sixth fractions. The activities of the third, as well as of the seventh, were surprisingly stimulated.

In the third group of experiments, hormone-sensitive lipases in some other organs were compared with that of adipose tissue and their heterogeneity examined. The presence of hormone-sensitive lipase in skeletal and heart muscle,

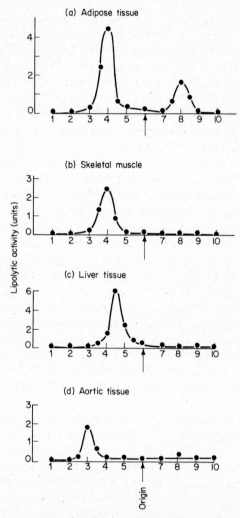

Figure 1. Separation of rat adipose tissue (a), skeletal muscle (b), liver (c), and aortic tissue (d) lipases by zone electrophoresis in starch medium. Substrate, 2-naphthol laurate. Abscissa: the width of strips of starch medium (cm); Ordinate: lipolytic activity in units. The results are the sum of 16(a), 9(b), 14(c) and 10(d) experiments, respectively.

Table 1. Effects of hormones and cyclic $3',5'$-AMP on lipase activity of cathodic (I) and anodic (II) fractions separated by zone electrophoresis in starch medium from rat adipose tissue homogenate.

Preincubation medium	Lipolytic activity I	II	N I	II	Statist. signif. p ⟨ against contr. I	II
Control medium	100.0 ± 10.1	100.0 ± 9.8	15	15	–	–
+ ACTH (Corticotrophin Spofa), 1 U/ml	307.2 ± 31.4	127.0 ± 11.3	11	11	0.001	0.05
+ Synacthen Ciba, 0.01 µg/ml	249.9 ± 20.1	106.5 ± 10.1	7	8	0.001	NS
+ Norepinephrine, 2 µg/ml	188.7 ± 22.8	301.6 ± 24.1	15	15	0.01	0.001
+ Epinephrine, 2 µg/ml	136.5 ± 12.7	199.2 ± 22.4	10	8	0.02	0.01
+ TSH Spofa, 0.1 U/ml	203.4 ± 24.4	181.3 ± 19.2	7	8	0.01	0.01
+ Glucagon (Eli Lilly Co.), 10 µg/ml	215.2 ± 25.8	190.3 ± 16.8	8	7	0.01	0.01
+ Insulin (Novo), 1 mU/ml	80.0 ± 6.9	156.5 ± 10.9	14	12	0.02	0.001
+ ACTH, 1 U/ml + insulin, 1 mU/ml	300.9 ± 32.2	134.8 ± 11.1	9	10	0.001	0.02
+ Norepinephrine, 2 µg/ml + insulin, 1 mU/ml	195.2 ± 21.1	298.7 ± 28.0	10	12	0.01	0.001
+ Cyclic $3',5'$-AMP, 2×10^{-5}M	159.2 ± 13.3	99.2 ± 10.8	14	15	0.001	NS

Substrate, 2-naphthol laurate. Lipolytic activity is expressed in per cent of control values, mean values ± S.E. are given. NS = not significant.

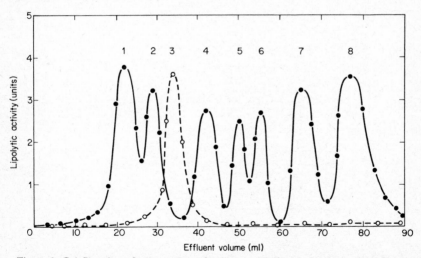

Figure 2. Gel filtration of adipose tissue fractions on Sephadex G-200. Cathodic fraction (full line) and anodic fraction (dotted line), obtained after electrophoresis of adipose tissue homogenate in starch medium (see Fig. 1), were separately applied on the column of Sephadex G-200 (see "Methods"). Abscissa: effluent volume in ml; Ordinate: lipolytic activity in units. Substrate, 2-naphthol laurate. The results are the means of 12 experiments.

and in aortic tissue, have been reported elsewhere [15, 16]. Evidence for this hormone-sensitive enzyme in liver tissue, however, is lacking. The effects of individual lipolytic hormones, as well as of insulin, on lipase activity of liver tissue homogenate, by comparison with those of muscle, aortic and adipose tissue, are given in Table 2. Although the stimulatory effects of the lipolytic hormones were found in all tissues compared, the effects of insulin differed.

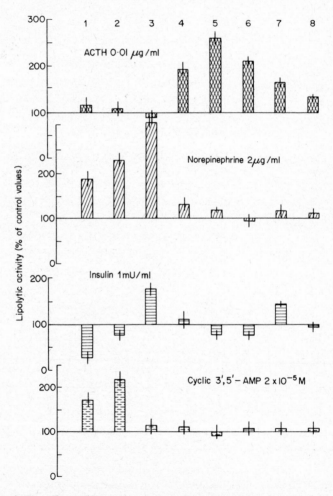

Figure 3. The effects of hormones on rat adipose tissue lipase fractions isolated by zone electrophoresis in starch medium and following gel filtration on Sephadex G-200 (see Fig. 2). Abscissa: individual fractions 1-8; ordinate: lipolytic activity expressed in per cent of control values. Substrate, 2-naphthol laurate. The results are the means of 12 experiments.

Table 2. The effects of hormones and cyclic $3',5'$-AMP *in vitro* on lipase activity of homogenates of rat adipose tissue, skeletal muscle, liver and aortic tissue.

Preincubation medium	Lipolytic activity			
	Adipose tissue	Skeletal muscle	Liver	Aortic tissue
(A) Substrate: 2-naphthol laurate				
Control medium	0.41 ± 0.05 (17)	0.52 ± 0.07 (14)	62.0 ± 7.2 (15)	0.53 ± 0.04 (12)
+ ACTH (Corticotrophin Spofa), 1 U/ml	1.82 ± 0.21 (15) $p < 0.001$	0.78 ± 0.09 (9) $p < 0.05$	110.9 ± 13.2 (12) $p < 0.01$	1.10 ± 0.09 (10) $p < 0.001$
+ Norepinephrine, 2 μg/ml	1.40 ± 0.11 (16) $p < 0.001$	1.69 ± 0.13 (12) $p < 0.001$	174.3 ± 19.0 (15) $p < 0.001$	1.62 ± 0.12 (9) $p < 0.001$
+ Epinephrine, 2 μg/ml	1.02 ± 0.08 (12) $p < 0.001$	1.11 ± 0.09 (10) $p < 0.001$	125.6 ± 10.4 (10) $p < 0.001$	1.08 ± 0.10 (11) $p < 0.001$
+ TSH Spofa, 0.1 U/ml	1.98 ± 0.12 (7) $p < 0.001$	0.82 ± 0.07 (7) $p < 0.01$	181.1 ± 15.2 (8) $p < 0.001$	1.87 ± 0.15 (9) $p < 0.001$
+ Glucagon (Eli Lilly Co.), 15 μg/ml	1.71 ± 0.2 (8) $p < 0.001$	0.77 ± 0.08 (7) $p < 0.05$	170.7 ± 19.0 (8) $p < 0.001$	2.19 ± 0.24 (7) $p < 0.001$
+ Insulin (Novo), 1 mU/ml	0.31 ± 0.03 (14) $p < 0.01$	0.41 ± 0.03 (12) $p < 0.01$	189.9 ± 21.0 (15) $p < 0.001$	2.01 ± 0.19 (12) $p < 0.001$
+ ACTH, 1 U/ml + insulin, 1 mU/ml	1.04 ± 0.13 (9) $p < 0.01$	0.64 ± 0.06 (10) $p < 0.001$	202.5 ± 18.7 (14) $p < 0.001$	2.63 ± 0.28 (10) $p < 0.001$
+ Norepinephrine, 2 μg per ml 3 insulin, 1 mU/ml	0.78 ± 0.09 (12) $p < 0.01$	0.99 ± 0.08 (8) $p < 0.001$	225.7 ± 21.5 (10) $p < 0.001$	3.02 ± 0.29 (10) $p < 0.001$
+ Cyclic $3',5'$-AMP, 2×10^{-5}M	0.79 ± 0.08 (12) $p < 0.001$	0.87 ± 0.09 (9) $p < 0.01$	112.1 ± 12.6 (11) $p < 0.01$	0.89 ± 0.09 (8) $p < 0.01$
(B) Substrate: mixture of triglycerides C_{14}-C_{18}				
Control medium	0.85 ± 0.08 (10)	1.50 ± 0.11 (11)	26.6 ± 2.2 (8)	2.01 ± 0.19 (14)
+ ACTH (Corticotrophin Spofa), 1 U/ml	3.62 ± 0.31 (7) $p < 0.001$	1.81 ± 0.15 (10) NS	54.1 ± 6.1 (8) $p < 0.01$	4.22 ± 0.40 (10) $p < 0.001$
+ Norepinephrine, 2 μg/ml	2.47 ± 0.27 (6) $p < 0.01$	4.71 ± 0.40 (10) $p < 0.001$	75.7 ± 6.2 (8) $p < 0.001$	6.74 ± 0.58 (11) $p < 0.001$
+ Epinephrine, 2 μg/ml	1.95 ± 0.21 (7) $p < 0.01$	3.25 ± 0.31 (8) $p < 0.001$	59.2 ± 5.8 (7) $p < 0.01$	4.02 ± 0.39 (10) $p < 0.001$
+ TSH Spofa, 0.1 U/ml	4.21 ± 0.33 (6) $p < 0.001$	2.26 ± 0.18 (7) $p < 0.01$	83.2 ± 7.0 (5) $p < 0.01$	6.89 ± 0.59 (7) $p < 0.001$
+ Glucagon (Eli Lilly Co.), 15 μ/ml	3.11 ± 0.31 (6) $p < 0.001$	2.34 ± 0.19 (10) $p < 0.01$	79.8 ± 8.1 (6) $p < 0.01$	8.27 ± 0.75 (8) $p < 0.001$

Table 2—*continued*

Preincubation medium	Adipose tissue	Lipolytic activity Skeletal muscle	Liver	Aortic tissue
+ Insulin (Novo), 1 mU/ml	0.71 ± 0.08 (8) NS	1.22 ± 0.10 (10) $p < 0.01$	68.9 ± 5.9 (8) $p < 0.001$	7.99 ± 0.81 (12) $p < 0.001$
+ ACTH, 1 U/ml + insulin, 1 mU/ml	1.41 ± 0.10 (7) $p < 0.01$	1.70 ± 0.18 (9) NS	89.1 ± 9.7 (7) $p < 0.001$	9.75 ± 0.90 (10) $p < 0.001$
+ Norepinephrine, 2 μg/ml + insulin, 1 mU/ml	1.23 ± 0.11 (6) $p < 0.02$	2.25 ± 0.21 (6) $p < 0.02$	97.9 ± 10.1 (7) $p < 0.001$	10.22 ± 0.92 (11) $p < 0.001$
+ Cyclic 3′,5′-AMP, 2 x 10^{-5}M	1.54 ± 0.12 (6) $p < 0.01$	2.39 ± 0.20 (7) $p < 0.01$	45.7 ± 4.7 (7) $p < 0.01$	3.87 ± 0.44 (9) $p < 0.01$

Lipolytic activity is expressed in units per g wet weight of tissue, mean values ± S.E. are given, the number of experiments is in brackets. Statistical significance is evaluated against control of the respective tissue. NS = not significant.

Table 3. Effects of inhibitors on hormone-sensitive lipases of rat adipose tissue, skeletal muscle, liver and aortic tissue.

Incubation medium	Adipose tissue	Lipolytic activity Skeletal muscle	Liver	Aortic tissue
Control medium	100.0 ± 11.9 (15)	100.0 ± 8.9 (10)	100.0 ± 12.1 (10)	100.0 ± 11.2 (14)
+ Iodoacetic acid, 10^{-5}M	58.1 ± 6.2 (14) $p < 0.001$	64.5 ± 5.8 (9) $p < 0.001$	75.2 ± 8.1 (9) $p < 0.02$	125.2 ± 6.9 (12) $p < 0.01$
+ 2-Propanol, 1%	56.0 ± 7.1 (12) $p < 0.001$	85.5 ± 9.7 (6) $p < 0.01$	110.3 ± 9.8 (10) $p < 0.01$	123.1 ± 10.1 (11) $p < 0.05$
+ NaF, 2 x 10^{-2}M	28.4 ± 6.9 (10) $p < 0.001$	70.1 ± 8.0 (7) $p < 0.01$	56.7 ± 7.2 (9) $p < 0.001$	54.3 ± 6.8 (10) $p < 0.001$
+ E 600, 10^{-5}M	15.9 ± 4.8 (7) $p < 0.001$	51.4 ± 4.3 (5) $p < 0.001$	76.9 ± 8.3 (7) $p < 0.05$	52.2 ± 8.4 (12) $p < 0.001$
+ Eserine, 10^{-5}M	83.9 ± 9.2 (8) NS	115.1 ± 12.7 (4) NS	109.5 ± 10.1 (6) NS	98.1 ± 11.5 (7) NS
+ Protamine sulphate, 0.3 mg/ml	105.8 ± 8.2 (6) NS	111.8 ± 15.6 (7) NS	98.7 ± 14.8 (8) NS	103.0 ± 13.3 (8) NS

Substrate, 2-naphthol. Lipolytic activity is expressed in per cent of control values, mean values ± S.E. are given. The number of experiments is in brackets. Statistical significance is evaluated against control of the respective tissue. NS = not significant.

Whereas in adipose and muscle tissues, insulin itself slightly lowered the lipase activity and, when applied simultaneously with ACTH or norepinephrine, decreased their lipolytic effects, in aortic and liver tissues it had not only the clearcut stimulatory effect on lipase activity but it also potentiated the lipolytic effects of other hormones. This was found not only in the experiments with the synthetic substrate 2-naphthol laurate (Table 2A), but also with a mixture of triglycerides (Table 2B) or with tristearin [15, 16]. The behaviour of hormone-sensitive lipases in homogenates of individual tissues in the presence of some inhibitors is summarized in Table 3. The nature of the inhibition of hormone-sensitive lipase of rat adipose tissue confirmed the results of Vaughan and other authors; on the other hand, the characteristics of analogous enzymes from other tissues differed. Although the electrophoretic mobilities of cathodic fractions of individual lipases, demonstrated in Fig. 1, differed only slightly from that of adipose tissue (the anodic fraction, however, was found only in adipose tissue), the inhibition properties resembled those of whole homogenates. It was of interest that the properties (e.g. localization, number and hormonal effects) of individual fractions of muscle hormone-sensitive lipase (after separation on starch medium electrophoresis and following gel filtration on Sephadex G-200) were practically identical with those of adipose tissue. Only the third, seventh and eighth fraction failed and the sixth fraction had slightly different mobility. The properties of liver lipase fractions were, however, quite different. The detailed analysis will be reported elsewhere.

DISCUSSION

Although the physiological interpretation of the role of hormone-sensitive lipases in individual tissues is not the subject of this paper, the above-mentioned experiments based on the use of well-defined synthetic substrates lead to the conclusion that hormone-sensitive lipase represents the key enzyme in lipid metabolism, not only in adipose tissue but also in some other tissues. In muscle tissue its participation in the regulation of lipid metabolism has already been postulated by Garland and Randle [23] and Challoner and Steinberg [24]. Direct evidence for hormone-sensitive lipase in muscle, and especially in liver tissue, was delayed by the fact that, when endogenous substrate is used, both classical methods (the estimation of free fatty acids, and of glycerol) do not test the end-product since both are rapidly reutilized in synthetic reactions. The high activity of glycerokinase in liver has long been known; in adipose tissue and heart muscle it is much lower and was therefore demonstrated only recently by more sensitive radiometric methods [25].

The different effects of hormones and inhibitors on "hetero-enzymes" [26] in the tissues compared may be explained by the differing participation of individual subfractions in each tissue. The surprising and unexpected separation of cathodic fractions into a number of multiple forms of lipase with different molecular weights reveals that more lipases participate in the tissue hormone-sensitive lipase spectrum, but their classification requires further experimental work. It could not be excluded that there is some kind of interaction between molecules of enzyme and particles of Sephadex. It can be seen that the last fraction is somewhat delayed, perhaps due to adsorption. The effects of inhibitors on individual fractions [27], not examined in this paper, revealed that only the first fraction was inhibited with iodoacetic acid and 2-propanol, which are considered to be specific inhibitors of hormone-sensitive lipase, whereas the fractions with lower molecular weights were not affected or were activated. The most surprising was the fact that the effects of individual hormones, as well as of cyclic $3',5'$-AMP, were demonstrable when enzyme fractions from Sephadex G-200 were briefly incubated with hormones, and that these effects clearly differed. A number of authors, on the basis of experiments with α- and β-adrenolytics, which blocked not only the lipolytic effects of catecholamines, but also those of polypeptide hormones, proposed a common mode of action for both groups of hormones—the elevation of the cyclic $3',5'$-AMP level. The identity of receptors for both catecholamines and polypeptide hormones, however, is still uncertain. Schwandt *et al.* [28], from experiments with nicotinic acid and with the β-adrenergic blocking agent propranolol, concluded that catecholamines act on lipolysis by a mechanism other than that for the polypeptide hormones. Stock and Westermann [29] found that β-adrenergic blocking agent Kö 592 seems to block norepinephrine competitively and the ACTH effect non-competitively. The elegant analyses of Lech and Calvert [30] provided some evidence for differentiation of ACTH and norepinephrine lipolytic receptors in adipose cells as well. The possibility that ACTH and epinephrine stimulate different lipases in adipose tissue is also supported by the results of Hollet [22]. She found that the centrifugal properties of lipase stimulated by ACTH and epinephrine were quite different. From this point of view our experiments supplement the foregoing results.

It is of special interest that in our experiments cyclic $3',5'$-AMP, which is believed to be the mediator of the action of lipolytic hormones in intact tissue, was capable of stimulating the activity of only two fractions. These results suggest that the lipolytic effects of the respective hormones are mediated not only through the released cyclic $3',5'$-AMP. The resulting effect may involve additional different mechanisms of action of the respective hormone. The experiments with isolated fractions indicated that the direct interaction of hormone with enzyme may be one of these mechanisms. According to this idea, individual lipase fractions may represent a part of the cellular receptor field of the respective hormone.

The behaviour of insulin was also remarkable. Whereas in the whole homo-genate insulin slightly inhibited lipase activity, the inhibitory effect on adipose tissue fractions was demonstrable especially in the first, and also in the second, fifth and sixth fractions. The third fraction, as well as the seventh, surprisingly were stimulated. It is necessary to emphasize that Novo insulin used in these experiments did not contain traces of glucagon [31]. It is probable that the total effect of insulin on the lipase activity of the whole homogenate or intact adipose tissue is the sum of its opposing effects on isolated lipase fractions from which the first one predominates. The detailed mechanism and the physiological significance of the lipolytic effect of insulin in liver and aortic tissue will be discussed in a subsequent paper.

ACKNOWLEDGEMENTS

The authors wish to acknowledge the skilful technical assistance of Miss Zdena Stará.

REFERENCES

1. Rizack, M. A., *J. biol. Chem.* **236** (1961) 657.
2. Hollenberg, C. H., Raben, M. S. and Astwood, E. B., *Endocrinology* **86** (1961) 589.
3. Vaughan, M. and Steinberg, D., *J. Lipid Res.* **4** (1963) 193.
4. Rodbell, M., *J. biol. Chem.* **239** (1964) 375.
5. Rizack, M. A., *J biol. Chem.* **239** (1964) 392.
6. Vaughan, M., Berger, J. E. and Steinberg, D., *J. biol. Chem.* **239** (1964) 401.
7. Strand, O., Vaughan, M. and Steinberg, D., *J. Lipid Res.* **5** (1964) 554.
8. Vaughan, M., Steinberg, D., Lieberman, F. and Stanley, S., *Life Sci.* **4** (1965) 1077.
9. Gorin, E. and Shafrir, E., *Biochim. biophys. Acta* **137** (1967) 189.
10. Björntorp, P., *J. Lipid Res.* **7** (1966) 621.
11. Björntorp, P., *Life Sci.* **6** (1967) 367.
12. Sutherland, E. W., Øye, I. and Butcher, R. W., *Recent Prog. Horm. Res.* **20** (1965) 623.
13. Butcher, R. W., *Pharmac. Rev.* **18** (1966) 237.
14. Palkovič, M. and Macho, L., *Endocr. experimentalis* **1** (1967) 12.
15. Chmelař, M. and Chmelařová, M., *Experientia* (1969) in press.
16. Chmelař, M. and Chmelařová, M., Abstracts of the Third International Symposium on Drugs Affecting Lipid Metabolism, Milan, 1968.
17. Chmelař, M. and Chmelařová, M., Abstracts of the Fourth Conference of Comparative Endocrinologists, Karlsbad, 1967.
18. Nachlas, M. M. and Seligman, A. M., *J. biol. Chem.* **181** (1949) 343.
19. Lambert, M. and Neish, A. C., *Can. J. Res.* **28** (1950) 83.
20. Kunkel, H. G. and Slater, R. J., *J. clin. Invest.* **31** (1952) 677.
21. Andrews, P., *Biochem. J.* **96** (1965) 595.
22. Hollet, C. R., *Biochem. biophys. Res. Commun.* **15** (1964) 575.
23. Garland, P. B. and Randle, P. J., *Nature, Lond.* **199** (1963) 381.
24. Challoner, D. R. and Steinberg, D., *Nature, Lond.* **205** (1965) 602.
25. Robinson, J. and Newsholme, E. A., *Biochem. J.* **901** (1966) 41.
26. Wieland, T. and Pfleiderer, G., *Angew. Chem.* (Int. Edn.) **1** (1962) 169.

27. Chmelař, M. and Chmelařová, M., paper delivered at the 5th FEBS Meeting, Prague, 1968.
28. Schwandt, P., Hartmann, Th. and Karl, H. J., *Z. ges. exp. Med.* **143** (1967) 79.
29. Stock, K. and Westermann, E., *Life Sci.* **5** (1966) 1667.
30. Lech, J. J. and Calvert, D. N., *Life Sci.* **6** (1967) 833.
31. Wenkeová, J., *Physiologia bohemoslov.* (1969) in press.

FEBS Symposium, Volume 18, 1970, pp. 227-239

The Heterogeneity of Human Alkaline Phosphatase

D. W. MOSS

University of Edinburgh, Department of Clinical Chemistry, The Royal Infirmary, Edinburgh, Scotland

Alkaline phosphatase (orthophosphoric monoester phosphohydrolase, EC 3.1.3.1) is widely distributed throughout the animal kingdom and among the tissues of particular species. It is an enzyme of low specificity, acting on a wide range of phosphate esters: for many years, the balance of opinion favoured the view that its action was confined to orthophosphate monoesters [1], but recent work has confirmed that inorganic [2, 3] and organic [4-6] pyrophosphates are substrates of the enzyme, though less readily attacked under optimum conditions than orthophosphates. Diesters are, however, apparently not hydrolysed. Alkaline phosphatases from various sources have been shown to be activated by divalent metal ions, notably magnesium, and to contain zinc as a constituent of the enzyme molecule [7-9].

The exact role of alkaline phosphatase in metabolism has proven difficult to assign. The closest relationship between the presence of the enzyme and a physiological event is to be found in calcification processes, and in man the primary pathological consequence of a reduced activity of the enzyme (hypophosphatasia) is defective bone mineralization. The occurrence of alkaline phosphatase in abundance in small intestine, kidney, placenta and liver suggests an involvement in transport processes in these tissues.

Determinations of serum alkaline phosphatase activity are of practical clinical importance in the investigation and management of hepatobiliary and bone diseases, and in consequence considerable interest has focused on the nature and properties of the enzyme and on the possible existence of tissue-specific variants or isoenzymes. As with other enzymes of wide distribution, the ubiquity of alkaline phosphatase also provides an opportunity to study the effects of genetic variation on the enzyme as it occurs in different organs and tissues.

Although distinctions between alkaline phosphatases from different sources were noted as long ago as 1937 [10], a more intensive investigation of alkaline phosphatase isoenzymes has resulted from the introduction of refined methods of fractionation of protein mixtures on both an analytical and a preparative scale, notably gel electrophoresis and ion-exchange chromatography. However, the extensive purification of alkaline phosphatase from human tissues in quantities sufficient for its characterization at a molecular level has proved difficult. In spite of this it is possible to conclude that tissue-specific isoenzymes of human alkaline phosphatase do exist, the distinctions between the several enzymes being in some cases quite marked, and the evidence for this conclusion is reviewed in the present paper.

DIFFERENTIATION OF HUMAN TISSUE ALKALINE PHOSPHATASES

Catalytic properties

The substrate specificity of human liver, bone and intestinal alkaline phosphatases has recently been re-examined and their action on pyrophosphate substrates has been confirmed [3, 5, 6]. The pyrophosphatase activity of human placental phosphatase has also been demonstrated [11]. Purified phosphatases from human bone, liver and small intestine show differences in their relative rates of release of inorganic phosphate from a range of pyrophosphate substrates (Table 1), while intestinal phosphatase differs from other human phosphatases in its specificity when acting on orthophosphate substrates, notably in its rapid hydrolysis of adenosine 5'-phosphate compared with that of p-nitrophenyl phosphate or β-glycerophosphate ([12] and Table 1).

Human placental alkaline phosphatase also shows different relative rates of hydrolysis of orthophosphate esters when compared with phosphatases from other human tissues [13].

The measurement of Michaelis constants for alkaline orthophosphatase activity has resulted in somewhat conflicting reports, partly because of the use of impure enzyme preparations in earlier work and partly due to the variations introduced by the marked dependence of K_m on pH in the range from pH 8 to pH 10, approximately, in which measurements are usually made. However, when allowance is made for this effect (e.g. by plotting initial reaction velocity at optimum pH against substrate concentration [14]), small but reproducible differences in K_m values for alkaline phosphatases from several human tissues have been shown [15]. Values of pK_e derived from plots of pK_m against pH also vary between purified human bone, liver and intestinal alkaline phosphatases [16]. The differences in pK_e for these three enzymes are more pronounced when inorganic pyrophosphate is the substrate, and values of pH 7.5 (liver), 8.1 (bone) and 8.4 (intestinal) have been reported [16].

Alkaline phosphatases from bacterial and animal sources incorporate radioactive phosphorus at acid pH, with phosphorylation of a serine residue at or near the active centre of the enzyme [17, 18]. This effect has been utilized to obtain estimates of the catalytic-centre activity of bovine milk [19] and human liver

Table 1. Relative rates of release of inorganic phosphate from ortho- and pyro-phosphate esters by purified human alkaline phosphatases.

Enzyme source:	Bone		Liver		Intestine	
Substrate:	Without Mg^{2+}	With Mg^{2+}	Without Mg^{2+}	With Mg^{2+}	Without Mg^{2+}	With Mg^{2+}
pNPP	22	100	45	100	77	100
AMP	18	69	35	55	95	125
ADP	15	33	22	19	150	50
ATP	5	7	10	5	115	15
UMP	18	54	45	60	45	50
UDP	11	34	30	20	130	40
UTP	2	13	15	11	75	10
TPP	8	17	50	40	110	50

Substrate concentration 2 mM in each case, at optimum pH for that substrate, with and without addition of 10 mM Mg^{2+}. The rates are referred to hydrolysis of 2 mM pNPP + 10 mM Mg^{2+} as 100, for each enzyme [5, 6].

Abbreviations: pNPP, p-nitrophenyl phosphate; TPP, thiamine pyrophosphate.

and intestinal [20] alkaline phosphatases. The values so obtained are of similar magnitude for the three enzymes when allowances are made for differences in experimental conditions, and are of the order of 5000-6000 s^{-1} at 37°C. The dissociation constants for the phosphorylated intestinal and milk phosphatases are also similar, but that for liver phosphatase seems significantly smaller [20]. Phosphorylation by radioactive inorganic pyrophosphate has also been demonstrated [21].

The use of inhibitors to differentiate isoenzymes of alkaline phosphatase has resulted in a clear distinction between, on the one hand, intestinal and placental phosphatases, and on the other, phosphatases from liver, bone and kidney. Bodansky [10] observed that bile acids inhibit non-intestinal, but not intestinal, alkaline phosphatases, while Fishman and his colleagues have demonstrated stereospecific inhibition of intestinal and placental phosphatases by L-phenylalanine, alkaline phosphatases from other tissues being virtually unaffected by this inhibitor [22, 23].

Studies on the catalytic properties and inhibition characteristics of human alkaline phosphatases thus indicate a clear distinction between the intestinal and placental enzymes and those from non-intestinal, non-placental sources, while further suggesting, though less conclusively, that the enzymes in the latter category are not themselves identical.

Molecular constitution

Molecular weights of the order of 100,000 have been reported on the basis of ultracentrifuge data, e.g. for the calf-intestinal enzyme [8]. Gel-filtration techniques [24-26] have indicated higher values, in the range 130,000-220,000 (Fig. 1), the most marked difference in size being observed in the case of the alkaline phosphatase of human urine which appears to have a molecular weight about half that of kidney phosphatase [27].

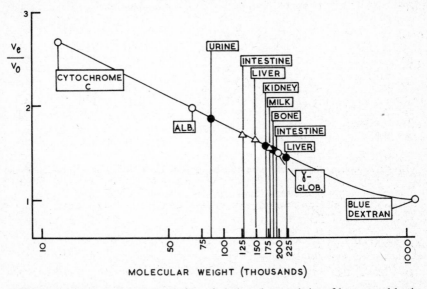

Figure 1. Diagrammatic representation of elution characteristics of human and bovine alkaline phosphatases on "Sephadex" G-200 gel filtration. The ratio of elution volume of enzyme (V_e) to void volume of column (V_o) is plotted against molecular weight. (●) Human phosphatases; (△) bovine phosphatases; (○), reference macromolecules. (For references, see text.)

Comparisons of the mobilities of alkaline phosphatases on starch gel electrophoresis have played a central role in focusing attention on the existence of phosphatase isoenzymes. While human alkaline phosphatases do not give sharp, homogeneous zones on this medium, it is now agreed that tissue-specific patterns can consistently be distinguished [28-30]. (The question of the heterogeneity of enzyme activity within a single tissue extract is discussed in a later section.) These patterns are shown in Fig. 2.

The differences in migration between intestinal alkaline phosphatase and phosphatases from other sources is accentuated if the enzymes are incubated with neuraminidase before electrophoresis [31]. The mobility of intestinal phosphatase is not affected by this treatment but the phosphatases of bone, liver, kidney and placenta are retarded [6, 32-34]. That the effect of neuraminidase is due to a release of sialic acid from the enzyme molecule has been verified in the case of placental phosphatase by Ghosh *et al.* [34], while Engström had earlier shown that purified calf-intestinal phosphatase contained no detectable sialic acid [8]. The catalytic activity of the alkaline phosphatases is unaffected by incubation with neuraminidase.

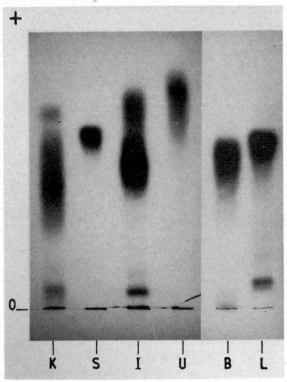

Figure 2. Zones of alkaline phosphatase activity in extracts of human tissues separated on horizontal starch gel electrophoresis at pH 8.6. O, origin; K, kidney; S, normal blood serum; I, small intestine; U, normal urine; B, bone; L, liver (anode at top). The heterogeneous character of the main zones can be seen, and also the occurrence of minor, slow-moving zones in the tissue extracts.

As with other examples of isoenzymes, attempts have been made to discriminate between alkaline phosphatase variants by comparison of their stabilities under different conditions. Differences in the rates of inactivation of human alkaline phosphatases on heating at 55°C were reported by Moss and King [30], while Neale *et al.* [35] subsequently demonstrated the resistance of placental phosphatase to prolonged heating at 70°C. Apart from the placental enzyme, the differences in stability between phosphatases from other tissues are small and experiments have therefore been made to determine whether the relative stabilities of the enzymes (particularly those from bone and liver) are reproduced in mixtures of the enzymes in serum [23, 36] and of the purified phosphatases (R. H. Eaton and D. W. Moss, unpublished; Fig. 3). These studies indicate that the differences in stability are observed consistently.

Other denaturing agents which have been applied to the differentiation of alkaline phosphatases include treatment with concentrated urea solutions and

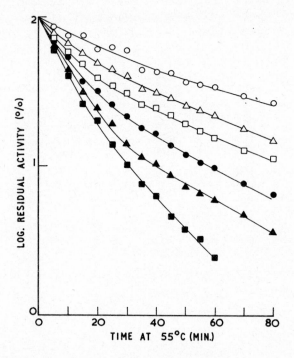

Figure 3. Inactivation of purified liver (O) and bone (■) alkaline phosphatases by heating at 55°C and pH 7.7 (tris buffer), and of mixtures of different proportions of the two enzymes. (△) 80% liver + 20% bone; (□) 60% liver + 40% bone; (●) 40% liver + 60% bone; (▲) 20% liver + 80% bone. (R. H. Eaton and D. W. Moss, unpublished data.)

exposure to acid pH. Treatment with urea confirms the resistance to denaturation of placental phosphatase [37] and shows that, while less stable than the placental enzyme, intestinal phosphatase is more resistant to urea than are the phosphatases of bone, liver or kidney [37, 38]. Bone phosphatase is also slightly less stable to urea than the liver enzyme [6]. Inactivation by acid also shows the greater stability of human intestinal phosphatase compared with liver phosphatase, which is again rather more resistant than the bone enzyme [6, 39].

When purified alkaline phosphatases from human tissues are used to produce antisera, further differentiation of the individual isoenzymes is obtained. Three

Table 2. Differences between three main categories of human alkaline phosphatases (see text for references).

	Placental phosphatase	Intestinal phosphatase	Non-placental, non-intestinal (liver, bone, etc.) phosphatases
I. Catalytic properties			
(i) Substrate specificity	Hydrolyses β-glycerophosphate and phenyl phosphate at similar rates (others more active on phenyl phosphate).	Hydrolyses AMP and PP_i faster with respect to p-nitrophenyl phosphate than non-intestinal enzymes.	Similar relative rates of hydrolysis of various substrates by enzymes within this group.
(ii) Inhibition			
(a) by bile acids	–	Not inhibited	Inhibited
(b) by L-phenylalanine	Inhibited	Inhibited	Not inhibited
II. Molecular properties			
(i) Resistance to denaturation			
(a) Time for half-inactivation at 55°C	Completely stable	60 min	Varies slightly for different phosphatases (bone 5-10, liver 30-40 min)
(b) Urea conc. for irreversible inactivation	8M	6-7M	Varies with source of enzyme (liver 3M, bone and kidney < 3M).
(c) Exposure to low pH	–	More stable than non-intestinal enzymes	Liver phosphatase more stable than bone phosphatase
(ii) Effect of neuraminidase on electrophoretic mobility	Retarded	Not retarded	Retarded
(iii) Immunochemical characteristics	Antigenically distinct (partial cross-reaction with intestinal enzyme)	Antigenically distinct (partial cross-reaction with placental enzyme)	Antigenically distinct from placental and intestinal enzymes. Antiserum to liver phosphatase does not cross-react with bone and kidney phosphatases.

classes of phosphatase, firstly, a group composed of the enzymes from bone, liver and spleen together with the major kidney component; secondly, intestinal phosphatase; and thirdly, the placental enzyme, were recognized as a result of earlier work [40, 41], with cross-reactions occurring between the groups. More recently, it has been shown immunochemically that, within the non-placental, non-intestinal group, an antiserum to liver phosphatase did not react with the phosphatases from bone or kidney [42].

The incorporation of radioactive phosphorus at acid pH into alkaline phosphatase has been shown to be due to phosphorylation of a serine residue in the cases of the enzyme from *E. coli*, calf intestine and bovine liver [17, 18, 24]. Human alkaline phosphatases also bind phosphate under similar conditions [20], presumably by a similar mechanism, though this has not yet been confirmed. Attachment of the label is most probably at the active centre of the enzyme, so that analysis of the serine peptides obtained by partial degradation of the labelled enzyme allows conclusions to be drawn as to the amino acid sequence in this region of the molecule. Milstein [43] has shown the presence of the tripeptide Asp-Ser-Ala in *E. coli* alkaline phosphatase in this way, and Engström [44] has confirmed that this tripeptide also occurs in calf-intestinal phosphatase. Extension of the known amino acid sequences away from the reactive serine residue may reveal variations in phosphatase molecules from diverse sources.

Alkaline phosphatase from *E. coli*, which has a molecular weight of about 80,000, has been shown to be composed of two subunits [45]. The larger size of mammalian alkaline phosphatases makes it probable that these enzymes also have a subunit structure, although this is not as yet proved.

The molecular properties of human alkaline phosphatases further emphasize the clear distinction between placental and intestinal phosphatases, and those from other sources. The placental and intestinal enzymes are themselves distinct, with the sum of evidence again favouring heterogeneity within the non-intestinal, non-placental group.

Some of the major criteria by which these categories of alkaline phosphatase may be differentiated are summarized in Table 2.

HETEROGENEITY OF ALKALINE PHOSPHATASE WITHIN A SINGLE TISSUE

The diffuse nature of the main zones of alkaline phosphatase activity following starch gel electrophoresis of tissue extracts, or of purified enzyme preparations, has frequently been noted. In addition, minor zones of enzyme activity migrating with mobilities lower than those of the main bands are usually seen after electrophoresis of unfractionated tissue extracts. Heterogeneity within the main zones of activity, e.g. of intestinal or kidney extracts, seems to result from variations in net charge, since gel-filtration indicates that the components of these zones are homogeneous with respect to molecular size [26, 46]. The origin of this gradation of net charge may lie in the occurrence of enzyme molecules containing different numbers of sialic acid residues, since incubation with neuraminidase reduces the diffuse main zone of human kidney phosphatase to a single, more compact region [32]. However, this is unlikely to be the explanation in the case of intestinal phosphatase, which is resistant to the action of neuraminidase. Studies of the enzymic properties of different regions of the diffuse zones of intestinal [47] and kidney [48] phosphatases have shown that, within the respective zones, catalytically-different enzyme variants cannot be distinguished.

The minor zones of alkaline phosphatase with a low electrophoretic mobility which are seen in gel electrophoresis of a particular tissue extract are, similarly, indistinguishable in their stability to heating and in catalytic properties from the major enzyme band in the same extract [30]. When an extract of liver is treated with neuraminidase before electrophoresis, the minor enzyme zones, as well as the major band, are retarded [33]. The method of preparation of the phosphatase-containing extracts also has an effect on the appearance and relative proportions of subsidiary electrophoretic zones of alkaline phosphatase [49], while phosphatase zones of an intermediate mobility, possibly representing partially degraded enzyme, have been observed during purification of human intestinal, liver and bone phosphatases [6, 26].

The experimental evidence therefore favours the view that, within a single tissue or organ, there is only one type of alkaline phosphatase and that the heterogeneity which is observed probably results from modification of this single phosphatase to give rise to a population of molecules with a range of molecular charges but of equivalent enzyme activity, while the existence of minor zones may result from degradation of the primary enzyme type or its attachment to inactive proteins (perhaps as complexes with lipo-protein). In the placenta, however, a genetically-controlled polymorphism of alkaline phosphatase, which gives rise to differences between individuals, has been established [50, 51]. Robson and Harris [51] differentiated human placental phosphatase into at least six different phenotypes by starch gel electrophoresis under various conditions, the phenotypes being determined by the genotype of the foetus. The observed polymorphism can largely be accounted for by postulating three relatively common autosomal allelic genes. No functional differences between the respective polymorphic forms have been established so far, and, in view of the apparent failure of electrophoretic heterogeneity to give rise to a variation in catalytic properties in other tissues [47, 48], marked catalytic differences between the placental enzyme variants are perhaps not to be expected.

THE ORIGIN OF ALKALINE PHOSPHATASE VARIATION

It is clear that while there are basic similarities in the alkaline phosphatases from different sources, for example, in the probable presence of an active serine residue in each and, among mammalian phosphatases, a general similarity in molecular size and catalytic activity, there are marked differences between human placental and non-placental phosphatases, and between intestinal and non-intestinal enzymes. Intestinal phosphatases from different species resemble each other more closely than each of them resembles phosphatases from other tissues of its own species. The question as to whether these different classes of phosphatases have fundamentally distinct structures, and are therefore the products of separate structural genes, can probably be answered with certainty for placental phosphatase, the evidence favouring a specific placental phosphatase gene or genes. A distinct genetic control of intestinal phosphatase probably also operates.

Within the non-placental, non-intestinal category of alkaline phosphatases the differences between the enzymes composing the group are less clear-cut, and the possibility arises that variations in properties within the group are the result of environmental factors, related to different types of cell, operating on a fundamentally similar enzyme molecule. The genetically-determined condition of hypophosphatasia [52], in which low or absent serum alkaline phosphatase is associated with defective bone mineralization, might be expected to throw some light on this point. Reports of these cases have suggested that the alkaline phosphatase activity of all tissues is depressed, implying perhaps a common genetic control, but in at least one case normal levels of phosphatase in liver and duodenal juice were found [53], and normal or increased intestinal alkaline phosphatase activity was detected in biopsy specimens in a recent series of cases [54]; this result would favour independent genetic control of bone and intestinal enzymes. It is not certain, however, that the underlying defect in hypophosphatasia is a genetically-determined inability to produce alkaline phosphatase, since a fragment of bone taken from a hypophosphatasic patient gave rise in tissue culture to osteoblast-like cells, rich in alkaline phosphatase [53].

The observed heterogeneity of alkaline phosphatase within a single tissue is also difficult to interpret, except for the placenta in which a genetic basis for the polymorphism has been demonstrated. The part apparently played by sialic acid residues in determining this heterogeneity in certain tissues (e.g. kidney) may suggest that a population of enzyme molecules arises by attachment of these residues in varying numbers to the enzyme protein. However, the possibility of variation in the protein core of the molecule cannot be dismissed, and intestinal alkaline phosphatase, which shows considerable heterogeneity, does not appear to contain sialic acid, at least not in a combination accessible to the action of neuraminidase.

The existence of tissue-specific isoenzymes of alkaline phosphatase prompts speculation as to how far catalytically-similar enzymes prepared from the different tissues of one animal species may be expected to be identical, and to what extent variation may be probable, particularly with the considerable degree of heterogeneity shown by the alkaline phosphatases. Some enzymes of wide occurrence in various tissues of the body have been shown to be closely similar or identical when the enzyme from one tissue is compared with its homologue from a different source. Isoenzyme 1 of lactate dehydrogenase from heart appears to be identical with isoenzyme 1 from other tissues of the same species, for example, while homologous creatine kinases appear to be closely similar, and this is true also of glyceraldehyde-3-phosphate dehydrogenases. Furthermore, the respective isoenzymes of these enzymes are themselves homogeneous. Compared with human alkaline phosphatases, these enzymes have lower molecular weights (according to more recent estimates of the molecular weights of the phosphatases) and it may be that in smaller enzyme molecules there are fewer opportunities for modifications in the protein structure to arise without adversely affecting the enzymic properties. In comparison with alkaline phosphatases these enzymes have clearly definable roles in metabolism and much more restricted catalytic specificities; any mutation in the enzyme structure which involved even a small impairment of catalytic efficiency could therefore be expected to result in considerable adverse pressure of natural selection tending towards its deletion, so that homogeneity amongst the molecules of more specific enzymes might seem the more probable situation. Alkaline phosphatases do not appear to occupy positions in metabolism which make them essential to life, and their specificity is low. It seems possible that considerable structural modification of the phosphatases could take place without so altering their properties that the mutant forms would be subjected to selection pressures strongly favouring the predominance of one type and the exclusion of others. Consequently, a heterogeneous population of enzyme molecules might arise and subsequently be perpetuated.

The variant forms of alkaline phosphatase which are seen in man may have originated in the following way: the basic alkaline phosphatase molecule may have been derived by modification of a prototype, active serine-containing, hydrolytic enzyme, the ancestor also of other "serine" hydrolases. Further modification of the gene determining this phosphatase into three or more types, expressed to different extents in different tissues, perhaps resulted in the emergence of tissue-specific phosphatase groups, i.e. placental, intestinal, etc. (Elucidation of the amino acid sequences of the isoenzymes would enable a choice to be made between this hypothesis and an alternative, convergent evolution from separate precursors.) Modification and multiplication of the tissue-specific phosphatase genes may then have led to the heterogeneity of the enzymes within each tissue, if this does indeed reflect the existence of a population of genetically-distinct molecules.

From the point of view of predicting which enzymes might profitably be studied in order to disclose organ-specific isoenzymes of use in diagnosis, arguments of the kind set out above suggest that the objects of such investigations should be enzymes of low specificity and of fairly high molecular weights, in which the possibilities of modifications associated with particular tissues are perhaps strongest. Exceptions to this kind of generalization arise in those situations in which variations in the pattern of metabolism in one tissue, as compared with another, are associated with corresponding adaptations of key enzymes: the correlation of the relative abundance of lactate dehydrogenase isoenzymes 1 and 5 with the balance of aerobic and anaerobic metabolism is a striking example of this functional enzyme adaptation.

REFERENCES

1. Dixon, M. and Webb, E. C., "Enzymes", 2nd edn., Longmans, Green and Co., London, 1964, p. 223.
2. Cox, R. P. and Griffin, M. J., *Lancet* ii (1965) 1018.
3. Moss, D. W., Eaton, R. H., Smith, J. K. and Whitby, L. G., *Biochem. J.* 102 (1967) 53.
4. Fernley, H. N. and Walker, P. G., *Biochem. J.* 104 (1967) 1011.
5. Eaton, R. H. and Moss, D. W., *Biochem. J.* 105 (1967) 1307.
6. Eaton, R. H. and Moss, D. W., *Enzymologia* 35 (1968) 31.
7. Mathies, J. C., *J. biol. Chem.* 233 (1958) 1121.
8. Engström, L., *Biochim. biophys. Acta* 52 (1961) 36.
9. Plocke, D. J., Levinthal, C. and Vallee, B. L., *Biochemistry* 1 (1962) 373.
10. Bodansky, O., *J. biol. Chem* 118 (1937) 341.
11. Sussman, H. H. and Laga, E., *Biochim. biophys. Acta* 151 (1968) 281.
12. Landau, W. and Schlamowitz, M., *Archs Biochem. Biophys.* 95 (1961) 474.
13. Ahmed, Z. and King, E. J., *Biochim. biophys. Acta* 45 (1960) 581.
14. Motzok, I., *Biochem. J.* 72 (1959) 169.
15. Moss, D. W., Campbell, D. M., Anagnostou-Karakas, E. and King, E. J., *Biochem. J.* 81 (1961) 441.
16. Eaton, R. H. and Moss, D. W., *Enzymologia* 35 (1968) 168.
17. Schwartz, J. H. and Lipmann, F., *Proc. natn. Acad. Sci. U.S.A.* 47 (1961) 1996.
18. Engström, L., *Biochim. biophys. Acta* 52 (1961) 49.
19. Barman, T. E. and Gutfreund, H., *Biochem. J.* 101 (1966) 460.
20. Moss, D. W., Eaton, R. H. and Scutt, P. B., *Biochim. biophys. Acta* 154 (1968) 609.
21. Fernley, H. N. and Bisaz, S., *Biochem. J.* 107 (1968) 279.
22. Fishman, W. H., Green, S. and Inglis, N. I., *Biochim. biophys. Acta* 62 (1962) 363.
23. Fishman, W. H. and Ghosh, N. K., *Adv. clin. Chem.* 10 (1967) 255.

24. Engström, L., *Biochim. biophys. Acta* **92** (1964) 71.
25. Andrews, P., *Biochem. J.* **96** (1965) 595.
26. Smith, J. K., Eaton, R. H., Whitby, L. G. and Moss, D. W., *Analyt. Biochem.* **23** (1968) 84.
27. Butterworth, P. J., Moss, D. W., Pitkanen, E. and Pringle, A., *Clinica chim. Acta* **11** (1965) 220.
28. Chiandussi, L., Greene, S. F. and Sherlock, S., *Clin. Sci.* **22** (1962) 425.
29. Hodson, A. W., Latner, A. L. and Raine, L., *Clinica chim. Acta* **7** (1962) 255.
30. Moss, D. W. and King, E. J., *Biochem. J.* **84** (1962) 192.
31. Robinson, J. C. and Pierce, J. E., *Nature, Lond.* **204** (1964) 472.
32. Butterworth, P. J. and Moss, D. W., *Nature, Lond.* **209** (1966) 805.
33. Moss, D. W., Eaton, R. H., Smith, J. K. and Whitby, L. G., *Biochem. J.* **98** (1966) 32C.
34. Ghosh, N. K., Goldman, S. S. and Fishman, W. H., *Enzymologia* **33** (1967) 113.
35. Neale, F. C., Clubb, J. S., Hotchkis, D. and Posen, S., *J. clin. Path.* **18** (1965) 359.
36. Posen, S., Neale, F. C. and Clubb, J. S., *Ann. intern. Med.* **62** (1965) 1234.
37. Birkett, D. J., Conyers, R. A. J., Neale, F. C., Posen, S. and Brudenell-Woods, J., *Proc. Austral. Assoc. Clin. Biochem.* **1** (1965) 175.
38. Butterworth, P. J. and Moss, D. W., *Enzymologia* **32** (1967) 269.
39. Scutt, P. B. and Moss, D. W., *Biochem. J.* **105** (1967) 43P.
40. Schlamowitz, M. and Bodansky, O., *J. biol. Chem.* **234** (1959) 1433.
41. Boyer, S. H., *Ann. N. Y. Acad. Sci.* **103** (1963) 938.
42. Sussman, H. H., Small, P. A. and Cotlove, E., *J. biol. Chem.* **243** (1968) 160.
43. Milstein, C., *Biochem. J.* **92** (1964) 410.
44. Engström, L., *Biochim. biophys. Acta* **92** (1964) 79.
45. Levinthal, C., Signer, E. R. and Fetherolf, K., *Proc. natn. Acad. Sci. U.S.A.* **48** (1962) 1230.
46. Moss, D. W., *Nature, Lond.* **200** (1963) 1206.
47. Moss, D. W., *Biochem. J.* **94** (1965) 458.
48. Butterworth, P. J., *Biochem. J.* **96** (1965) 74P.
49. Moss, D. W., *Nature, Lond.* **193** (1962) 981.
50. Boyer, S. H., *Science, N.Y.* **134** (1961) 1002.
51. Robson, E. B. and Harris, H., *Nature, Lond.* **207** (1965) 1257.
52. Fraser, D., *Am. J. Med.* **22** (1957) 730.
53. Scaglione, P. R. and Lucey, J. F., *J. Diseases Childr.* **92** (1956) 493.
54. Danovitch, S. H., Baer, P. N. and Laster, L., *New Engl. J. Med.* **278** (1968) 1253.

FEBS Symposium, Volume 18, 1970, pp. 241-247

Inorganic Pyrophosphate Phosphohydrolase in Boar Seminal Plasma

K. J. HRUŠKA

Department of Biochemistry, Faculty of Veterinary Science, Brno, Czechoslovakia

Many metabolic processes lead to formation of inorganic pyrophosphate, which is hydrolyzed in tissues by inorganic pyrophosphate phosphohydrolase [1, 2]. The significance and properties of this enzyme from microsomes of rat liver and kidney cells have been widely studied [3-14], and the results indicate that the pyrophosphate cleavage is catalyzed by an enzymatic system which is not specific but which also catalyzes phosphate transfer from pyrophosphate to hexose, and hydrolytic cleavage of other phosphate esters and polyphosphates. On the other hand, the pyrophosphatase that has been purified from the erythrocytes is very specific and hydrolyzes only pyrophosphate [15]. The available data on the properties of pyrophosphatases from different sources are of some interest and led us to study inorganic pyrophosphatase of seminal plasma, hitherto examined only to a limited extent [16-20]. Histochemical methods pointed to a very high activity of inorganic pyrophosphatase in the epithelial cells of the seminal vesicles and in their secretion. In the rat, mouse, and guinea-pig, pyrophosphatase is not found in all accessory glands where fructose is produced and, therefore, the participation of this enzyme in fructose formation does not seem to be very important [21]. Moreover, we could not find a correlation between the fructose level and pyrophosphatase activity in bull seminal plasma [22]. An additional finding, the disappearance of pyrophosphatase activity in the secretion of the sex glands following castration of rabbits, requires further investigation, since testosterone application did not increase the activity to the initial value, although the fructose level reached it during several days [22]. The present study is devoted to some properties of the inorganic pyrophosphatase of the seminal plasma of the boar.

Enzymes. Inorganic pyrophosphatase: inorganic pyrophosphate phosphohydrolase (EC 3.6.1.1); alkaline phosphatase: orthophosphoric monester phosphohydrolase (EC 3.1.3.1).

MATERIALS AND METHODS

Seminal plasma was obtained by centrifugation of boar semen (18,000 x g, 20 min, 4°C). Only semen with high pyrophosphatase activity was used. All experiments were repeated at least twice and results were similar with the plasma of different boars.

The activity of inorganic pyrophosphatase was estimated after 20 min incubation (30°C) of appropriately diluted enzyme (x 100) in tris-HCl buffer (pH 8.3, I = 0.05), in the presence of inorganic pyrophosphate and magnesium, each at a final concentration of 3 mM. Inorganic phosphate was estimated in the whole reaction mixture (0.5 ml) by the method of Baginski et al. [23]. The reaction was initiated by addition of preheated substrate and stopped by addition of trichloroacetic acid with ascorbic acid as the first step for estimation of phosphate. The results are expressed either as extinctions or in μmoles of inorganic phosphate liberated per ml of enzyme preparation during 1 min. The same method was used when glucose-6-phosphate or fructose-1,6-diphosphate served as substrates. Blanks included the enzyme incubated under the same conditions without substrate, which was added after deproteination.

For the alkaline phosphatase the method of Neumann and Van Vreedendaal [24] was used and proteins were estimated either by absorption at 280 nm or by the method of Lowry et al. [25].

Proteins were separated by gel filtration on Sephadex G-100 or G-200 on columns K 25/45 (Pharmacia, Uppsala). Further details are described in the legends to figures.

RESULTS AND DISCUSSION

The activity of inorganic pyrophosphate phosphohydrolase in boar seminal plasma is relatively high. For further study we tried to purify this enzyme, but neither butanol extraction nor ammonium sulphate fractionation were successful. However, by gel filtration on a column of Sephadex G-200, we have obtained all pyrophosphatase in the first smaller protein fraction (Fig. 1). This fraction, with a specific activity increased 5-10 times, was used for further study.

After gel filtration, and during the other purification experiments, we also estimated the activity of alkaline phosphatase with p-nitrophenylphosphate as substrate. In all instances the changes in both activities were very similar. Figure 2 shows the results of purification of inorganic pyrophosphatase by comparison with those for alkaline phosphatase. The elution patterns of both phosphatases after gel filtration are the same and, furthermore, the peak of activity is capable not only of hydrolyzing inorganic pyrophosphate and p-nitrophenylphosphate, but also glucose-6-phosphate and fructose-1,6-diphosphate. Consequently, further study of the specificity of boar seminal phosphatases seems to be necessary.

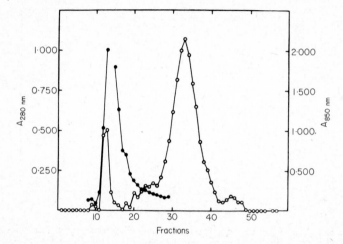

Figure 1. Gel filtration of boar seminal plasma on Sephadex G-200. Column 2.5 x 35 cm, elution with tris-HCl buffer (pH 7.4, I = 0.05) with 0.5M NaCl (20 ml/hod.), fractionation at 10 min intervals. Fresh seminal plasma was dialyzed against buffer and then concentrated (x 2-3), sample volume 2 ml. In each fraction, protein (open circles) and pyrophosphatase activity (solid circles) were estimated as described in the text.

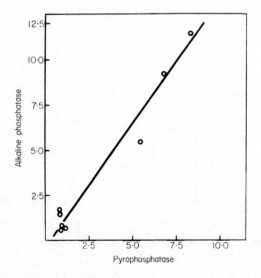

Figure 2. Interdependence during purification of pyrophosphatase and alkaline phosphatase activities of boar seminal plasma (coefficient of correlation, 0.98).

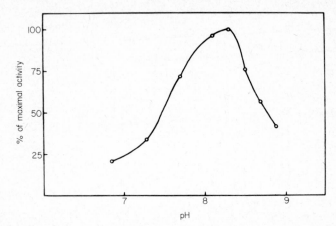

Figure 3. Effect of pH on the activity of boar seminal pyrophosphatase. To 0.6 ml of tris-HCl buffer (I = 0.1), 0.2 ml of the enzyme diluted in saline and 0.2 ml pyrophosphate with magnesium (final concentration 3 mM and 1.2 mM, respectively) were added. Other conditions were as described in "Methods".

The enzymatic hydrolysis of pyrophosphate by boar seminal plasma has an optimal pH of 8.3 (Fig. 3). This value is similar to that of blood serum pyrophosphatase [26], but differs from the lower optimal pH for the same enzyme from liver microsomes (pH 5) [7], *B. megaterium* (pH 7) [27], rat heart (pH 7-7.3) [28], and erythrocytes (pH 7.7) [15].

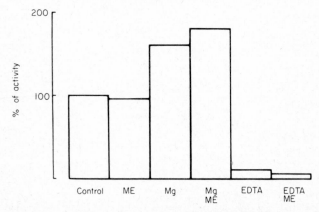

Figure 4. Effect of incubation with 1 mM magnesium (Mg), 1 mM ethylenediaminetetraacetic acid (EDTA), and 10 mM mercaptoethanol (ME) on the activity of boar seminal pyrophosphatase. After 20 min, pyrophosphate (3 mM) was added to the mixture and the reaction stopped at 5-min intervals. The pyrophosphate hydrolysis was linear during the whole period (40 min). All conditions are as described in "Methods". Concentrations are expressed as final values in the whole reaction mixture.

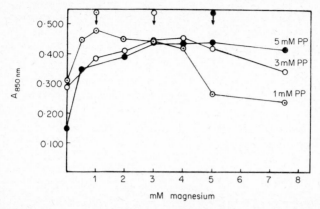

Figure 5. Effect of magnesium on the activity of boar seminal pyrophosphatase at three substrate concentrations. All concentrations are expressed as final concentrations of pyrophosphate and magnesium. Arrows indicate the values at equimolar concentrations of magnesium and substrate.

Inorganic pyrophosphatase from boar seminal plasma is activated by magnesium and inhibited by ethylenediaminetetraacetic acid in 1 mM concentration. On the other hand, mercaptoethanol in 10 mM concentration does not influence the activity either in the presence of magnesium or ethylenediaminetetraacetic acid (Fig. 4).

The ratio of molar concentrations of substrate and magnesium is very important, as for some other pyrophosphatases [29]. The activity at different substrate concentrations is the same when the molar ratio of magnesium to pyrophosphate is about 1.0. At higher values of this ratio the activity of pyrophosphatase decreases (Fig. 5). However, it must be noted that our results on rabbit seminal plasma [30], and the results of other authors on bull seminal plasma, did not point to a higher activation of pyrophosphatase by magnesium [16-18]. Moreover, it can be seen from Fig. 5 that, in the absence of magnesium in the reaction mixture, the higher concentration of pyrophosphate has an inhibitory effect.

The molecular weight of boar seminal pyrophosphatase was estimated by gel filtration. From a column of Sephadex G-100, the enzyme was eluted in the same fractions as a blue dextran 2000 (Pharmacia, Uppsala), so that the molecular weight of the enzyme is at least 150,000. Our earlier separation of boar seminal plasma by thin layer gel filtration on Sephadex G-200 indicated that a protein with pyrophosphatase mobility had a molecular weight of about 200,000 (unpublished results). This value is higher that that of pyrophosphatase from *B. megaterium,* which is 58,000 [27].

By agar gel electrophoresis [31], the inorganic pyrophosphatase from rabbit seminal plasma was separated into two isoenzymes [32]. One of these exhibited cathodic, the other anodic, mobility. The activity of both isoenzymes was approximately the same. We were unable to separate boar seminal pyrophosphatase, either by electrophoresis or gel filtration, into two or more fractions.

The foregoing results point to the high activity of inorganic pyrophosphate phosphohydrolase in boar seminal plasma and to some differences of this enzyme from others of this type. Further study is necessary for an understanding of the significance and properties of this characteristic product of the accessory gland of the male genital tract and for extending our knowledge of the seminal vesicles function.

ACKOWLEDGEMENT

The supply of the boar semen by the Division of Artificial Insemination (Dr. J. Kozumplík, CSc, Head), and the technical assistance of Mrs. J. Korytárová and Mrs. A. Krejzlová, are gratefully acknowledged.

REFERENCES

1. Stetten, D., Jr., *Am. J. Med.* **28** (1960) 867.
2. Rafter, G. W., *J. biol. Chem.* **235** (1960) 2475.
3. Stetten, M. R., *J. biol. Chem.* **239** (1964) 3576.
4. Stetten, M. R. and Taft, H. L., *J. biol. Chem.* **239** (1964) 4041.
5. Stetten, M. R., *J. biol. Chem.* **240** (1965) 2248.
6. Stetten, M. R. and Burnett, F. F., *Biochim. biophys. Acta* **128** (1966) 344.
7. Stetten, M. R. and Burnett, F. F., *Biochim. biophys. Acta* **132** (1967) 138.
8. Fisher, C. J. and Stetten, M. R., *Biochim. biophys. Acta* **121** (1966) 102.
9. Stetten, M. R. and Rounbehler, D., *J. biol. Chem.* **243** (1968) 1823.
10. Nordlie, R. C. and Arion, W. J., *J. biol. Chem.* **239** (1964) 1680.
11. Nordlie, R. C. and Soodsma, J. F., *J. biol. Chem.* **241** (1966) 1719.
12. Cox, R. P., Gilbert, P., Jr. and Griffin, M. J., *Biochem. J.* **105** (1967) 155.
13. Nordlie, R. C., Arion, W. J., Hanson, T. L., Gilsdorf, J. R. and Horne, R. N., *J. biol. Chem.* **243** (1968) 1140.
14. Nordlie, R. C., Gilsdorf, J. R., Horne, R. N. and Paur, R. J., *Biochim. biophys. Acta* **158** (1968) 157.
15. Pynes, G. D. and Younathan, E. S., *J. biol. Chem.* **242** (1967) 2119.
16. Heppel, L. A. and Hilmoe, R. J., *J. biol. Chem.* **192** (1951) 87.
17. Zhivkov, V., *Izv. Inst. srav. Patol. dom. Zhiv. Sof.* **8** (1960) 177.
18. Buruiana, L. M. and Hadarag, E., *Naturwissenschaften* **50** (1963) 478.
19. Buruiana, L. M. and Hadarag, E., *Naturwissenschaften* **51** (1964) 494.
20. Buruiana, L. M. and Hadarag, E., *Lucr. stiint. Inst. agron. 'N. Balcescu' Bucuresti Ser. C.* **8** (1965) 165.
21. Litošová, I., Holman, J. and Hruška, K. J., *Acta. histochem.* **32** (1969) 178.
22. Tamirat, H. Y., Thesis, Faculty of Veterinary Science, Brno, 1967.
23. Baginski, E. S., Foà, P. P. and Zak, B., *Clin. Chem.* **13** (1967) 326.
24. Neumann, H. and Van Vreedendaal, M., *Clinica chim. Acta* **17** (1967) 183.
25. Lowry, O. H., Roseborough, N. J., Farr, A. L. and Randall, R. J., *J. biol. Chem.* **193** (1951) 265.

26. Pitkänen, E., *Scand. J. clin. Lab. Invest.* **12** (1960) 143.
27. Tono, H. and Kornberg, A., *J. Bact.* **93** (1967) 1819.
28. Soodsma, J. F. and Nordlie, R. C., *Biochim. biophys. Acta* **122** (1966) 510.
29. Bloch-Frankenthal, L., *Biochem. J.* **57** (1954) 87.
30. Hruška, K. J., Thesis, Faculty of Veterinary Science, Brno, 1965.
31. Hruška, K. J., *Sb. vys. Sk. zeměd. les. Fac. Brne–B–Spisy fak. vet.* **13** (1965) 103.
32. Petráš, K., Thesis, Faculty of Veterinary Science, Brno, 1967.

FEBS Symposium, Volume 18, 1970, pp. 249-255

Further Characterization of Porcine α-Amylases

G. MARCHIS-MOUREN, L. PASÉRO and P. COZZONE

Institut de Chimie Biologique, Faculté des Sciences,
Marseille, France

ISOLATION OF TWO ISOAMYLASES

Early chromatographic studies [1] performed to determine the enzyme composition of porcine pancreatic juice indicated the presence of two α-amylases. In addition, disc electrophoresis of total juice revealed two components [2]. Amylase has been purified in large amounts by selective precipitation of the enzyme as a substrate complex with glycogen [3]. Pure porcine amylase analysed by disc electrophoresis gave two bands, while rat pancreatic amylase obtained by the same purification procedure at identical specific activity gave only one band (Fig. 1). However, the glycogen purification technique was not used further because of difficulties encountered in elimination of the co-precipitated polysaccharides. Total porcine amylase was then purified from porcine homogenate (Table 1). The first two steps of the procedure are identical to the technique of Fischer and Bernfeld [4]. Final purification of amylase was achieved by chromatography on a column of DEAE-cellulose.

Analysis of the eluate showed that amylolytic activity is associated with two peaks (Fig. 2a), both of identical specific activity. When peaks I and II were submitted to re-chromatography under the same conditions, each gave only one peak at the expected position (Figs. 2b and c). Analysis by gel electrophoresis of peaks I and II gave one distinctive band for each, while the mixture of both gave

Table 1. Purification of the two porcine amylases.

Purification steps		Yield (%)	Specific activity
Aqueous extract		100	100
Acetone precipitate (49-65%)		71	230
Ammonium sulphate precipitate (40% saturation)		65	400
Chromatography on DEAE-cellulose	Peak I:	40%	
	Peak II:	25%	550
	Total:	65%	

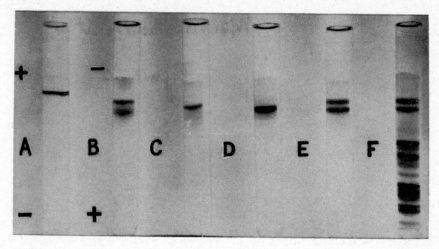

Figure 1. Results of disc electrophoresis in polyacrylamide gel. Electrophoresis was carried out at 6°C, for Expt. A in an acidic buffer containing 15.6 g β-alanine and 4.0 ml acetic acid per l; for Expts. B to F in a basic buffer containing 6.0 g Tris and 28.8 g glycine per l. The concentration of the gel was 7.5%. A and B: purified rat and porcine amylase, respectively. C and D: porcine amylases I and II, respectively. E: Mixture of porcine amylases I and II. F: Anionic fraction of porcine pancreatic juice; the two upper bands are amylases I and II.

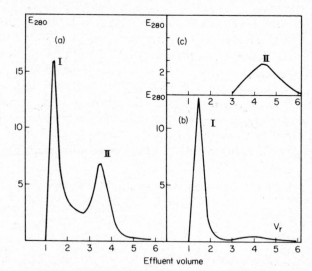

Figure 2. Separation of porcine amylases I and II on DEAE-cellulose. Left: Chromatography of an amylase preparation (2×10^5 units) on a 1.6 cm × 30 cm column (void volume, 34 ml) equilibrated and eluted with a 5 mM phosphate buffer at pH 8.0. Right: Re-chromatography of Peaks I and II, respectively, under the same conditions on a 0.9 cm × 14 cm column.

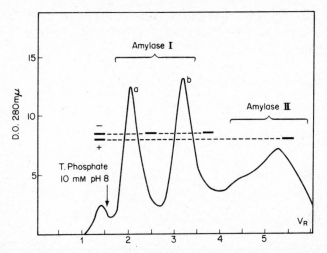

Figure 3. Separation of porcine amylases I and II on DEAE-cellulose. The conditions are the same as in Fig. 2 except that 10 mM phosphate (pH 8) was added as indicated. The results of disc electrophoresis analysis are given at the right of each peak and referred to the total pure amylase pattern (left).

two bands (Fig. 1). In agreement with its behaviour on DEAE-cellulose, peak II, the more acidic component, migrates further towards the anode. In some experiments peak I was split into two peaks named Ia and Ib (Fig. 3), each giving only one band on disc electrophoresis. The reason for the splitting of peak I is not understood; chromatographic heterogeneity of amylase in two or four active fractions separated on DEAE-cellulose has also been reported by Szabó and Straub [5]. Peaks I and II have been referred to as amylases I and II, and differences other than electrophoretic mobilities and chromatographic behaviour have been investigated.

CHEMICAL CHARACTERIZATION

(a) End groups

C-terminal leucine residue was determined by the hydrazinolysis technique for both porcine amylases, as against lysine previously found in rat amylase. The search for an α-NH$_2$ terminal group was carried out using the technique of Sanger; no ether- or water-soluble DNP amino acid could be identified in amylases I and II. These negative results indicate that the α-NH$_2$ terminal group is blocked. Characterization of an acetyl group at this position was then attempted either chemically [2] or by physical methods.

The acetyl group was characterized as an acetyl hydrazide [6] and as the DNP-acetylhydrazide [7] in amylases I and II. However, such determinations seemed to us rather poor since both compounds have been identified only by their positions in two-dimensional chromatography. Better characterization is hopefully in progress, using nuclear magnetic resonance, since the protons of the acetyl group give a very specific resonance signal [8].

Table 2. Amino acid composition of porcine amylase I.

	(1)	(2)
Aspartic acid + asparagine	17.73	66-67
Threonine	5.84	24-25
Serine	7.17	34
Glutamic acid + glutamine	9.86	34
Proline	5.27	23
Glycine	7.89	52-53
Alanine	5.21	29
1/2 Cystine	2.42	10
Valine	9.73	41-42
Methionine	2.21	7
Isoleucine	5.54	21
Leucine	6.09	23
Tyrosine	6.57	18
Phenylalanine	7.09	21
Lysine	5.63	19
Histidine	2.45	8
Arginine	9.38	27

(1) Grams of amino acid in 100 g of protein.
(2) Moles of amino acid in one mole of amylase I, assuming 50,000 as molecular weight.

(b) Amino acid composition

Significant differences are not to be expected between molecules of 50,000 molecular weight or more. The amino acid composition of amylase I (Table 2) is given in per cent since some controversy exists as to the exact molecular weight (M.W.) of amylase. Assuming a M.W. = 50,000, the composition indicates five possible SS bridges; of interest for subsequent structural work is the number of methionines 7-8. For several amino acids, large differences prevail between our results and those published by Caldwell *et al.* [9].

(c) Cyanogen bromide cleavage

In an attempt to get a better insight into the primary structures of amylases I and II, preparative cleavage of the protein molecules was achieved. Amylase was dissolved in 70% trifluoroacetic acid, containing 5×10^{-4}M of the reducing agent dithioerythritol, at a concentration of 1%. Cyanogen bromide was added in 60-fold molar excess and the reaction carried out at 25°C for 24 h. After a 10-fold dilution with water, volatile products were evaporated and the concentrated solution lyophilized. The extent of cleavage was estimated from amino acid analysis of the digest: under our condition all methionines were degraded and transformed into equivalent molar amounts of homoserine lactone.

(d) Purification of the cyanogen bromide peptides.

Eight to nine peptides were expected from the methionine content of amylases. Most of the peptides are water-insoluble and have been dissolved either in 6M urea or in 50% acetic acid. Aminoethylation by ethlylenimine did not favour solubilization and impaired further characterization and isolation. Total peptides of the digest were dissolved in 50% acetic acid and their fractionation was achieved by gel filtration on a Sephadex G-75 fine column equilibrated with the same solvent (Fig. 4). Six peaks and an identical pattern were obtained from digests of amylases I and II. As indicated in Fig. 4, analysis of the peptides of the digest, and of each of the six peaks, was achieved by disc electrophoresis in 6M urea at pH 4.4. The two electrophoretic patterns of the cyanogen bromide digest revealed nine bands of various intensities (Figs. 4 and 5) named a-i.

The two patterns appeared to be identical except for one band (h), the colour of which was significantly more intense in the case of amylase I. Electrophoretic analysis of each peak of the chromatogram (I-VI) was also performed at acidic pH, the results being given in Fig. 4. The first two peaks each contain at least two peptides, while peak III is contaminated by peak II components; peaks IV and V appear homogeneous. Electrophoresis of peak VI did not give any band at

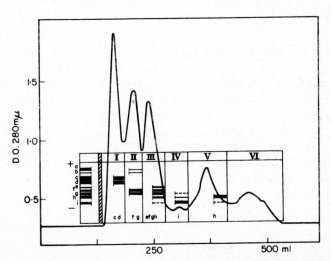

Figure 4. Fractionation of the cyanogen bromide digest peptides on a 2.5 cm x 100 cm Sephadex G-75 column. Total disc electrophoretic pattern of the peptides is given on the left. Disc electrophoresis analysis of each fraction (I-VI) is also given under each corresponding peak.

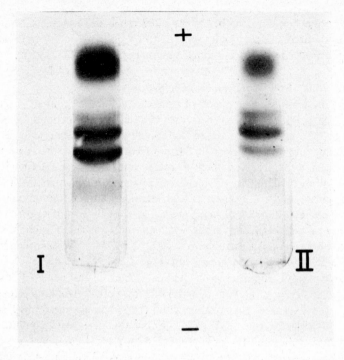

Figure 5. Results of disc electrophoresis in polyacrylamide gel. Electrophoresis was carried out in 6M urea. The electrophoresis buffer was the same as in Fig. 1. Note the variation in the staining of the h-band (second from the bottom).

any migration pH. Peak VI might be either a small peptide unstained by aniline black and blue or a non-peptidic substance absorbing at 280 mμ.

Further purification of cyanogen bromide peptides of amylases I and II is in progress. At the moment amylases I and II appear similar since they can only be distinguished by their respective electrophoretic mobility and chromatographic behaviour. Surprisingly enough the peptide mobility pattern appeared identical. The most likely difference at the molecular level is expected from the h-peptides, the variation observed in staining suggesting differences in amino acid compositions and sequences.

REFERENCES

1. Marchis-Mouren, G., Charles, M., Ben Abdeljlil, A. and Desnuelle, P., *Biochim. biophys. Acta* **50** (1961) 186.
2. Marchis-Mouren, G. and Paséro, L., *Biochim. biophys. Acta* **140** (1967) 366.
3. Loyter, A. and Schramm, M., *Biochim. biophys. Acta* **65** (1962) 200.
4. Fischer, E. H. and Bernfeld, P., *Helv. chim. Acta* **31** (1948) 1831.

5. Szabó, M. T. and Straub, F. B., *Acta Biochim. biophys. Acad. Sci. Hung.* 1 (1966) 379.
6. Narita, K., *Biochim. biophys. Acta* 28 (1958) 184.
7. Phillips, D. M. P., *Biochem. J.* 86 (1963) 397.
8. Cozzone, P. and Marchis-Mouren, G., Journées de Résonance Magnétique Nucléaire, Marseille, 1968.
9. Caldwell, M. L., Dickey, E. S., Hanrahan, V. M., Kung, H. C., Kung, J. T. and Misko, M., *J. Am. chem. Soc.* 76 (1954) 143.

FEBS Symposium, Volume 18, 1970, pp. 257-262

Two Forms and Two Chains of Pancreatic Amylase

F. B. STRAUB, M. T. SZABÓ and T. DÉVÉNYI

Institute of Medical Chemistry, University Medical School,
Budapest University, and Institute of Biochemistry, Hungarian
Academy of Sciences, Budapest, Hungary

It has been observed in our laboratory [1] and by Marchis-Mouren and Pasero [2] that pig pancreatic amylase may be resolved by column chromatography into two fractions with practically identical enzyme activities. Using a column of DEAE-Sephadex we were able to separate the two forms from 3-times crystallized pig pancreatic amylase. This is shown in Fig. 1. The protein is adsorbed at pH 7.5 at a very low ionic strength (0.005M Tris, pH 7.5 + 0.001M $CaCl_2$), and eluted with a Tris-chloride gradient. Elution of both forms is complete at a concentration of 0.03M Tris-chloride. The small difference in specific activity of the two peaks is accidental. As the two fractions, I and II, are clearly separated by this technique, we have prepared them in larger quantities. The eluted fractions of each form have been pooled, dialyzed against concentrated ammonium sulfate, and the precipitated protein dissolved in 0.01M Tris buffer and dialyzed against the same buffer. Both forms crystallized from these solutions; the crystals are consequently of two different types, as shown in Fig. 2.

We have observed that the ratio of the two forms changes during preparation, and from one preparation to the other. This is partly due to the better (apparent) solubility of amylase I. We have also studied the ratio of these two forms in extracts of individual organs. A piece of pancreas was homogenized with a solution containing 3% butanol and 0.01M $CaCl_2$ (pH 7.5). The homogenate was kept at 37°C for 30 min and cleared by centrifugation. The main protein component of such a supernatant is amylase. From the gel electrophoretic patterns of different extracts which we have studied, we find that, about four times more frequently, form II is more abundant than form I, while in the others the ratio is reversed in favour of form I. We believe it is as yet premature to conclude that we are dealing with genetically-determined isoenzymes. It is not yet finally settled whether amylase is produced as the active enzyme or whether it has a zymogen precursor. In the latter case the different

α-Amylase (pig pancreatic), EC α-1,4-glucan-4-glucanohydrolase.

ratios of the two forms might be the result of a difference in the activating mechanism.

The second part of this contribution is devoted to our studies on the structure of amylase. These were carried out by using either unfractionated amylase, or form I or II after chromatographic separation. As no differences have yet been observed, we describe the results obtained with unfractionated amylase.

We find a molecular weight of 53,000 for crystalline pig pancreatic amylase. The fingerprint of the tryptic digest of the performic acid-oxidized enzyme

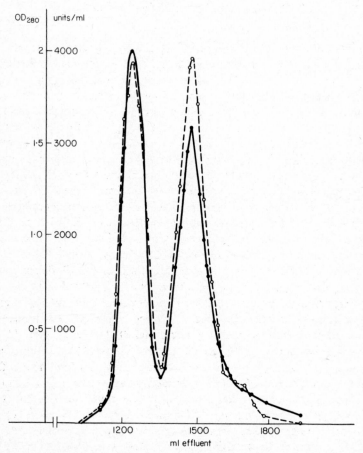

Figure 1. Chromatography of crystalline pig pancreatic amylase on DEAE-Sephadex (40 cm × 3 cm column). Gradient elution between 0.005M and 0.035M Tris buffer, pH 7.5, at constant concentration of $CaCl_2$ (10^{-3}M).
●——● OD_{280}; ○——○ enzyme activity.

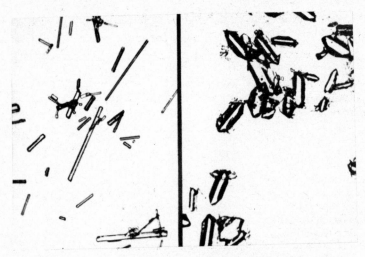

Figure 2. Amylase crystals. Left: amylase I; right: amylase II.

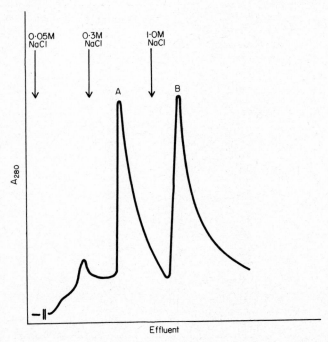

Figure 3. DEAE-cellulose column chromatography of fully reduced and carboxy-methylated amylase in 8M urea, pH 8.5.

Table 1. Distribution of cysteine residues in amylase A and B fractions before and after reduction.

	D.P.M./mg	A/B
^{14}C-Carboxymethylation *after* reduction	A 76,388 B 21,445	3.5
	A 160,777 B 67,556	2.4
	A 551,724 B 178,294	3.0
^{14}C-Carboxymethylation *before* reduction	A 39,421 B 29,379	1.3

Table 2. Amino acid composition of pig pancreatic amylase and of chains separated after oxidation.

	Amylase = 53,000 Mol. wt.	Fraction A = 31,000 Mol. wt.	Fraction B = 22,000 Mol. wt.
Lys	17	8	6
His	8	5	3
Arg	24	14	10
Cys	8-10	5-6	2-3
Asp	61	40	26
Thr	18	11	8
Ser	26	14	11
Glu	33	22	15
Pro	20	10	8
Gly	43	30	20
Ala	28	16	12
Val	33	20	15
Met	10	?	?
Ile	20	12	9
Leu	23	13	10
Tyr	16	9	6
Phe	22	13	9
Try	10	6	3
Total:	420	248	173
A + B:	421		

shows about forty peptide spots, corresponding to the number of lysine and arginine residues.

When the crystalline enzyme is fully reduced with mercaptoethanol in the presence of 8M urea and then carboxymethylated or, if the enzyme is treated with alkaline sulfite in the presence of copper ions in 8M urea, it gives two different fractions after chromatography on DEAE-cellulose (Fig. 3). Chromatography was performed in 8M urea at pH 8.5 in the presence of 0.1% sodium dodecylsulfate. The yield of both fractions A and B is 30-40% of the input material (OD_{280}).

The pooled fractions A and B were dialyzed, treated with performic acid and taken up in 0.1M Tris buffer, pH 8.5. The molecular weights were determined by sedimentation. After correcting for the presence of sodium dodecylsulfate, the molecular weights of fractions A and B were calculated to be 32,000 and 21,000, respectively. It appears that amylase is composed of two chains (A and B) linked by disulfide bonds.

We have therefore studied the distribution of cysteine residues in chains A and B. It is known that amylase contains two SH-groups which are, however, masked in the native molecule, becoming available to reagents only in the presence of concentrated urea. These SH-groups react with ^{14}C-labelled bromoacetate in the presence of 8M urea. This product was then fully reduced with mercaptoethanol, and the newly appearing SH-groups blocked with cold bromoacetate. The separated A and B chains contained about equal radioactivity (Table 1). In another type of experiment we first reduced amylase in 8M urea by mercaptoethanol, and then reacted all the SH-groups with labelled bromoacetate. In this case the A chain had about three times greater radioactivity than chain B (Table 1).

Our amino acid analysis and molecular weight figures show that amylase contains two SH-groups and four disulfide bridges. The distribution of radioactivity suggests that each chain contains one SH-group, while chain A has two, chain B only one intrachain bridge, and there is one interchain disulfide bond.

The amino acid analysis of amylase is shown in Table 2. There are some definite differences in the composition of chains A and B as far as tyrosine, tryptophan and cysteine residues are concerned. The rather similar amino acid composition of the chains is in accordance with our preliminary results from fingerprint analysis. It appears that amylase is composed of two chains of similar, but not identical, composition.

Note added in proof. Using the method of Edelhoch (*Biochemistry* **6** (1967) 1948) we have found a considerably higher amount of tryptophan than indicated in Table 2.

REFERENCES

1. Szabó, M. T. and Straub, F. B., *Acta Biochim. biophys. Acad. Sci. Hung.* **1** (1966) 379.
2. Marchis-Mouren, G. and Paséro, L., *Biochim. biophys. Acta* **140** (1967) 366.

FEBS Symposium, Volume 18, 1970, pp. 263-268

Biochemical and Immunochemical Properties of Neutral and Acid β-Galactosidases of the Small Intestine of the Rat

J. KRAML, O. KOLDOVSKÝ, A. HERINGOVÁ, V. JIRSOVÁ
and K. KÁCL

*First Department of Medical Chemistry, Charles University;
Laboratory of Developmental Nutrition, Czechoslovak Academy
of Sciences; and Institute of Mother and Child Care, Prague,
Czechoslovakia*

Chromatography and gel filtration have revealed at least two β-galactosidases in the mucosa of the small intestine of the rat [1, 2]. These differ in their pH optimum and affinity to some substrates. One of the enzymes appeared to be mainly a hetero-β-galactosidase and had a pH optimum at 3-4. We shall term this enzyme "acid" β-galactosidase. The other appeared to be chiefly a disaccharidase and had a pH optimum at 5-6. We shall call it "neutral" β-galactosidase. The neutral enzyme is bound in the brush-border [3] and can be solubilized by treatment with papain or trypsin [2, 4]. On the other hand, a part of the acid enzyme is solubilized after homogenization without papain or trypsin treatment.

In the present paper some physico-chemical and immunochemical characteristics of both galactosidases are examined. Homogenates of the mucosa of the small intestine were prepared from rats of both sexes, aged 10-13 days, of the Wistar Konárovice strain. The mucosa was scraped off and homogenized for 1-2 min in a glass homogenizer with a Teflon piston in the cold. Solubilization of mucosal homogenates with papain was performed according to Asp and Dahlqvist [2]. The subsequent procedure for isolation of enzyme (Sephadex G-200 column), and the determination of activities using o-nitrophenyl-β-D-galactoside (ONPG), were performed as described by Asp and Dahlqvist [2].

The left-hand side of Fig. 1 shows a typical picture of β-galactosidase activities on a Sephadex G-200 chromatogram. Activity was measured at pH and/or 5.5. The first major peak represents the neutral, the second the acid enzyme. The neutral enzyme has practically no activity at pH 3.5, whereas the acid enzyme at pH 5.5 is about 40% as active as at pH 3.5. The right-hand side of Fig. 1 presents the molecular weight determinations for both enzymes in a calibrated Sephadex G-200 column. Proteins used as standards were: cytochrome *c*, human hemoglobin, bovine serum albumin, human serum γ-globulin, beef liver catalase, *E.*

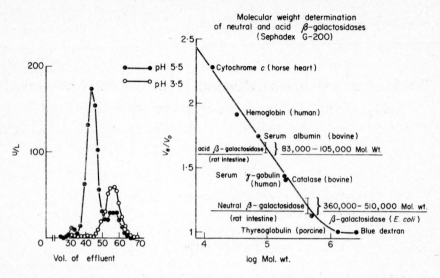

Figure 1. Sephadex G-200 chromatography of neutral and acid β-galactosidases (substrate, ONPG) of the small intestine of the rat (left) and determination of their minimal molecular weights by gel filtration (right).

coli β-galactosidase, and hog thyroglobulin. Blue dextran was used for determination of void volume. Protein content in the effluent was estimated by absorption at 280 mμ. β-Galactosidases were determined from their enzymatic activities. The ratio V_e/V_o (where V_e is the elution volume of the respective protein and V_o represents the void volume) was plotted against the log of molecular weight. The elution volumes represent the mean values of at least three estimations. Apparent molecular weight values reported by Andrews [5] were employed for the calibration. The minimal mol. wt. of the acid β-galactosidase ranged between 83,000 and 105,000, whereas for the neutral enzyme the range was between 360,000 and 510,000. The elution volume of the neutral enzyme did not differ much from that of *E. coli* β-galactosidase.

Figure 2 shows inactivation with urea and heat inactivation at 50°C for both the neutral and acid intestinal galactosidases. The neutral enzyme, with the higher molecular weight, was more readily inactivated with urea and by heating (between 40-53°C). The left-hand side of Fig. 2 shows that this enzyme was inactivated almost completely by 2-3M urea in the incubation mixture at 37°C, whereas the acid enzyme still retained the greater part of its activity. Both enzymes were relatively labile during thermal inactivation (right-hand side of Fig. 2). Whereas bovine serum albumin at a concentration of 2 mg/ml had no protective effect for the neutral enzyme, it partially protected the acid enzyme against heat inactivation. This effect may be explained by the protection of the SH-groups of the acid enzyme. Koldovský *et al.* [6] showed that the acid

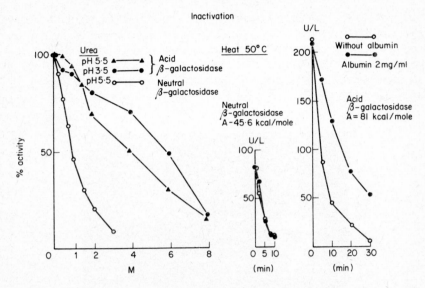

Figure 2. Inactivation of the neutral and acid β-galactosidases. Left: with urea; right: by heat (50°C).

enzyme, unlike the neutral one, may be inhibited with *p*-chloromercuribenzoate. Heat inactivation followed first-order kinetics. The first-order rate constants were higher for the neutral enzyme. The Arrhenius plot revealed an activation energy for inactivation of 45.6 kcal/mole for the neutral enzyme and 81 kcal/mole for the acid one.

	K_M (mM)		K_i (mM)
	o-Nitrophenyl-β-D-galactoside (1-16 mM)	Galactono-(1→4)-lactone (0.195-25 mM)	Sodium galactonate (0.012-3.125 mM)
Neutral β-galactosidase (pH 5.5)	14.35 11.75	1.1	No inhibition
Acid β-galactosidase (pH 3.5)	0.50 0.40	0.15	0.17

Figure 3. Competitive inhibition (K_i) of the two β-galactosidases by galactono-(1→4)-lactone and sodium galactonate. Substrate, ONPG.

Figure 3 shows the mean K_M values with 1-16 mM ONPG for both the neutral and acid enzymes. These are in accord with the values reported by Asp and Dahlqvist [2] and also with those found in whole homogenate by Koldovský *et al.* [6]. On the right-hand side of the figure, K_i values for competitive inhibition are given. The acid enzyme was inhibited by galactono-$(1 \rightarrow 4)$-lactone (0.2-25 mM) and sodium galactonate (0.012-3.125 mM), whereas the neutral enzyme

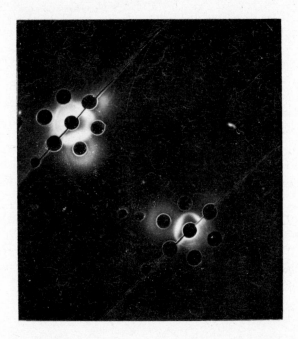

Figure 4. Ouchterlony test for acid β-galactosidase as antigen precipitated with two homologous rabbit antisera (AS 3 and AS 4 placed in the middle holes). Left: AS 3; right: AS 4. The middle section represents specific staining for enzyme activity with 6-bromo-2-naphthyl β-galactoside and fast blue BB. Outer sections (upper left and lower right): staining for protein with amido black. Two concentrations of enzyme antigen were used (about 0.1 and 0.2 mg protein/ml).

was not inhibited by sodium galactonate in the range of concentrations used. Thus the neutral enzyme seems to be more specific both for substrates and inhibitors.

Immunochemical properties of both enzymes were also compared. Rabbits were immunized with the proteins from the enzymatically-active peaks after separation on Sephadex G-200. The antigens were adsorbed on aluminium hydroxide gel. The neutral enzyme was applied to three rabbits in a total amount of 2-3 mg protein for one animal at 2-4-week intervals, but no anti-β-galactosidase activity for this enzyme was found in any of the three antisera. Two animals were given the acid enzyme in five injections, 12 mg protein each. Precipitating antisera were obtained, described here as antiserum nos. 3 and 4. The Ouchterlony plate (Fig. 4) shows double staining both for protein with amido black and, in the middle section, for enzyme activity in the precipitate after a simultaneous azo-coupling reaction, with 6-bromo-2-naphthyl β-galactoside and fast blue BB. Antisera nos. 3 and 4 were used, the former showing a stronger precipitate at two concentrations of the acid enzyme (about 0.1 and 0.2 mg protein/ml). Incubation was conducted in the refrigerator.

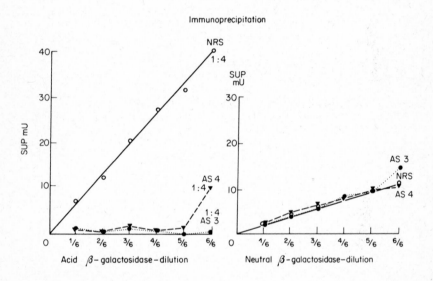

Figure 5. Quantitative immunoprecipitation of acid and neutral β-galactosidases with antisera against the acid enzyme. Enzyme activity above the immunoprecipitate was measured and compared with the same dilution of enzyme in normal rabbit serum (NRS).

Figure 5 shows, on the left-hand side, a typical immunoprecipitation experiment with 0.2 ml of the 1:4 diluted antisera no. 3 and/or 4, and 0.05 ml of varying concentrations of the acid enzyme as the antigen, and to the right an experiment with the undiluted antisera (0.2 ml), and 0.1 ml of varying concentrations of the neutral enzyme as the antigen. Incubation proceeded in the refrigerator for 24 h. The enzymatic activity of the supernatant above the immunoprecipitate was measured and compared with the corresponding dilution of the antigen in normal rabbit serum. From the right-hand side of the figure, it is apparent that the antisera against the acid enzyme do not precipitate the neutral enzyme. Thus no cross-reaction between the two enzymes was apparent.

ACKNOWLEDGEMENTS

We wish to thank Dr. Miloš Ledvina for his help in the determination of the molecular weights and to Dr. Spížek (Institute of Microbiology, Czechoslovak Academy of Sciences) for supplying us with the purified $E.$ $coli$ β-galactosidase. Parts of Figs. 1, 2 and 3 are reproduced in a modified version from the $Biochemical$ $Journal$ [7] with the kind permission of the journal.

REFERENCES

1. Furth, A. J. and Robinson, D., $Biochem.$ $J.$ **97** (1965) 59.
2. Asp, N. G. and Dahlqvist, A., $Biochem.$ $J.$ **106** (1968) 841.
3. Koldovský, O., Noack, R., Schenk, G., Jirsová, V., Heringová, A., Braná, H., Chytil, F. and Friedrich, M., $Biochem.$ $J.$ **96** (1965) 492.
4. Dahlqvist, A., Bull, B. and Thomson, D. L., $Archs$ $Biochem.$ $Biophys.$ **109** (1965) 159.
5. Andrews, P., $Biochem.$ $J.$ **96** (1965) 595.
6. Koldovský, O., Asp, N. G. and Dahlqvist, A., $Analyt.$ $Biochem.$ **27** (1968) 409.
7. Asp, N. G. and Dahlqvist, A., $Biochem.$ $J.$ **110** (1968) 143.
8. Kraml, J., Koldovský, O., Heringová, A., Jirsová, V., Kácl, K., Ledvina, M. and Pelichová, H., $Biochem.$ $J.$ **114** (1969) 621.

FEBS Symposium, Volume 18, 1970, pp. 269-279

Comparative Aspects of Creatine Kinase Isoenzymes

H. M. EPPENBERGER, M. E. EPPENBERGER and A. SCHOLL

*Institute of Biochemistry, University of Neuchâtel, Neuchâtel,
and Zoological Institute, University of Berne, Berne,
Switzerland*

Around 1963, the enzyme creatine kinase was added to the rapidly increasing group of enzymes which can appear in multiple forms. The soluble fractions obtained from different tissues generally show one to three electrophoretically-distinguishable creatine kinase isoenzymes. Depending upon conditions, one may observe also a fourth, and a fifth, enzymatically-active band. In addition to these cytoplasmatic creatine kinases, one additional form exists in mitochondria from several tissues [1, 2]. Recent data support also the existence of another particulate, but non-mitochondrial enzyme, in heart [3]. The data presented in this survey are derived from the soluble forms of creatine kinase.

Evidence has been presented that the catalytically-active enzyme consists of two subunits [4]. One dimer, present mainly in skeletal muscle, consists of two monomers of one type, subsequently called the MM-type. Another, present mainly in brain and nervous tissue, consists of monomers of a second type subsequently called the BB-type.

Yue *et al.* [5] recently presented some evidence that at least one of the enzymes, the MM-type, consists of two non-covalently bound polypeptide chains of similar molecular weight. Work under way in this laboratory should provide an answer to this question for the BB-enzyme. Figure 1 is a combination of all isoenzyme patterns from crude tissue extracts of thirteen different vertebrate species. In addition to relative differences in migration rates between species, major qualitative differences also occur in the isoenzyme distribution pattern. The brain extract always shows the fast migrating enzyme only, whereas the skeletal muscle always has the slow type. The enzyme found in heart varies from one species to another, as exemplified in this figure. From the dimer picture, one may anticipate an intermediate type as being the hybrid of the two extreme forms, and indeed, this hybrid occurs in several species. In other cases, it might readily appear as the result of artificial hybridization between the two M- and B-type monomers. However, the final explanation for the occurrence of four

Enzyme. Creatine kinase or ATP: creatine phosphotransferase (EC 2.7.3.2).

bands in the sparrow, or five bands in the duck, requires a more elaborate hypothesis. Physical and chemical studies of purified isoenzymes obtained from the chicken and rabbit provide evidence for a similar mol. wt. for all three types, namely 81,000; however, the enzyme proteins behave quite differently in their kinetic activity and they have markedly different amino acid compositions [4, 6]. One can therefore draw the conclusion that two genes are responsible for the synthesis of the two subunits.

With the appearance of more isoenzymes, as in the case of several avian species [7], some additional factor must be considered. Can we anticipate the

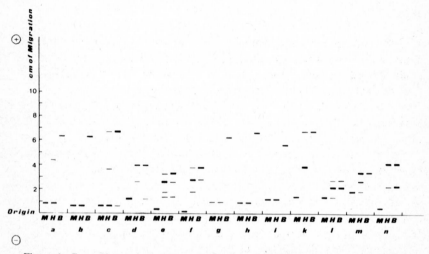

Figure 1. Comparison of the mobilities of the creatine kinase isoenzymes of different vertebrate species. Species compared are: a, human; b, rabbit; c, rat; d, chick; e, duck; f, sparrow; g, snake; h, lizard; i, waran; k, alligator; 1, turtle; m, frog; n, mackerel. The ordinate gives the electrophoretic migration in cm on an agarose-gel, whereas the abscissa represents the starting points of the samples. Electrophoresis conditions: veronal buffer, pH 8.6, 0.06M in the trays, 0.03M in the gel; 4°C; 75 min. For activity staining procedure see ref. 6. M, skeletal muscle; H, heart muscle; B, brain.

possible effects of a mutation in the structural genes responsible for the synthesis of creatine kinase subunits, or could other mechanisms be involved?

The characterization of the states of an enzyme during early ontogeny could possibly provide some indications. The enzyme activity can indeed change in many ways, both in activity as well as in isoenzyme patterns, as shown for creatine kinase [8]. Figure 2 indicates how the isoenzyme pattern in rat skeletal muscle changes with age, and how it can be correlated with the general development of muscle tissue. The BB-type enzyme is the first catalytically-active form. Whether the MM-type is also present in some inactive form, for instance as M-type monomer, is an important question, as yet unsolved.

The subsequent appearance of the active hybrid, which consists of one M- and one B-subunit, may be helpful in trying to answer the question and studies in this direction are currently under way. An interesting point is the timing of this transition from BB- to MM-type enzyme in relationship to the onset of muscular activity. Myofibril formation, and the appearance of phosphocreatine in large amounts, coincide with the appearance of the MM-creatine kinase (Fig. 2). A quite similar behaviour has been described for other isoenzymes, e.g. lactate dehydrogenase.

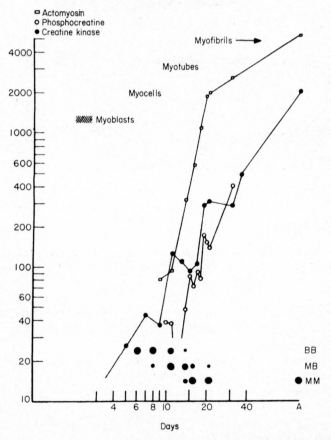

Figure 2. Comparison of developmental changes of actomyosin [9], phosphocreatine [10] and creatine kinase [11] in chicken muscle. Actomyosin in mg protein/g muscle, phosphocreatine in µg phosphate/g wet weight, creatine kinase in µmoles/min/g wet weight. At the top of the figure the morphological stages, and at the bottom the isoenzyme patterns, are shown. (Reproduced from Eppenberger *et al.* [8], with permission.)

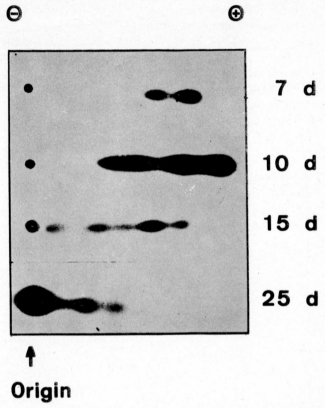

Figure 3. Isoenzyme pattern of creatine kinase during embryonic development of skeletal muscle from duck; d, days after incubation. Agarose gel electrophoresis as described in Fig. 1.

During the investigation of a large number of avian species, we frequently observed the additional bands referred to. They are never absent in the order Passeriformes [7].

In the light of earlier studies on the developmental features of the creatine kinase isoenzymes, it becomes very interesting to know if the two brain bands would also be present at early embryonic stages, or if the usual pattern known for mammalian species or chicken prevails.

The patterns in Figs. 3-5 show, indeed, a shift from one to another form very similar to that observed for rat or chicken tissues, but one also notes the emergence of an additional brain band, even in very early stages. This suggests a genetic determinant and raises the question as to whether these new bands are actual isoenzymes, or whether they could be different from the three previously described forms of creatine kinase.

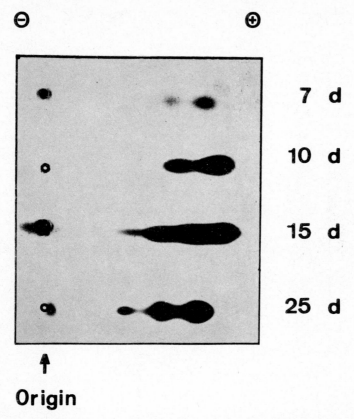

$\ominus$ $\oplus$

7 d

10 d

15 d

25 d

↑
Origin

Figure 4. Isoenzyme pattern of creatine kinase during the embryonic development of heart muscle from duck.

The two brain forms subsequently called BB and (BB)$_a$, have been purified and compared [12]. The specific activities and the catalytic behaviour of both fractions were identical. Dissociation experiments in 6M guanidine-HCl, followed by reassociation to the active dimer forms, have been made. The results are given in Fig. 6. When the pure (BB)$_a$-enzyme was dissociated in 6M guanidine-HCl and reconstituted by dialysis, a mixture of BB and (BB)$_a$ was obtained, showing that (BB)$_a$ was indeed convertible into the normal brain-type enzyme. It was also possible to initiate formation of a hybrid MB-enzyme by dissociating in the same solution pure rabbit MM-enzyme and pure sparrow (BB)$_a$-enzyme. Besides MB, under these conditions, BB also appeared in considerable amounts. The hybrid enzyme MB obtained in this case always migrated to an intermediate position with respect to BB and MM. However, when the hybridization was performed with crude extracts, one could also identify a hybrid of the MB$_a$-form. This leads

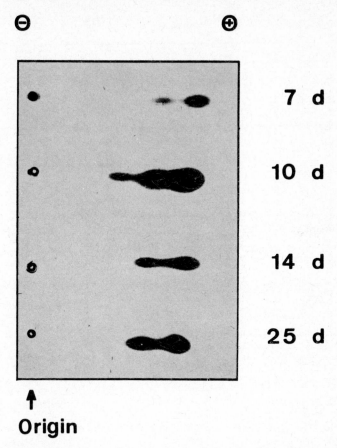

Figure 5. Isoenzyme pattern of creatine kinase during the embryonic development of brain tissue from duck.

to the type of distribution found in the duck, namely with five bands. However, a conversion from BB_a to $(BB)_a$, or a formation of a hybrid between BB and $(BB)_a$, could never be observed.

Summarizing these observations, one might assume that the brain-type enzyme is structurally less organized, and can therefore give rise to slightly different molecular forms. It is quite possible that this represents a state of structural conformers, as described in the literature [13]. In this case, the two subunits of the B-forms would not differ in their primary structure, but only in their sterical conformation.

This means that in the BB-form of creatine kinase, a precise conformation might not be as critical in maintaining the catalytic activity as it would be in the other isoenzymes of the MM-type.

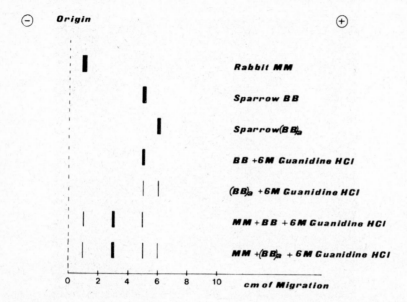

Figure 6. Reversible dissociation of several pure creatine kinases in 6M guanidine-HCl, 0.1M tris-buffer and 0.1M 2-mercaptoethanol. Tracings of starch gel electrophoresis at pH 8.6, tris-borate buffer, during 16 h, 4°C. Reassociation and hybridization of subunits by overnight dialysis against 0.1M tris and 0.1M 2-mercaptoethanol. Initial concentrations of enzyme proteins are identical.

Differences in the stability between MM- and BB-types of creatine kinase may also be observed with the chicken enzyme. Whereas the MM-enzyme did not give rise to additional bands after several kinds of treatment, one could obtain such bands after similar treatment of the BB-type enzyme. One example is provided by Fig. 7. Occasionally, one could even observe the appearance of an additional band during the purification of BB-type enzyme from chicken hearts.

Inhibiting the two SH-groups which directly or indirectly form part of the active site of the enzyme, we made another peculiar observation: a total inhibition of the chicken BB-enzyme could never be obtained, using either iodoacetate or iodoacetamide; only a maximum of about 70% inhibition was achieved, whereas the MM-enzyme could be completely inhibited under similar conditions [6]. This observation has been recently confirmed [15]. We have gained more evidence that conformational changes of the protein may be involved in the appearance of these additional enzymatically-active bands. Antibodies prepared against the sparrow BB-form show a single line of precipitation on immunoelectrophoresis, when opposed to homologous antigen as well as with the (BB)a-protein. No coalescence of the precipitation lines occurs. Figure 8 shows the result of a microcomplement fixation experiment following the method described by Levine and Wasserman [16].

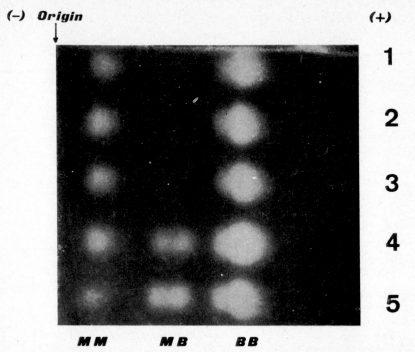

Figure 7. Photograph of starch gel electrophoresis (pH 8.6) of a mixture of pure MM- and BB-enzyme from chicken. Effect of multiple fast freezes and slow thaws on hybridization and appearance of additional bands. The numbers indicate how many freezing-thawing operations took place. All protein samples were in 1M NaCl + 0.1M Na₃PO₄. For detailed method see ref. 14.

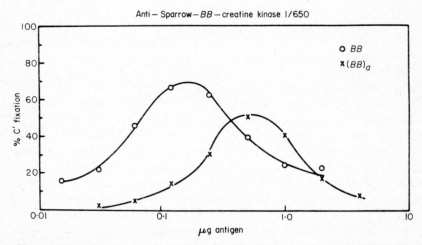

Figure 8. Quantitative microcomplement fixation. Reactions of an antibody, prepared against sparrow BB-enzyme, with BB- and (BB)ₐ-antigen. Method according to ref. 16.

The slight lateral shift, and the decrease in maximum fixation of the (BB)$_a$-form, suggest a reduced availability of the majority of immunologically-active sites on the protein. This may be caused either by an increased distance between complementary antigenic sites, or by certain sites becoming unavailable as a result of a modification in the structure of the protein. The same kind of

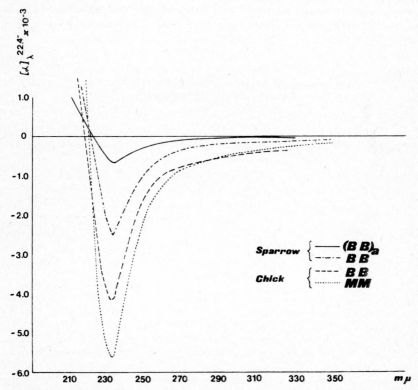

Figure 9. ORD spectrum of sparrow BB-type creatine kinases and chicken BB- and MM-type creatine kinases. Drawings corrected for protein concentrations and refraction. The measurements were performed with a Cary 60 recording spectropolarimeter with a sensitivity of ± 0.0005° in the wavelength range 600-200 mμ. Protein concentrations were of the order of 0.25 mg/ml.

lateral shift, and of decrease in maximum fixation, can be obtained by treating the BB-enzyme with 2-4M urea. This experiment suggests that (BB)$_a$ could be an "opened up" form of the molecule. The change in electrophoretic mobility can be caused by the exposure of some previously buried charged groups.

Again, this seems to confirm the assumption that a precise conformation is not too critical for the occurrence of full catalytic activity. Another considerable difference between BB and (BB)$_a$ is provided by optical rotatory dispersion studies. Figure 9 gives a survey of such measurements. A comparison of the Cotton effects of the two sparrow enzymes, and of the MM- and BB-enzymes for chicken, provide further evidence in favour of a decrease in the level of structural organization from chick-MM down to sparrow-(BB)$_a$. The reduced mean rotations for these four proteins have been computed in Table 1. Detailed results will be published in a forthcoming paper.

In conclusion one can say that the observed multiple electrophoretic forms of creatine kinase are very probably not the result of only one mechanism but several.

Table 1. Reduced mean residue rotation [m′] at 233 mμ of creatine kinases from different sources.

Species	Enzyme type	[m′] 233
Chick	M M	− 4900
	B B	− 3640
Sparrow	B B	− 2200
	(B B)$_a$	− 644

REFERENCES

1. Jacobs, H., Heldt, H. W. and Klingenberg, M., *Biochem. biophys. Res. Commun.* **16** (1964) 516.
2. Kleine, T., *Nature, Lond.* **207** (1965) 1393.
3. Keto, A. I. and Doherty, M. D., *Biochim. biophys. Acta* **151** (1968) 721.
4. Eppenberger, H. M., Dawson, D. M. and Kaplan, N. O., *J. biol. Chem.* **242** (1967) 204.
5. Yue, R. H., Palmieri, R. H., Olson, O. E. and Kuby, S. A., *Biochemistry* **6** (1967) 3204.
6. Dawson, D. M., Eppenberger, H. M. and Kaplan, N. O., *J. biol. Chem.* **242** (1967) 210.
7. Eppenberger, M. E., Eppenberger, H. M. and Kaplan, N. O., *Nature, Lond.* **214** (1967) 239.
8. Eppenberger, H. M., Eppenberger, M. E., Richterich, R. and Aebi, H., *Devl Biol.* **10** (1964) 1.
9. Csapo, A. and Herrmann, H., *Am. J. Physiol.* **165** (1951) 701.
10. Herrmann, H. and Cox, W. M., *Am. J. Physiol.* **165** (1951) 711.
11. Eppenberger, H. M., Fellenberg, R., Richterich, R. and Aebi, H., *Enzymol. Biol. Clin.* **2** (1962/63) 139.
12. Eppenberger, H. M., in "Homologous Enzymes and Biochemical Evolution" (edited by N. Thoai and J. Roche), Gordon and Breach, New York, 1968, pp. 231-242.

13. Kitto, G. B., Wassarman, P. M. and Kaplan, N. O., *Proc. natn. Acad. Sci. U.S.A.* **56** (1966) 578.
14. Chilson, O. P., Costello, L. A. and Kaplan, N. O., *Biochemistry* **4** (1965) 271.
15. Hooton, B. T., Proc. British Biophysical Society, Kings College, London, May 9, 1968.
16. Wasserman, E. and Levine, L., *J. Immun.* **87** (1961) 290.

FEBS Symposium, Volume 18, 1970, pp. 281-289

Isoenzymes of Soluble Mitochondrial Monoamine Oxidase

M. B. H. YOUDIM, G. G. S. COLLINS and M. SANDLER

*Bernhard Baron Memorial Research Laboratories and
Institute of Obstetrics and Gynaecology, Queen
Charlotte's Maternity Hospital, London,
England*

A strong body of evidence has now accumulated to suggest that mitochondrial monoamine oxidase (monoamine: O_2 oxidoreductase (deaminating) EC 1.4.3.4) (MAO) exists in multiple forms [1], although much of it is based on data obtained using impure particle-bound preparations of the enzyme. In recent years, mitochondrial MAO has been solubilized and purified and some of its properties have been described [2-7]. Youdim and Sandler [8] used poly-acrylamide gel electrophoresis to separate at least three bands of activity from solubilized rat liver and two from human placental MAO; these results have been partially confirmed by Kim and D'Iorio [9] and Ragland [10]. Very recently, Collins *et al.* [11] have separated solubilized human and rat liver mitochondrial MAO into four isoenzymes, each possessing a differing substrate specificity and heat inactivation pattern. We should like to report some further properties of these isoenzymes with regard to their pH optima, substrate specificities and the effects of heat and various inhibitors on their respective activities.

METHODS

Preparation of isoenzymes of mitochondrial MAO

Soluble mitochondrial MAO from human placenta, and rat and human liver, was prepared using the method described by Youdim and Sourkes [3] as adapted by Youdim and Sandler [12]. Soluble enzyme from rat and human brain was prepared using the method described by Youdim and Sourkes [3], as adapted by mitochondrial extract in the presence of tris buffer (0.05M, pH 8.6) and a

Abbreviations. MAO: monoamine oxidase; RMAO: rat monoamine oxidase; HMAO: human monoamine oxidase.

Enzymes. Monoamine oxidase (monoamine: O_2 oxidoreductase (deaminating) EC 1.4.3.4).

non-ionic detergent (unpublished results). A 5% polyacrylamide gel was made by dissolving 5 g of 'Cyanogum 41' (British Drug Houses Ltd.) in 98 ml tris/ hydrochloric acid buffer (0.05M, pH 8.6) and filtering to remove insoluble matter; 1 ml each of freshly prepared solutions of 2-dimethyl-amino-ethyl cyanide (10% w/v) and ammonium persulphate (10% w/v) were added to aid gelling, and the mixture was poured into glass tubes (5 x 75 mm), care being taken to exclude air bubbles. Samples of the enzyme under study were mixed with Sephadex G-200 (approx. 40 ml/g) and 0.1 ml was introduced at the cathodal or the anodal end of a polyacrylamide column. Disc electrophoresis was performed using a continuous tris/hydrochloric acid buffer system (0.05M, pH 8.6) for 3-5 h at room temperature with a constant current of 6 mA per tube. Eight columns were run simultaneously. The gels were then removed from the columns and tested for enzyme activity and the presence of protein.

ESTIMATION OF MAO ACTIVITY

Two methods were used to test for MAO activity:

1. Gels were incubated with a freshly prepared solution containing tryptamine hydrochloride (15 mg), sodium sulphate (9 mg) and tetrazolium nitro BT (5 mg) dissolved in 10 ml phosphate buffer (0.05M, pH 7.4) [13]. Incubation was carried out at 37°C for 30-60 min; MAO activity appeared as blue-mauve bands bearing a fixed relationship to, or coinciding with, a number of yellowish bands of protein (Naphthalene Black 12B-staining) which also migrated towards the anode.

2. After electrophoresis, the bands of enzyme activity whose location had already been established (see above), were separated from the gels using a sharp razor, homogenized in phosphate buffer (0.05M, pH 7.4) and centrifuged at 500 x g for 10 min. The supernatant was retained for protein estimation [14] and for enzyme assay with [14C] tyramine and [14C] tryptamine [15], kynuramine [16] and benzylamine [17].

Phosphate buffer (0.05M, in a pH range 6.0-7.7) and boric acid/borate buffer (0.05M, in a pH range 8.1-9.1) were used for the pH-activity studies of the isoenzymes with kynuramine as substrate [16]; in these experiments, the gel segments were homogenized in water. Heat inactivation studies were carried out at 45°C on the buffered enzyme (0.05M, pH 7.4) and activity was measured after 10, 20 and 40 min heating, using kynuramine as substrate [16].

RESULTS

Soluble mitochondrial MAO preparations submitted to polyacrylamide gel electrophoresis developed varying numbers of bands of enzyme activity, depending on the tissue of origin (Fig. 1). Thus, three bands from human and rat liver preparations and two from rat brain, two from human placenta and one from human cerebellum migrated towards the anode.

Activity was also observed at the origin with the cerebellar and hepatic enzymes; when the latter were subjected to electrophoresis with reversed polarity, one band migrated towards the cathode. The rat and human liver mitochondrial isoenzymes migrating to the anode have been designated $RMAO_{1-4}$ and $HMAO_{1-4}$, respectively, as described by Collins *et al.* [11], whilst the bands migrating towards the cathode have been allocated the symbols $HMAO_5$ and $RMAO_5$.

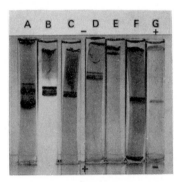

Figure 1. Polyacrylamide gel disc electrophoretic patterns of isoenzymes of mitochondrial monoamine oxidase prepared from various tissues. (A) Rat liver; (B) human liver; (C) human placenta; (D) rat brain; (E) human cerebellum (all cathode to anode); (F) rat liver (anode to cathode); (G) human liver (anode to cathode).

Thermal inactivation of hepatic isoenzymes

The effect of exposing rat and human liver isoenzymes to heat, using kynuramine as substrate, is shown in Table 1. $RMAO_1$, $RMAO_2$ and $RMAO_3$ are heat-labile whereas $RMAO_4$ and $RMAO_5$ are relatively heat-stable. In contrast $HMAO_{1-4}$ are heat-stable whereas $HMAO_5$ is heat-labile.

Effect of pH on hepatic and cerebellar isoenzyme activity

Figs. 2A and B show pH-activity curves of rat and human liver isoenzymes using kynuramine as substrate; pH values for optimum activity of $RMAO_{1-5}$ are 8.1-8.7, 7.7, 8.1, 8.7 and 6.7, respectively. The corresponding values for $HMAO_{1-5}$ are 8.3, 8.3, 8.7, 7.5 and 7.1, respectively. The pH optimum for the two human cerebellar enzymes is 8.3.

Effect of MAO inhibitors on hepatic isoenzyme activity

The effect of various drugs, known to be MAO inhibitors, on the activity of $HMAO_{1-5}$, using kynuramine as substrate, is shown in Table 2. There are marked

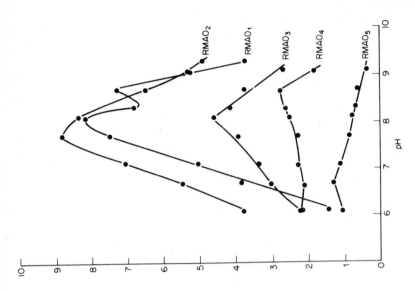

Figure 2B. pH-Activity curves of isoenzymes of rat (RMAO) liver mitochondrial monoamine oxidase, using kynuramine as substrate.

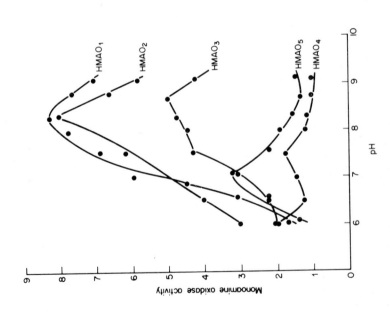

Figure 2A. pH-Activity curves of isoenzymes of human (HMAO) liver mitochondrial monoamine oxidase, using kynuramine as substrate.

Table 1. Thermal inactivation of hepatic monoamine oxidase isoenzymes.

Isoenzymes separated electrophoretically were heated at 45°C in 0.05M phosphate buffer, pH 7.4, for various times as shown. Figures represent percentage activity compared with control. Some of these data have been published by Collins *et al.* [11].

Enzyme	Rat Time (min)			Human Time (min)		
	10	20	40	10	20	40
MAO_1	48.5	35.3	18.6	87.2	82.7	84.2
MAO_2	32.6	18.2	10.0	98.4	86.9	85.3
MAO_3	55.8	34.8	26.5	91.7	86.5	80.8
MAO_4	76.8	71.4	72.4	92.1	92.1	88.4
MAO_5	91.3	83.0	52.0	40.6	29.5	20.5

differences in sensitivity. In contrast to the other isoenzymes, $HMAO_3$ and $HMAO_5$ are resistant to all inhibitors tested apart from M & B 9302 and harmaline, at least at the concentrations employed. $HMAO_1$ and $HMAO_2$ are more sensitive to pargyline than $HMAO_4$, whereas the latter is more sensitive to iproniazid, pheniprazine, M & B 9302 and harmaline. Preliminary results, using [^{14}C] tyramine and [^{14}C] tryptamine as substrates, show that $HMAO_4$ is relatively resistant to iproniazid, even at a concentration of 10^{-4}M. The rat liver isoenzymes exhibit similar sensitivities towards iproniazid and pheniprazine, as do the human isoenzymes (in preparation).

Table 2. The effect of inhibitors on isoenzymes of human liver mitochondrial monoamine oxidase.

Each enzyme preparation was preincubated for 15 min with inhibitor before the addition of substrate (kynuramine, final concentration 2×10^{-5}M). Each figure represents percentage inhibition.

Inhibitor	Concentration (M)	$HMAO_1$	$HMAO_2$	$HMAO_3$	$HMAO_4$	$HMAO_5$
Pargyline	2×10^{-8}	26	17	0	10	0
	4×10^{-8}	42	49	0	15	0
Tranylcypromine	2×10^{-7}	53	31	10	36	17
Iproniazid	2×10^{-6}	28	20	6	45	0
Pheniprazine	2×10^{-6}	36	36	20	72	21
M & B 9302	5×10^{-6}	69	75	77	90	52
Harmaline	1.5×10^{-4}	20	4	58	50	61
	3.3×10^{-4}	65	60	100	100	94

Table 3. The oxidative deamination of various substrates by the five electrophoretically-separated bands of activity of solubilized human (H) and rat (R) liver mitochondrial monoamine oxidase (MAO_{1-5}).

Each figure represents mμM deaminated product formed per 10 min per mg protein and is the mean ± S.E.M. of 3-7 experiments. Some of these data have been published by Collins et al. [11].

		$HMAO_1$	$HMAO_2$	$HMAO_3$	$HMAO_4$	$HMAO_5$
Human	Tryptamine	8.03 ± 1.20	14.40 ± 1.60	2.44 ± 0.71	0.51 ± 0.08	—
	Tyramine	26.50 ± 4.40	38.70 ± 4.50	3.75 ± 0.85	0.55 ± 0.04	—
	Kynuramine	54.60 ± 4.78	114.20 ± 10.44	13.10 ± 2.55	4.51 ± 0.88	2.22 ± 0.67
	Benzylamine	1.55 ± 0.47	2.07 ± 0.61	0.027 ± 0.01	0.37 ± 0.09	—
		$RMAO_1$	$RMAO_2$	$RMAO_3$	$RMAO_4$	$RMAO_5$
Rat	Tryptamine	5.40 ± 0.65	5.18 ± 0.40	5.48 ± 0.43	0.73 ± 0.27	—
	Tyramine	10.20 ± 0.41	6.37 ± 0.24	1.10 ± 0.11	0.56 ± 0.07	—
	Kynuramine	21.20 ± 3.2	11.60 ± 2.4	2.41 ± 1.0	1.33 ± 0.3	3.22 ± 0.79

Substrate specificity of hepatic isoenzymes

Table 3 shows the activity of rat and human hepatic isoenzymes per mg protein with various substrates. Of the human liver isoenzymes, $HMAO_2$ possessed greatest activity towards all substrates investigated; kynuramine was more, and benzylamine less actively deaminated than other substrates. In addition, tyramine was a better substrate than tryptamine for $HMAO_1$ and $HMAO_2$, whereas both amines were deaminated to a similar extent by $HMAO_3$ and $HMAO_4$. The rat enzyme, however, manifested a different pattern. $RMAO_1$ was almost twice as active as $RMAO_2$ against kynuramine and tyramine, although both possessed similar activity against tryptamine. $RMAO_3$ showed greatest activity towards tryptamine, whereas $RMAO_1$ and $RMAO_2$ deaminated kynuramine at a faster rate than the other substrates tested.

DISCUSSION

The data presented in this paper provide direct support for the presence of more than one form of MAO in the mitochondria. The bands of MAO activity separated by disc electrophoresis exhibit a number of physiochemical properties which differ with respect to substrate preference, sensitivity to inhibitors, pH-optimum and thermal stability. These data correlate well with indirect evidence for multiple forms of the enzyme, recently reviewed by Squires [1].

The possibility that the various bands are due to incomplete solubilization of enzyme, and that the observed phenomena stem from the action of one enzyme bound to parts of cell membranes of various sizes, has already been discussed elsewhere [9, 11]. Suffice it to say that the bands of activity obtained are always consistent in the same tissue; their number differs from one tissue to another. If we are observing an artefact, it is a highly reproducible one and the individual bands, when eluted and re-run, migrate to similar positions. Indications are that the hepatic isoenzymes, as we feel it justifiable to call them, are represented in many different tissues but that their proportions differ quantitatively from one tissue to another. The nomenclature we have adopted for them is very much a *pro tempore* measure. Any definitive classification must await more detailed information on their properties and distribution.

From the preliminary data reported here, human liver mitochondrial MAO isoenzymes appear to be not too dissimilar from those of the rat, although there are distinct differences in thermal inactivation, substrate specificity and pH-activity curve. All the human bands are relatively stable at 45°C except $HMAO_5$, but only $RMAO_4$ and $RMAO_5$ have a similar stability. It may be relevant that Youdim and Sourkes [18] and Nagatsu and Yagi [19] found a heat-stable MAO in rat liver mitochondria resistant to MAO inhibitors. Squires [1] has recently confirmed these results in mouse liver and intestinal MAO.

Although the enzymes were capable of oxidizing all substrates tested, their preferences for some were greater than for others. The data obtained in this area of investigation are still preliminary; work is in progress to study a number of other important MAO substrates, particularly the catecholamines and 5-hydroxytryptamine.

The picture is even more difficult to interpret when one considers the effect of MAO inhibitors. With the limited number of compounds and concentrations tested, it is not yet possible to draw any conclusions although the present data would seem to indicate that hydrazine and non-hydrazine inhibitors act differently. It would be of interest if $HMAO_3$ and $HMAO_5$, which were particularly sensitive to harmaline and, therefore, presumably to harmine and other β-carboline derivatives [20], are eventually found to be relatively specific for 5-hydroxytryptamine as substrate. Gorkin et al. [20] have shown that this group of inhibitors can prevent the deamination of 5-hydroxytryptamine in mitochondria from certain species at a lower concentration than when tyramine or tryptamine are employed as substrates.

ACKNOWLEDGEMENTS

We are grateful to the Medical Research Council and Hoechst Pharmaceuticals Ltd. who respectively defrayed the salaries of G.G.S.C. and M.B.H.Y. M & B 9302 was kindly made available to us by Dr. D. R. Maxwell, May & Baker Ltd.

REFERENCES

1. Squires, R. F., *Biochem. Pharmac.* **17** (1968) 1401.
2. Nara, S., Gomes, B. and Yasunobu, K. T., *J. biol. Chem.* **241** (1966) 2774.
3. Youdim, M. B. H. and Sourkes, T. L., *Can. J. Biochem.* **44** (1966) 1397.
4. Erwin, V. G. and Hellerman, L., *J. biol. Chem.* **242** (1967) 4230.
5. Tipton, K. F., *Eur. J. Biochem.* **4** (1968) 103.
6. Barbato, L. M. and Abood, L. G., *Biochim. biophys. Acta* **67** (1963) 531.
7. Gabay, S. and Valcourt, A. J., *Biochim. biophys. Acta* **159** (1968) 440.
8. Youdim, M. B. H. and Sandler, M., *Biochem. J.* **105** (1967) 43P.
9. Kim, H. C. and D'Iorio, A., *Can. J. Biochem.* **46** (1968) 295.
10. Ragland, J. B., *Biochim. biophys. Res. Commun.* **31** (1968) 203.
11. Collins, G. G. S., Youdim, M. B. H. and Sandler, M., *FEBS Letters* **1** (1968) 215.
12. Youdim, M. B. H. and Sandler, M., *Biochim. appl.* **14** (1968) Suppl. 1, p. 175.
13. Glenner, G. C., Burtner, H. J. and Brown, G. W., *J. Histochem. Cytochem.* **5** (1957) 591.
14. Lowry, O. H., Roseborough, N. J., Farr, A. L. and Randall, R. J., *J. biol. Chem.* **193** (1951) 265.
15. Robinson, D. S., Lovenberg, W., Keiser, H. and Sjoerdsma, A., *Biochem. Pharmac.* **17** (1968) 109.
16. Kraml, M., *Biochem. Pharmac.* **14** (1965) 1684.

17. Tabor, C. W., Tabor, M. and Rosenthal, S. M., *J. biol. Chem.* **208** (1954) 645.
18. Youdim, M. B. H. and Sourkes, T. L., *Can. J. Biochem.* **43** (1965) 1305.
19. Nagatsu, T., and Yagi, K., *J. Biochem., Tokyo* **58** (1965) 302.
20. Gorkin, V. Z., Tat'yanenko, L. V., Krasnokutskaya, D. M., Pronina, E. V. and Yakhontov, L. N., *Biokhimiya* **32** (1967) 510.

FEBS Symposium, Volume 18, 1970, pp. 291-296

Isoenzymes in Mammalian Liver Catalases:
A Controversial Subject

H.-G. HEIDRICH

*Max-Planck-Institut fur Eiweiss- und Lederforschung,
München, West Germany*

One of the most persistent difficulties in protein structure analysis is heterogeneity. In the case of enzymes the heterogeneity, apart from impurities, can be caused by isoenzymes. That is, the different isoenzymes vary from each other in their primary structure. This variation is produced by different genetical determination of subunits and by combination of these subunits to the fully active enzyme hybrids. As a consequence, it is impossible to convert isoenzymes into one another by chemical reactions.

Beef liver catalase is known to be built from four subunits. According to Schroeder *et al.* [1], these subunits are identical, as shown by a comparison of the tryptic peptides of this enzyme and by advanced primary structure analysis. For this reason it should be impossible to find isoenzymes in beef liver catalase. However, contradictory results have been published [2] showing that rat liver catalase is heterogeneous and that the heterogeneity in mammalian liver catalases is caused by catalase isoenzymes. Only one of these two conflicting postulates can be correct. The experiments described below were performed to resolve these discrepancies.

Experimental procedures and results

Different commmercial beef liver catalases were examined by acrylamide gel electrophoresis and were shown to be heterogeneous and to vary in their heterogeneity [3]. The most heterogeneous catalase preparation showed five bands. Elution of the electrophoresis gel, followed by activity tests according to Bergmeyer [4], demonstrated that all of the enzyme fractions were active catalases with almost identical specific activity (cat f-values around 30,000). As

Enzyme. Beef liver catalase-hydrogen peroxide:hydrogen peroxide oxidoreductase (EC 1.11.1.6).

shown in Fig. 1, separation of these catalase fractions by column chromatography on DEAE-Sephadex resulted in two peaks (I and III) with a shoulder (II) in between.

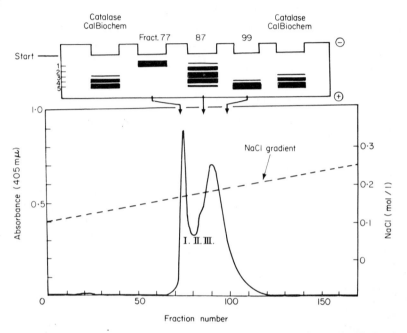

Figure 1. Column chromatography of 40 mg of beef liver catalase (CalBiochem) on DEAE-Sephadex A-50 (Column: 200 cm x 1 cm, 20°C). Gradient: first vessel 275 ml Tris 0.05M, NaCl 0.1M, pH 7.8 (HCl); second vessel 275 ml Tris 0.05M, NaCl 0.3M, pH 7.8 (HCl). Elution: 2 ml/h/fraction and acrylamide gel electrophoresis patterns of some column fractions (electrophoresis in 7.5% gel and Tris/boric acid buffer, pH 8.4, according to Heidrich [3]).

Electrophoretic examination of the fractions indicated that peak I contained a homogeneous material which must have been formed from the original enzyme during the chromatographic procedure. The shoulder II exhibited a fraction with five bands. The protein of peak III was found to be almost homogeneous. When the peak I fraction was re-chromatographed in the same system, it emerged from the column at the expected position. During re-chromatography of the peak III fraction, conversion of the chromatographed material to that of peak I could be observed with regularity.

In order to determine whether this alteration during the chromatographic process was due to an oxidation-reduction process, the different commercial beef liver catalase preparations were treated with Cleland's reagent or oxidized with oxygen. Figure 2 shows the results, in which the reduction product is band 1 while the oxidation product is mainly band 5. Apparently bands 1 and 5 are

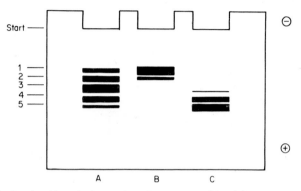

Figure 2. Acrylamide gel electrophoresis patterns of beef liver catalase. (A) Catalase Worthington; (B) same, reduced with Cleland's reagent 10 mg/100 ml Tris/0.05M, pH 7.8 (HCl), at 1°C for 12 h; (C) same as (A), oxidized with oxygen in the same buffer as (B) at 20°C for 48 h. Electrophoresis according to Heidrich [3].

definite forms of the enzyme and represent reduced and oxidized states respectively, whereas bands 2, 3 and 4 are hybrids of 1 and 5.

The oxidized and reduced forms of the catalase were compared as to their chemical and physical properties. No differences were found in their amino acid composition. Their molecular weights, as well as their absorption and ORD spectra, were identical. The only difference found was in their specific activities. The reduced form was a little more active than the oxidized form (cat f-value 33,000 red., 29,000 ox.), although this difference was not very striking.

It was necessary to determine whether the five fractions of beef liver catalase are also present *in vivo* or if they are artifacts formed during isolation of the enzyme. Liver catalase is known to be stored in the living cell in a particle fraction called peroxisomes [5]. Therefore a light mitochondrial fraction containing the peroxisomes was isolated from beef and rat liver homogenates by normal centrifugation techniques [6]. This fraction was then purified by using the Free Flow Electrophoresis III according to Hannig (90 mA, 100 kV, 3-4°C, triethanolamine-acetate-EDTA-sucrose buffer, pH 7.4). The peroxisome-enriched fractions obtained showed very little catalase activity, which indicated that the peroxisome membranes were intact and that the enzyme was still in its "containers". Part of the peroxisome fraction was then mixed with Triton X-100 and immediately put on disc electrophoresis. Simultaneously, another part of the peroxisome fraction was sonicated, mixed with Triton and then run on disc electrophoresis. After the run the gels were stained specifically for catalase [7]. Figure 3 indicates that the catalase carefully released from the peroxisomes during electrophoresis produces only one homogeneous band. It corresponds to the catalase fraction which was obtained by reducing the commercial prepara- tions with Cleland's reagent. On the other hand, the catalase released by

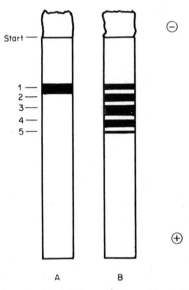

Figure 3. Disc electrophoresis of beef liver catalase released from peroxisomes: (A) by Triton X-100; (B) by sonification ($\lambda = 7\mu$ at $0°C$ for 3×10 s). Electrophoresis using an upper gel according to Heidrich [3] and activity test of Thorup *et al.* [7].

sonicating the particles showed five bands. The same five bands could be obtained when a beef or rat liver homogenate was centrifuged and the clear supernatant fluid run on disc electrophoresis. Results from similar homogenate supernatant electrophoresis runs have led investigators to attribute heterogeneity of catalase to "isoenzymes" of this protein [2].

DISCUSSION

Two facts are established by the experiments described above:

1. The heterogeneity of beef liver catalase is not due to impurities or to isoenzymes.

2. It is proved by the peroxisome experiment that liver catalase exists *in vivo* in only one form. This native form is altered during the process of isolation— perhaps by oxygen. If this is true, one may assume two forms of liver catalase. One is the reduced state, which is the native form; and the other is the oxidized form, which is an artifact. Since catalase has four subunits, it is possible to construct three additional hybrids from these two forms. Consequently, five

active catalase fractions can be postulated. This is a possible explanation for the five electrophoretically-separated bands.

There is presently no definitive evidence as to the nature of the alteration in the oxidized catalase. One speculation is that S-S bridges are formed. Figure 4 gives a schematic explanation for this reaction and the possible structures of the five "oximers". (A) depicts a mechanism in which equilibrium with oxygen forms S-S bridges within the subunits. This explanation must be considered as improbable because it was shown [8] that all of the cysteine residues in beef liver catalase can be titrated with p-chloromercuribenzoate after mild acid

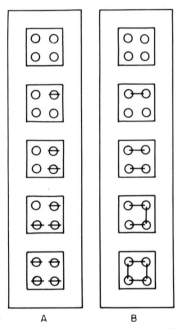

A   B

Figure 4. Speculation on the structure of the five "oximers" found in beef liver catalase. (A) S-S bridges within the subunits; (B) S-S bridges between the subunits.

denaturation at pH 3. (B) represents an alternative explanation of the conversion mechanism in which S-S bridges are formed between the subunits. Some evidence to support this speculation has been reported [8, 9]. However, since these authors were not aware of the heterogeneity of the catalases used, i.e. whether it was in the native or the altered state, their results may be revised by new investigations now in progress. Another explanation for the enzyme heterogeneity was given a few years ago by Kitto *et al.* [10], who attributed it to "conformers" of the enzyme molecule. This explanation must be examined to determine if it is the source of the different liver catalase forms. Experiments to clarify a similar situation in erythrocyte catalase were made recently by Cantz *et al.* [11].

It is quite possible that in future a number of "isoenzymes" will be found to be isolation artifacts. To determine this, it may be necessary to analyse enzymes directly after controlled release from the cell particles in which they are synthesized or stored.

REFERENCES

1. Schroeder, W. A., Shelton, J. R., Shelton, J. B. and Olson, M. M., *Biochim. biophys. Acta* **89** (1964) 47.
2. Holmes, R. S. and Masters, C. J., *Archs Biochem. Biophys.* **109** (1965) 196.
3. Heidrich, H.-G., *Hoppe-Seyler's Z. physiol. Chem.* **349** (1968) 873.
4. Bergmeyer, H.-U., *Biochem. Z.* **327** (1955) 255.
5. De Duve, Ch. and Baudhin, P., *Phys. Rev.* **46** (1966) 323.
6. Sawant, P. L., Shibko, S., Kumta, U. S. and Tappel, A. L., *Biochim. biophys. Acta* **85** (1964) 82.
7. Thorup, O. A., Strole, W. B. and Leavell, B. S., *J. Lab. clin. Med.* **58** (1961) 122.
8. Samejima, T. and Tsi Yang, J., *J. biol. Chem.* **238** (1963) 3256.
9. Sund, H., Weber, K. and Moelbert, E., *Eur. J. Biochem.* **1** (1967) 400.
10. Kitto, G. B., Wassarman, P. M. and Kaplan, N. O., *Proc. natn. Acad. Sci. U.S.A.* **56** (1966) 578.
11. Cantz, M., Moerikofer-Zwez, St., Bossi, E., von Wartburg, J. P. and Aebi, H., *Experientia* **24** (1968) 119.

FEBS Symposium, Volume 18, 1970, pp. 297-303

Peroxidase Isoenzymes in Germinating Wheat

P. SEQUI, A. MARCHESINI and A. CHERSI

*Laboratorio Virus Biosintesi Vegetali del C.N.R. and
Istituto di Chimica Organica dell'Universita, Milan,
Italy*

Two main groups of isoperoxidases, an acidic and a basic group, were found in wheat by means of their chromatographic and electrophoretic behaviour. In wheat germ the most represented forms of peroxidase fell in the basic group; the acidic group accounted for a very low percentage of the total enzyme activity. After germination the isoenzymes of the acidic group increased both in total activity and in number.

The occurrence of molecular multiplicity in plant peroxidase has been found by many authors; two peroxidase isoenzymes have been isolated and purified from wheat germ [1-3]. In wheat seedlings the peroxidase activity has been fractionated, by chromatography on a carboxylic resin, into two groups of isoenzymes, an acidic and a basic group; several peroxidase isoenzymes were found in both the acidic and the basic group by means of polyacrylamide gel electrophoresis [4].

Isoenzyme electrophoretic patterns have been studied by Macko *et al.* [5] in early growth of wheat and some differences were found between the dormant and the germinated seed; the authors, however, investigated only acid proteins, migrating in alkaline buffers. The present paper is concerned with the modification of the peroxidase isoenzyme electrophoretic pattern of wheat during germination, and with attempts to isolate and purify all the wheat peroxidase isoenzymes.

MATERIALS AND METHODS

Commercial purified wheat germ was kindly supplied by Mulino Sesia, Vercelli, Italy. Wheat seedlings were grown for five days on silica sand with tap water and cut by scissors from carioxides.

The procedure for extraction and partial purification of wheat peroxidases is shown schematically in Fig. 1. The effect of cations on peroxidase activity [6-8] led us to use only NH_4^+ ions during the purification. Wheat germ was stirred for

Enzyme. Peroxidase (EC 1.11.1.7).

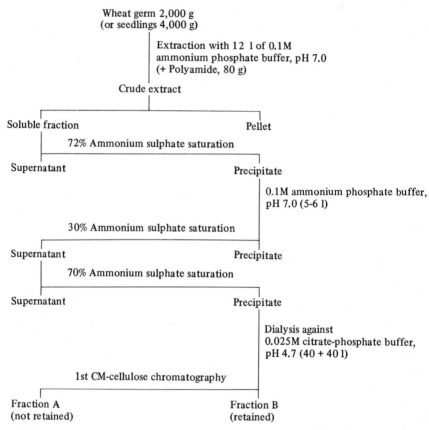

Figure 1. Extraction and partial purification of wheat peroxidase isoenzymes. About six million peroxidase units are present in the crude extract; 5 million units are recovered after dialysis.

one hour with 6 volumes by weight of 0.1M ammonium phosphate buffer, pH 7.0, at room temperature; wheat seedlings were homogenized for 3 min with 3 volumes by weight of the same buffer by means of a mechanical blender at 0-4°C and 20 g of polyamide were added to each kg of seedlings. The crude extract was centrifuged at 2,500 x g for one hour to obtain a "soluble fraction". Following the ammonium sulphate precipitations and dialysis, as reported in Fig. 1, the procedure yielded about 80% of the total enzyme activity originally present in the crude extract.

The enzyme activity was determined spectrophotometrically by measuring the rate of change in absorbancy at 460 mμ. Each reaction mixture for routine assay of peroxidase activity contained 2.75 ml of 0.05M ammonium citrate-phosphate buffer, pH 5.4, 0.1 ml of 0.1M H$_2$O$_2$, 0.05 ml of 0.5% o-dianisidine and 0.1 ml of enzyme. One peroxidase unit is arbitrarily taken as the quantity of enzyme which forms an amount of coloured reaction product corresponding to an optical density change of 1.00 in 1 min.

Polyacrylamide disc electrophoresis was performed with an alkaline buffer as described by Davis [9] and with an acidic buffer according to the method of Reisfeld et al. [10]. Peroxidase activities were measured with benzidine and H_2O_2 [11].

Carboxymethyl-cellulose (CM-cellulose) and acrylamide were purchased from Kodak (Rochester, U.S.A.); polyamide from Macherey, Nagel and Co. (Duren, Germany); Sephadex G-100 from Pharmacia (Uppsala, Sweden).

RESULTS AND DISCUSSION

The peroxidase isoenzyme pattern of wheat during germination was studied in the soluble fraction of the crude extract. As shown in Fig. 2, remarkable

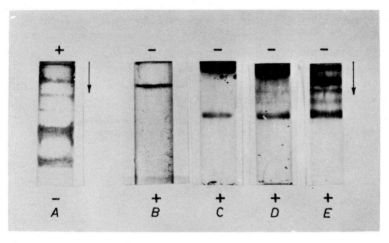

Figure 2. Polyacrylamide electrophoresis patterns of wheat peroxidase isoenzymes. A: cathodic isoperoxidase pattern in dormant and germinated seeds; B: anodic isoperoxidase pattern in dormant seed; C, D, E: anodic isoperoxidase patterns after 12, 24 and 48 h germination. Samples containing 150 μg of protein were applied to each gel for electrophoresis.

differences were found in the behaviour of the acidic and the basic groups of peroxidase isoenzymes. The isoenzymes of the basic group show on polyacrylamide electrophoresis a very characteristic pattern: their number and their electrophoretic behaviour do not change in the dormant and germinated seed of wheat. The isoenzymes of the acidic group, on the contrary, exhibit very different patterns during germination; the number of isoperoxidases increases gradually as a function of time.

Table 1. Acid and basic groups of wheat peroxidase isoenzymes before and after germination.

	Peroxidase activity*	
	Acid group†	Basic group‡
	%	%
Ungerminated wheat (germ)	1	75
Germinated wheat (seedlings)	10	60

* Percentage recoveries based on enzyme activity of crude extract as 100%.
† Fraction A of the first CM-cellulose chromatography.
‡ Fraction B of the first CM-cellulose chromatography.

We then attempted to isolate and purify the peroxidase isoenzymes. Initial chromatography on CM-cellulose columns (size 60 x 600 mm) gave a separation of two fractions: the first (fraction A) is not retained by the resin equilibrated in 0.025M ammonium citrate-phosphate buffer, pH 4.7; the second (fraction B) is eluted with 0.5M ammonium phosphate buffer, pH 7.0. As Table 1 shows, fraction A accounts for a considerable proportion of the total peroxidase activity in preparations from germinated wheat; in wheat germ, on the contrary, fraction A accounts for a very low percentage of the total peroxidase activity, as low as 1%. This finding agrees with the increase in the number of acid peroxidases during germination.

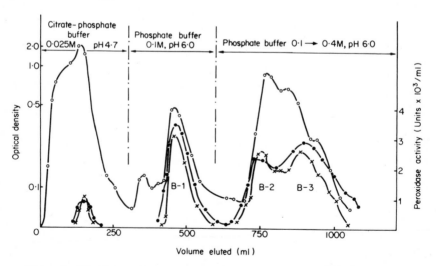

Figure 3. CM-cellulose chromatography of fraction B. The CM-cellulose column (size 25 x 480 mm) was equilibrated with 0.025M citrate-phosphate buffer, pH 4.7. ×——× Peroxidase activity; ●——● optical density at 400 mμ; ○——○ optical density at 280 mμ.

A second chromatography of the basic group of peroxidase isoenzymes (fraction B) resolves the enzyme activity into three main fractions, tentatively called B-1, B-2 and B-3 (Fig. 3).

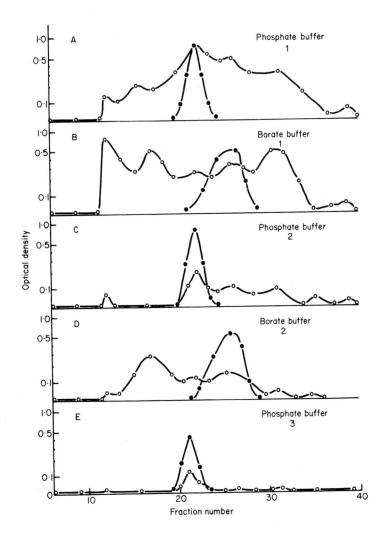

Figure 4. Gel filtrations of isoenzyme B-1 on Sephadex G-100. Column size 25 x 900 mm; each fraction 10.4 ml. Two identical preparations were chromatographed in experiments A and B. The purified preparation with phosphate buffer (A) was rechromatographed with borate buffer (D); the purified preparation with borate buffer (B) was rechromatographed with phosphate buffer (C). Both the preparations yield a protein of high purity in a third chromatography (E). ●———● Optical density at 400 mμ; ○———○ optical density at 280 mμ.

The purification of the three isoenzymes may be accomplished by means of further CM-cellulose chromatography, with linear concentration gradients; presently the isoenzymes B-1 and B-3 are being isolated and crystallized. It seems interesting to us, however, to refer to an alternative purification procedure for isoenzyme B-1 on Sephadex.

Isoperoxidase B-1 shows very different behaviour on gel filtration in phosphate or borate buffers. On standard Sephadex G-100 columns (size 25 x 900 mm; each fraction 10.4 ml) the isoenzyme B-1 is eluted in fraction 21 with 0.1M ammonium phosphate buffer, pH 6.3; *horseradish* peroxidase, in the same conditions, is eluted in fraction 18. With 0.1M borate buffer, pH 8.5, there is a slower migration of the isoenzyme B-1, which is eluted in fraction 24, and horseradish peroxidase in fraction 19.

Owing to this behaviour, by performing the gel filtrations with phosphate and borate buffers in turn, the purification of isoenzyme B-1 on Sephadex G-100 is easily improved. As shown in Fig. 4A and B, two identical preparations of the isoenzyme yield very different elution patterns with the two buffers. The preparation purified with phosphate buffer (Fig. 4A), rechromatographed with borate buffer (Fig. 4D) is further purified from faster migrating substances; the preparation obtained from borate buffer chromatography (Fig. 4B), by contrast, is further purified by phosphate buffer gel filtration (Fig. 4C) from substances which are eluted behind the isoenzyme. A third chromatography (Fig. 4E) yields isoperoxidase of high purity.

The borate buffer effect on gel filtration of isoenzyme B-1 does not correlate with the pH of the buffer. It disappears at higher ionic strengths; the isoperoxidase B-1 shows identical behaviour with phosphate buffers and, for example, with 0.1M borate buffer, pH 8.5 + 2M NaCl. These findings suggest that the effect of borate buffer may be due to interactions with carbohydrates present in the enzyme or with free hydroxyl groups of the Sephadex dextran gel.

No complete correspondence of properties has been found between peroxidase isoenzymes isolated by us and haemoproteins referred to in the literature. Our peroxidase B-1 shows strikingly similar properties to the "peroxidase 556" of Hagihara and co-workers [1] and the "wheat germ haemoprotein 550" of Wassermann and Burris [12], but its reduced spectrum, with an identical Soret maximum, shows a difference in the second characteristic maximum (548 mμ, compared to 566 and 550). No peroxidase isoenzymes with Soret maxima at 417 mμ, such as the "peroxidase 566" of Shin and Nakamura [3], were isolated by our procedure.

We are now purifying all the peroxidase isoenzymes in wheat, before and after germination, and initiating two important research projects:

(a) A study of the kinetic properties and substrate specificity of isolated isoenzymes; progress in this field is very important to ascertain physiological functions of each peroxidase isoenzyme in plants;

(b) The determination of protein composition and structure. Although plant peroxidases are being extensively studied, their structure is still unknown. Structural features of each peroxidase isoenzyme may provide insights into the mechanisms of control and biosynthesis of these enzymes.

REFERENCES

1. Hagihara, B., Tagawa, K., Morikawa, I., Shin, M. and Okonuki, K., *Nature, Lond.* **181** (1958) 1656.
2. Tagawa, K. and Shin, M., *J. Biochem., Tokyo* **46** (1959) 865.
3. Shin, M. and Nakamura, W., *J. Biochem., Toyko* **50** (1961) 500.
4. Lanzani, G. A., Marchesini, A., Galante, E., Manzocchi, L. A. and Sequi, P., *Enzymologia* **33** (1967) 361.
5. Macko, V., Honold, G. R. and Stahmann, M. A., *Phytochemistry* **6** (1967) 465.
6. Whitaker, J. R. and Tappel, A. L., *Biochim. biophys. Acta* **62** (1962) 310.
7. Fridovich, I., *J. biol. Chem.* **238** (1963) 3921.
8. Perez-Villasenor, J. and Whitaker, J. R., *Archs Biochem. Biophys.* **121** (1967) 541.
9. Davis, B. J., *Ann. N.Y. Acad. Sci.* **121** (1964) 404.
10. Reisfeld, R. A., Lewis, U. J. and Williams, D. E., *Nature, Lond.* **195** (1962) 281.
11. Giacomelli, M. and Cervigni, T., *Radiat. Bot.* **4** (1964) 395.
12. Wasserman, A. R. and Burris, R. H., *Phytochemistry* **4** (1965) 413.

FEBS Symposium, Volume 18, 1970, pp. 305-320

Ontogenetic Evolution and Pathological Modifications of Molecular Forms of some Isoenzymes

F. SCHAPIRA, J. C. DREYFUS and G. SCHAPIRA

24, rue du Faubourg Saint-Jacques, Paris 14e, France. *

INTRODUCTION

A knowledge of the multiple molecular forms of enzymes—currently named isoenzymes—has opened new perspectives in molecular pathology. But the term "isoenzymes" itself is not well defined. In this work we shall let "isoenzymes" (or more briefly, "isozymes") represent various enzymatic forms which catalyse, in one tissue, the same main reaction, and which differ in one or more characteristics, for example, kinetic constants, thermostability, response to stimulators and inhibitors, or even the action on a secondary substrate.

It is now well known that these molecular forms may vary according to tissue, species and age, and also in some pathological conditions. Our working hypothesis, suggested by our previous studies on hepatoma aldolase, was the analogy between these anomalies (especially in molecular diseases and in some cancerous tissues) and their forms at the initial stage of ontogenic evolution. In this perspective, we have studied three enzymes with multiple molecular forms: Lactic dehydrogenase (LDH), creatine kinase (CK), and aldolase. We shall consider first the modifications of aldolase and LDH in cancerous tissues.

ALDOLASE IN CANCEROUS TISSUES

We recall the studies of Hers [1] Rutter *et al.* [2, 3] and ourselves [4], which have shown that there are three types of aldolase in mammalian tissues.

(a) Aldolase A (muscle type) acts primarily on fructose-1,6-diphosphate (FDP) and slightly on fructose-1-phosphate (F-1-P). The ratio of its activity against FDP to its activity against F-1-P is greater than fifty.

Non-standard abbreviations. Lactic dehydrogenase, LDH; creatine kinase, CK; fructose-1,6-diphosphate, FDP; fructose-1-phosphate, F-1-P.

Enzymes. Lactic dehydrogenase or lactate-NAD oxidoreductase (EC 1.1.1.27); aldolase or ketose-1-phosphate aldehyde-lyase (EC 4.1.2.7); creatine kinase or ATP, creatine phosphotransferase (EC 2.7.3.2).

* Université de Paris, Groupe U15 de l'I.N.S.E.R.M., laboratoire associé au C.N.R.S.

(b) Aldolase B (liver type) has approximately the same activity against the two substrates (ratio: $\frac{FDP}{F\text{-}1\text{-}P} = 1$).

(c) The third aldolase, brain type, or aldolase C, is more recently known. Its activity ratio is around ten. Many other criteria differentiate aldolase iso-enzymes, e.g: competitive inhibition of aldolase A by ATP, which does not inhibit aldolase B [5].

Immunological properties: Antiserum anti-A (prepared from chicken with rabbit muscle aldolase) does not inhibit aldolase B (or very little). Conversely, antiserum anti-B does not inhibit aldolase A.

Electrophoresis is another method used to differentiate the three aldolases.

After migration (in the presence of β-mercaptoethanol), the isozymes are revealed by reduction of tetrazolium salts to blue formazan. The aldolase reaction is coupled with phosphoglyceraldehyde dehydrogenase reaction, in which NAD is reduced to NADH, which transfers its hydrogen to tetrazolium. At alkaline pH, aldolase C is the most anodic.

In mammalian tissues, the distribution of the three types varies according to the organ. But Hers and Joassin [6] and ourselves [7] have shown that this distribution is also different in foetal and adult livers.

We have found that in hepatomas (human and experimental) the isozymic distribution differs from the normal and is similar to the isozymic distribution of foetal liver [7, 8].

Table 1 shows the modification of aldolase activity ratio in hepatomas (rat or human). The mean aldolase activity ratio is 6.2 ± 1.6 as compared to 1.0 in normal liver. In human foetal liver, we have found the aldolase activity ratio (according to the foetal age) to vary between 2 and 3; in foetal rat, between 2 and 6. Therefore, we have found a modification of substrate specificity which is similar to the foetal ratio.

We have performed, with Nordmann [9], further experiments in order to determine if this modification is due to the presence of a mixture of two aldolases A and B, in cancerous and foetal liver.

Table 1. Mean aldolase activity ratios of normal and hepatoma livers.

Adult human liver	1.05
Adult rat liver	1.02
Foetal human liver	2.0 to 3.0
Foetal rat liver	2.0 to 6.0
Human hepatoma	6.2
Rat hepatoma (3'MDAB)	3.5

Table 2 shows the different aldolase activity ratios and also the behaviour of various aldolases in the presence of ATP and antiserum anti-A. Experiments were performed with crystalline muscle and liver aldolases from rabbit and with extracts of human tissues. Antiserum anti-A was prepared from chickens by repeated injections of rabbit muscle aldolase; it inhibited not only rabbit muscle aldolase, but also human muscle aldolase. It was almost completely inactive towards normal liver aldolase.

Table 2. Catalytic and immunologic properties of aldolases.

	FDP/F-1-P activity ratio	Inhibition of FDP activity by ATP	Inhibition of FDP activity by anti-muscle aldolase
Purified rabbit liver aldolase	1.0	< 5%	< 4%
Normal human liver extract	< 1.1	< 12%	< 15%
Purified rabbit muscle aldolase	46.0	65%	98%
Normal human muscle extract	> 35.0	> 45%	> 95%
Human foetal liver extract (8 to 16 weeks old)	2.0 to 3.0	25% to 40%	40% to 50%
Human hepatomas (mean)	6.2 ± 1.6	38% ± 6	53% ± 7

It is seen that ATP and antiserum anti-A both inhibit foetal liver and hepatoma aldolases. The mean percentage of inhibition is very similar for both aldolases (about 35% for ATP and 50% for antiserum).

Figure 1 shows a photograph of electrophoresis on cellulose acetate, followed by specific coloration of aldolase. The normal rat liver, as described by Rutter, shows a marked cathodic band, and a slight band near the origin. This last band is the unique one shown by pure muscle aldolase.

On the other hand, foetal and hepatoma aldolases show a reinforced band; this last one corresponds to aldolase A.

Consequently, it appears that in cancerous liver there is a maintenance, or perhaps an increase of aldolase A (predominant in foetal liver, and still present in feeble percentage in normal adult liver). It is the most differentiated, most specific, aldolase B, which disappears during the cancerous process.

In order to confirm this hypothesis, we have studied [10] an organ containing a different type of aldolase, that from mouse spleen, mainly type A, with an aldolase activity ratio of about nine.

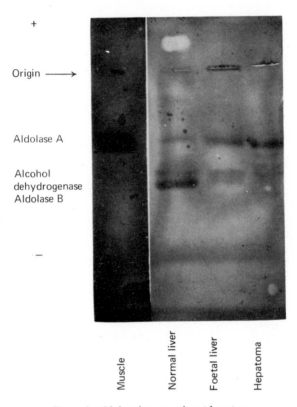

Figure 1. Aldolase isozymes in rat hepatoma.

Spleen reticulo sarcomas were induced in Swiss mice by chloramphenicol, and then transplanted. In this case, the mean aldolase activity ratio decreases (4.5 instead of 8.9). But, in this case also, its value is similar to the ratio in the foetal organ (as shown in Table 3).

More recently, our experiments on hepatomas were confirmed by the studies of Adelman *et al.* [11] who, in addition, compared the aldolase activity ratios in well differentiated and in poorly differentiated tumours.

Mention should also be made of the studies of Tanaka *et al.* [12] on pyruvate kinase. Normally, the activity of the muscle-type of this enzyme is very low in adult liver; on the contrary, this type is predominant in foetal liver. In experimental hepatoma, the muscle type becomes predominant.

In resumé, we have thus far proposed the hypothesis of the repression of the most differentiated and preponderant molecular form of enzymes in cancer.

We have also studied lactic dehydrogenase isozymes in this perspective.

Table 3. Normal and cancerous spleens (mice).

	Aldolases		
	F-1,6-P	F-1-P	Ratio
Normal adult	1.76 ± 0.055	0.195 ± 0.015	8.97 ± 0.5
Cancerous	0.759 ± 0.27	0.162 ± 0.009	4.51 ± 1.54
Foetal (mean of two experiments	2.24	0.47	4.7

LDH IN CANCEROUS TISSUES

It is not necessary to recall the investigations of Wieland and Pfleiderer [13], Vesell and Bearn [14], Markert and Moller [15, 16] and Cahn *et al.* [17] on LDH isoenzymes. There are two main types of LDH: M-type (occurring principally in muscle) and H-type (occurring principally in heart, but also in brain). Each subunit is of one of these two types. Different proportions of M and H exist in various tissues. The LDH molecule is tetrameric and the five tetramers (M4, M3Hl, M2H2, M1H3, H4) have been separated electrophoretically and chromatographically. They possess distinct physico-chemical and immunological properties, and subunits have different amino acid composition.

It has been shown that the LDH isoenzymic pattern varies not only according to species and organ, but also according to age [18, 19]. Many modifications of the LDH isoenzymic pattern were pointed out recently, especially by electrophoresis on starch gel, agar or cellulose acetate.

The electrophoresis separation is followed by specific coloration (with sodium lactate, NAD, and phenazine methosulphate: tetrazolium salts are reduced to blue formazan).

Starkweather and Schoch [20] have found that the distribution of the five isozymes tends to be uniform in cancers of various origins. Richterich and Burger [21] reported that the slow isozyme (M-type) is markedly increased in cancerous effusions.

The most extensive study on isozyme LDH in neoplastic diseases was carried out by Goldman *et al.* [22]. These authors found a shift in the pattern of isozymes in human neoplasms as compared, not only with normal controls, but also with benign tumours. These findings were confirmed on various tumours by Yasin and Bergel [23] and by Ishihara [24]. It is useful also to recall the findings of Hule [25], who showed that the activity of isoenzyme 5 (M-type) is higher in leukocytes of patients with myeloid leukemia.

According to the hypothesis of Goldman *et al.* [22], the shift towards an M-type is a shift towards a glycolytic pathway. They base this assumption especially on the behaviour of the two types with pyruvate: M-type is not inhibited by excess pyruvate; when this substrate is accumulated, lactate is

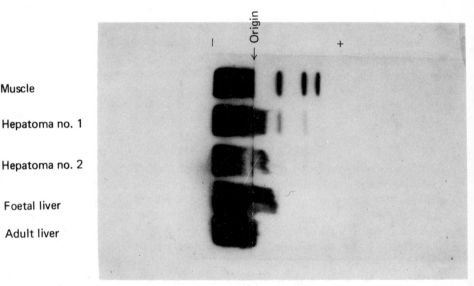

Figure 2. LDH isozymes in rat hepatoma.

formed. On the contrary, H-type is inhibited by the accumulation of pyruvate and the metabolism is oriented towards the citrate cycle.

However, Goldman *et al.* [22] have studied organs in which the adult form of lactic dehydrogenase isozymes is of the "H-type" or mixed; we may ask whether there would be a shift towards an H-type in organs with adult M-type and foetal H-type. Kline and Clayton [26] and Johnson and Kampschmidt [27] have demonstrated the appearance of one anodic isozyme in rat liver with experimental hepatoma. One of us has consequently studied with de Nechaud LDH isozymes in rat hepatoma and compared them with foetal rat liver [28]. The hepatomas were induced in the rats by 3′-methyl-dimethylamino azobenzene (3′-MDAB).

Figure 2 shows the isoenzymic pattern of LDH in normal liver, in foetal liver, in two hepatomas and, for comparison, in rat muscle.

The photograph confirms the predominance of M-subunits in normal liver:

only isozymes 5 and 4 are seen. Isozymes 3 and 2, very apparent in muscle, are only faintly visible: 1 is never visible. On the contrary, in foetal liver, these anodic bands are apparent, at the expense of the cathodic band. Consequently, the foetal liver contains more H subunits than the adult one.

In five hepatomas (of ten studied) we have found anodic, supplementary bands, with strengthening of isozymes 3 and 2 and the appearance of isozyme 1.

The comparative migration with muscle isozymes confirms that there are MH hybrids, and pure tetramer H.

Moreover, we see, in foetal liver as in hepatomas, an anomalous band, migrating between 4 and 3. We have no explanation for this variable band. We can only recall the analogous observations of Theret *et al.* [29] on the rat sarcomas and the supplementary forms described by Croisille [30] on chick embryo.

In any event, we believe that our experiments show that cancerous iso-enzymic modifications reflect tissue de-differentiation rather than a shift to a glycolytic metabolism.

LACTIC DEHYDROGENASE ISOZYMES OF PATHOLOGICAL MUSCLE

We recall that we have shown (with Demos) the ontogenic evolution of LDH in human muscle. On the other hand, Wieme and Herpol [31] and ourselves [19], have shown the absence of the cathodic band (characteristic of M-type) in the electrophoretic pattern of dystrophic muscle. Biopsies were performed in patients with progressive muscular dystrophy (Duchenne type, and also other types) and we have compared these with muscles of premature infants.

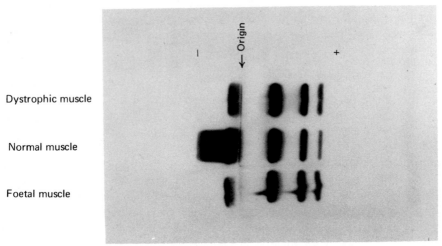

Figure 3. LDH isozymes of human muscle in muscular dystrophy.

Figure 3 shows that dystrophic as well as foetal muscles are of the H-type (brain or heart type), with a predominance of anodic bands.

Kaplan and Cahn [32] have found the same anomaly in muscle of chicken with hereditary muscular dystrophy, and Emery [33] and Wilkinson [34] in some muscles of female heterozygotes for muscular dystrophy.

But we have also demonstrated that this pattern is not limited to myogenic and hereditary diseases. We have found an analogous pattern in human dermatomyositis of myogenic origin and also in experimental neurogenic atrophy. Our experiments were performed on chicken atrophied by sciatic

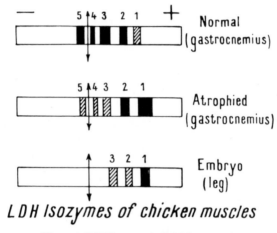

LDH Isozymes of chicken muscles

Figure 4. LDH isozymes of chicken muscle.

section [35]; Dawson *et al.* [36] have found analogous modifications in atrophied rabbit muscle and Lauryssens *et al.* [37] and Brody [38] in man with neurogenic atrophy.

Figure 4 shows the distribution in normal, atrophied, and embryonic chicken leg.

The anodic bands are strengthened at the expense of the slower bands, and the new pattern tends towards the embryonic pattern. The comparison between the LDH activities at high and low concentrations of pyruvate corroborates these findings (Table 4). Consequently, we find a foetal pattern (similar to the brain type) in pathological muscle.

Generally we have found that it is only in species having a foetal type different from the adult that there are modifications of the isozymic pattern in muscular diseases [39]. In mice with hereditary muscular dystrophy, and in rat gastrocnemius, atrophied by sciatic section, there are no modifications of isozymic distribution; in these species, the foetal pattern is identical with the adult.

Table 4

	LDH activity ratios $\dfrac{\text{Pyruvate } 10^{-4}\,\text{M}}{\text{Pyruvate } 3.3 \times 10^{-3}\,\text{M}}$
Human	
Normal muscle	0.4 to 0.6
Dystrophic muscle	0.8 to 1.0
Chicken	
Normal gastrocnemius	0.54
Atrophied gastrocnemius	0.90
Leg embryo	0.92

The "foetal-like" pattern seems to correspond to a de-differentiation process. It is known that the enzymatic equipment is different in the "white" muscles (fast contraction) and in the "red" muscles (slow contraction). The isozymic pattern is also different: white muscles (e.g. gastrocnemius) are of M-type; red muscles (e.g. soleus) are of H-type. After atrophy, white and red muscles tend towards a common isozymic pattern, as shown by Fig. 5.

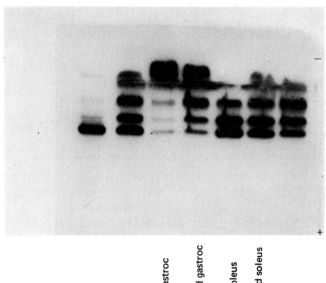

Figure 5. LDH isozymes of rabbit muscle.

We have also studied creatine kinase (CK) and aldolase isozymes in muscular diseases.

CREATINE KINASE ISOZYMES IN PATHOLOGICAL MUSCLE

The work of Dance and Watts [40] and Eppenberger *et al.* [41] has shown that CK is a dimer; each protomer may be "muscle type" (M) which is electrophoretically slow (at an alkaline pH), or "brain type" (B) which is fast; the hybrid "MB" exhibits electrophoretic migration on cellulose acetate or starch gel corresponding to an intermediate position.

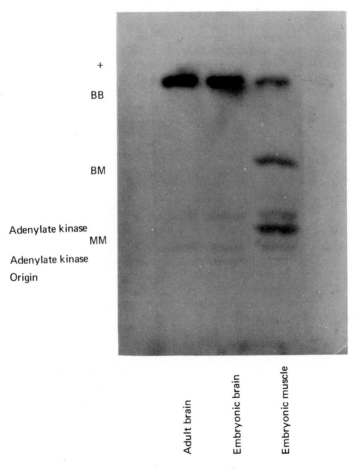

Figure 6. Creatine kinase isozymes of human brain and embryonic muscle.

After electrophoresis, creatine kinase isozymes are revealed by tetrazolium coloration. The creatine kinase reaction is coupled with the hexokinase and glucose-6-phosphate dehydrogenase reactions; the hydrogen ions come from the reduction of NADP, in the presence of glucose, creatine phosphate, ADP and Mg^{2+}; AMP is added in order to inhibit the adenylate kinase reaction.

Ontogenic evolution of creatine kinase isozymes varies according to the species. Eppenberger *et al.* [41] have shown that the chicken exhibits a very rapid evolution. The embryonic form, which is principally type "B" on the twelfth day, is identical with the adult type (M) from the eighteenth day on.

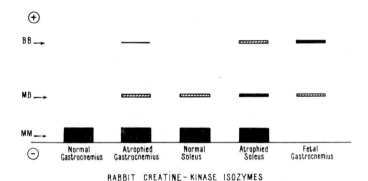

RABBIT CREATINE- KINASE ISOZYMES

Figure 7. Rabbit creatine kinase isozymes.

In mice, we have found no differences between the foetal and adult isozymic patterns; in both cases, a simple zone of migration, corresponding to the isozyme MM, is obtained.

In man, we have shown [42] that there is an ontogenic evolution, but, in order to demonstrate this, it was necessary to study embryos of less than five months. Under these conditions, we found not only the hybrid isozyme "BM" (slightly visible in some adult muscles) but also the isozyme "BB" (pure brain type) which normally is never visible (Fig. 6).

In rabbit, the ontogenic evolution is slow, and on the twenty-fifth day the muscle isozymes are still brain type.

We have studied the rabbit isozymic pattern of gastrocnemius (white muscle) and of soleus (red muscle), six weeks after sectioning of the sciatic nerve.

Figure 7 shows a scheme for the isozymic migration zones of atrophied muscles and of normal controlateral muscles in five experiments, as well as the electrophoretic migration of foetal rabbit muscle.

At a dilution where the enzymatic activities of normal and atrophied muscles would be comparable, we find a strengthening of the hybrid bands "BM" in atrophied gastrocnemius and soleus. Moreover, we find a supplementary band

"BB" in soleus, and we compare these patterns with the pattern of foetal muscle.

On the other hand, we have observed no modifications of CK isozymes in chicken and in mice with hereditary muscular dystrophy.

In man, we have studied muscle biopsies from patients with muscular dystrophy (Duchenne-type or atypical) and also two biopsies from patients with neurogenic atrophy.

In every case, electrophoresis on cellulose acetate or on starch gel showed a relative increase of the fast hybrid "BM" isozymes, and the appearance of "BB" isozymes, which are never visible under the same conditions in normal muscle.

ALDOLASE ISOZYMES IN PATHOLOGICAL MUSCLE

In chicken, we have confirmed the finding of Herskovitz *et al.* [43] which shows that the embryonic muscle contains primarily the aldolase "C" (brain type), the hybrid isozymes "A-C" appearing during the final stages of the incubation. Moreover, in foetal rabbit muscles, we have shown these isozymes to be of the "C" type.

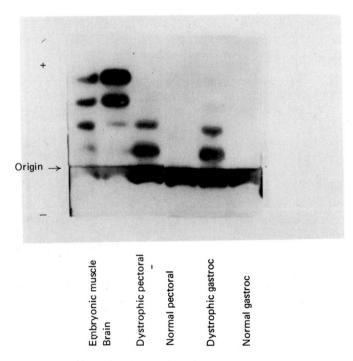

Figure 8. Aldolase isozymes of normal and dystrophic chicken muscles.

In chickens with hereditary muscular dystrophy we have studied the gastrocnemius and pectoral muscles. The photograph shows the results which we obtained on electrophoresis of five dystrophic and five normal chickens. We have compared them with the brain pattern, and also with embryonic muscle pattern [44] (Fig. 8).

The muscles of normal chickens show only one very strong cathodic band (pure type "A"). Embryonic muscle (on the thirteenth day of incubation) shows five bands, four anodic and a very feeble cathodic band. The comparison with the migration of a brain extract permits one to identify them as isozymes of the "C-type" (pure "C" and hybrids "A-C"). In dystrophic chickens the isozyme "A" is visible, but there are also supplementary, very abundant, isozymes which correspond most likely to the hybrids "A-C".

In this case also, this pattern (Fig. 8) is not specifically indicative of a hereditary, myogenic process. We have found an analogous modification in rabbit muscle, six weeks after sciatic nerve section. The photograph shows the electrophoretic pattern.

The rabbit brain shows on starch gel three or four rapid bands (Fig. 9). In foetal muscle, it is possible to distinguish these same anodic bands. In normal gastrocnemius, the slowest band is the only visible one, and it contrasts with

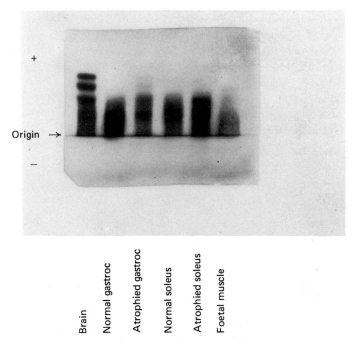

Figure 9. Aldolase isozymes in normal and atrophied rabbit muscles.

the atrophied gastrocnemius in which supplementary, anodic bands are seen. The difference between the normal and the atrophied patterns is less marked for soleus.

These findings recall our previous studies on the aldolase activity ratio $\dfrac{FDP}{F\text{-}1\text{-}P}$ in normal and atrophied muscles under the same conditions [43].

We should mention only that we have found a decrease of the aldolase activity ratio in atrophied muscle; we have noted that, in foetal leg muscle, this ratio is much lower than in adult [45]. Rutter has shown that the ratio $\dfrac{FDP}{F\text{-}1\text{-}P}$ is about ten in pure aldolase C. The presence of aldolase C in atrophied muscle consequently explains the lowering of the aldolase activity ratio in atrophied muscle.

CONCLUSIONS

In muscular diseases, we find for the three enzymes studied an increase (absolute or relative) of isozymes of the "brain type". This result, apparently unexpected, must be compared with the results obtained from the isozymic study of foetal or embryonic muscle, which shows a predominance of brain type.

In cancerous tissues, we have noted various isozymic modifications. Aldolase B (liver type) is decreased in hepatoma; aldolase A (muscle type) is decreased in spleen reticulo sarcoma. For LDH, it is generally the "M-type" which predominates in cancerous tissues, but we have found an increase of the H-type in some hepatomas.

The common feature of these modifications is their "foetal-like" pattern, which is an expression of the de-differentiation at the molecular level, characterized by the repression of the synthesis of the most specific forms of enzymes.

ACKOWLEDGEMENTS

This work was supported by a grant from the Muscular Dystrophy Associations of America Inc., and the National Cancer Institute.

REFERENCES

1. Hers, H. G., Le métabolisme du fructose, Arscia, Bruxelles (1957).
2. Blostein, R. and Rutter, W. J., *J. biol. Chem.* 238 (1963) 3280
3. Penhoet, E., Rajkumar, T. and Rutter, W. J., *Proc. natn. Acad. Sci. U.S.A.* 56 (1966) 1275.
4. Schapira, F., *Bull. Soc. Chim. biol.* 43 (1961) 1357-67.
5. Spolter, P. D., Adelman, R. C. and Weinhouse, S., *J. biol. Chem.* 240 (1965) 1327.

6. Hers, H. G. and Joassin, G., *Enzymol. Biol. Clin.* **1** (1961) 4.
7. Schapira, F., Schapira, G. and Dreyfus, J. C., *C. r. hebd. Séanc. Acad. Sci., Paris* **254** (1962) 3143.
8. Schapira, F., Dreyfus, J. C. and Schapira, G., *Nature, Lond.* **200** (1963) 995.
9. Nordmann, Y. and Schapira, F., *Eur. J. Cancer* **3** (1967) 247.
10. Schapira, F. and Tran Ba Loc, P., *C. r. hebd. Séanc. Acad. Sci., Paris* **260** (1965) 4856.
11. Adelman, R. C., Morris, H. P. and Weinhouse, S., *Cancer Res.* **21** (1967) 2408.
12. Tanaka, T., Harano, Y., Morimura, H. and Mori, R., *Biochem. biophys. Res. Commun.* **21** (1965) 55.
13. Wieland, T. and Pfleiderer, G., *Biochem. Z.* **329** (1957) 112.
14. Vesell, E. S. and Bearn, A. G., *J. clin. Invest.* **40** (1961) 586.
15. Markert, C. L. and Møller, F., *Proc. natn. Acad. Sci. U.S.A.* **45** (1959) 753.
16. Markert, C. L., *Science, N.Y.* **140** (1963) 1329.
17. Cahn, R. D., Kaplan, N. O., Levine, L. and Zwilling, E., *Science, N.Y.* **136** (1962) 962.
18. Pfleiderer, G. and Wachsmuth, E. D., *Biochem. Z.* **334** (1961) 185.
19. Dreyfus, J. C., Demos, J., Schapira, F. and Schapira, G., *C. r. hebd. Séanc. Acad. Sci., Paris* **254** (1962) 4384.
20. Starkweather, W. H. and Schoch, H. K., *Biochim. biophys. Acta* **62** (1962) 440.
21. Richterich, R. and Burger, A., *Enzymol. Biol. Clin.* **3** (1963) 65.
22. Goldman, R. D., Kaplan, N. O. and Hall, T. C., *Cancer Res.* **24** (1964) 389.
23. Yasin, R. and Bergel, F., *Eur. J. Cancer* **1** (1965) 203.
24. Ishihara, M., *Eur. J. Cancer* **3** (1968) 545.
25. Hule, V., *Clinica chim. Acta* **17** (1967) 349.
26. Kline, E. S. and Clayton, G. C., *Proc. Soc. exp. Biol. Med.* **117** (1964) 891.
27. Johnson, H. L. and Kampschmidt, R. F., *Proc. Soc. exp. Biol. Med.* **120** (1965) 557.
28. Schapira, F. and de Nechaud, B., *C. r. Séanc. Soc. Biol.* **162** (1968) 86.
29. Theret, C., Lalegerie, P. and Wicart, L., *C. r. Séanc. Soc. Biol.* **160** (1966) 2238.
30. Croisille, Y., *C. r. hebd. Séanc. Acad. Sci., Paris* **264** (1967) 348.
31. Wieme, R. J. and Herpol, H., *Nature, Lond.* **194** (1962) 287.
32. Kaplan, N. O. and Cahn, R. D., *Proc. natn. Acad. Sci. U.S.A.* **48** (1962) 2123.
33. Emery, A. E., *Nature, Lond.* **201** (1964) 1044.
34. Wilkinson, J. H., "Isoenzymes", Spon, London, 1965.
35. Schapira, F. and Dreyfus, J. C., *Enzymol. Biol. Clin.* **4** (1964) 23.
36. Dawson, D. M., Goodfriend, T. L. and Kaplan, N. O., *Science, N.Y.* **143** (1964) 929.
37. Lauryssens, N. G., Lauryssens, M. J. and Zondag, N. A., *Clinica chim. Acta* **9** (1964) 276.
38. Brody, I. A., *Neurology, Minneap.* **14** (1964) 1091.
39. Schapira, F. and Dreyfus, J. C., *Bull. Soc. Chim. biol.* **47** (1965) 2261.
40. Dance, N. and Watts, D. C., *Biochem. J.* **84** (1962) 114 P.
41. Eppenberger, H. M., Eppenberger, M., Richterich, R. and Aebi, H., *Dev. Biol.* **10** (1964) 1.

42. Schapira, F., Dreyfus, J. C. and Allard, D., *Clinica chim. Acta* **20** (1968) 439.
43. Herskovitz, J., Masters, C. J., Wassarman, P. M. and Kaplan, N. O., *Biochem. biophys. Res. Commun.* **26** (1967) 24.
44. Schapira, F., *C. r. hebd. Séanc. Acad. Sci., Paris* **264** (1967) 2654.
45. Schapira, F., *C. r. Séanc. Soc. Biol.* **159** (1965) 2189.

FEBS Symposium, Volume 18, 1970, pp. 321-328

Isoenzymes of Aldolase in Human Malignant Tumours

H. PANDOV and A. DIKOV

*Department of Biochemistry, Oncological Research
Institute, Sofia, Bulgaria*

The molecular and catalytic properties of aldolase have been the subject of numerous papers in recent years [1, 4, 15-17, 23]. The existence of three main forms of the enzyme—A, B and C, respectively—named "muscular", "hepatic" and "brain" aldolase has been demonstrated. The immunochemical, electrophoretic and chromatographic studies of Penhoet and collaborators [13] led to the identification of these three forms. The hybrid character of the intermediary fractions of aldolase was established by the same authors. Their electron microscope investigations [14] supported their hypothesis that the molecule of aldolase is not a three-chain model, but is composed of four similar subunits.

Further electrophoretic studies on the multiple fractions of the enzyme permitted a more detailed determination of the isoenzymes of aldolase [15, 24]. The isoenzyme patterns of numerous normal tissues were electrophoretically scanned and the changes in serum fractions determined in many pathological conditions [3, 6-10].

The purpose of our studies was to determine, by means of an improved method of fractionation, the tissue isoenzyme pattern of aldolase in various malignant tumours in humans.

MATERIALS AND METHODS*

The isoenzyme fractions and total aldolase activity were investigated in twenty-four human malignant tumours differing in histology and localization.

The tumour tissue, immediately after surgical removal, was thoroughly washed with physiological saline and stored at $-50°C$ until further examination.

Homogenation. The frozen material, under continuous addition of liquid nitrogen, was minced in a porcelain mortar. The homogenate was diluted with an

---

* The reagents utilized in this investigation had the following origins: Fructose-1,6-diphosphate-Na salt, glyceraldehyde-3-phosphate dehydrogenase, nicotinamide-dinucleotide, diaphorase, tris buffer (Boehringer, Mannheim); agarose (Serva, Heidelberg); *p*-nitrotetrazolium blue, DL-glyceraldehyde (Calbiochem, USA); EDTA, boric acid (Loba Chemie, Austria); sodium arsenate (BDH, England).

equal amount (w/v) of buffer solution and the final homogenation performed at
2°C for 10 min in a Waring blender. The homogenate was centrifuged at 0°C for
30 min at 40,000 r.p.m. (105,000 g, Ultracentrifuge Spinco L2). The supernatant
fraction was used for further investigation.

The measurement of total aldolase activity was carried out by the method of
Kulganek and Klaschka with dinitrophenylhydrazine [11]. The results are
expressed in μmoles/mg N/min.

Tissue nitrogen was determined according to the modified method of
Kieldahl-Nessler [19] by colorimetric recording.

The electrophoretic fractionation of aldolase isoenzymes was performed on
0.6% agarose gel, with 170, 150 and 4 mm plates. Zonal electrophoresis with
Aronson–Gronwall [2] tris buffer, pH 8.9 (tris, 0.5M; EDTA, 0.0205M; boric
acid, 0.075M) was run at a constant current −50 V and 15 mA (measured
directly at both ends of the gel) for 18 h at 2°C.

The aldolase activity of the isoenzymes was developed by pouring upon the
gel plate the substrate solution, followed by a 2 h incubation at 37°C. Clear
fractions were obtained with a substrate solution developed by Dikov [5],
having the following composition: fructose-1,6-diphosphate, 400 mg; NAD, 40
mg; sodium arsenate (0.1M), 1.5 ml; *p*-nitrotetrazolium blue, 10 mg; tris/
hydrochloric acid buffer (1M, pH 7.0), 3 ml; glyceraldehyde-3-phosphate
dehydrogenase 10 mg/ml, 0.5 ml; diaphorase 10 mg/ml, 0.1 ml; agarose 1%
(42°C), 15 ml; bidistilled water to 30 ml. After development the plates were
fixed in 10% acetic acid and air dried.

RESULTS

With the methods used, the number of electrophoretic bands, with manifested
aldolase activity, obtained by summing up the different fractions of the various
tumours, amounted to twelve.

Most of the fractions were anodic, one was located on the starting line and
only three migrated to the cathode. The slow-moving anodic fractions of the
investigated tumours, located near the starting line, were most frequently intensively stained. These fractions were not clearly differentiated in all tumours. In
some cases two or three fractions coalesced and formed a single wide band. The
fast-moving anodic fractions were usually less intense and, depending on the
investigated tumour, their number and position varied. They were located 3-8
cm from the origin.

Of the cathodic fractions there was one very intensively stained, which was
located near the starting line. The others were less stained and were located 2-3
cm from the origin.

The results obtained may be divided into the following seven groups:

1. *Ovarian tumours.* In comparison with the normal ovary, the total enzyme

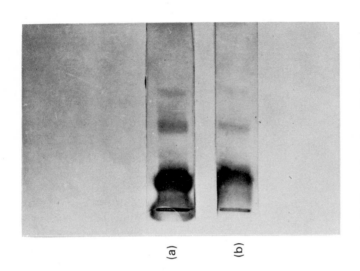

Figure 2. (a) Normal lung tissue; (b) bronchial cancer.

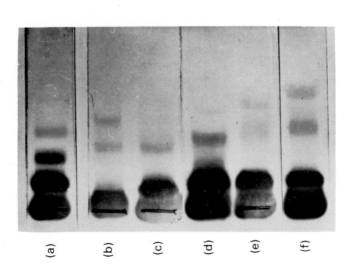

Figure 1. (a) Normal ovary; (b-f) ovarian carcinomas.

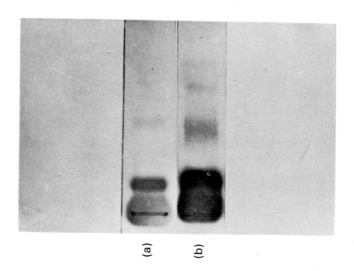

(a)

(b)

Figure 4. (a) Pigmented melanoma; (b) amelanotic melanoma.

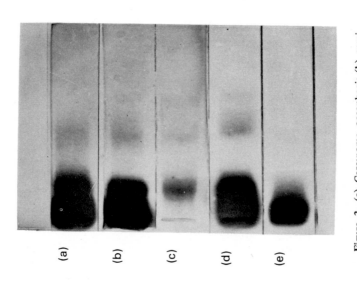

(a)

(b)

(c)

(d)

(e)

Figure 3. (a) Carcinoma oesophagi; (b) carcinoma solidum ventriculi; (c) carcinoma papilliferum ventriculi; (d) adenocarcinoma ventriculi; (e) adenocarcinoma signae.

activity was reduced by up to 5.7 to 11.2 μmoles/mg N/min. The isoenzyme fractions, depending on the histology, varied considerably in electrophoretic mobility by comparison with normal tissue (Fig. 1).

2. *Cancer of the lung.* Bronchial cancer was characterized by four to five not very intense fractions. Their electrophoretic mobilities did not differ from those of the normal lung tissue (Fig. 2).

3. *Tumours of the digestive tract.* In tumours originating from the mucosa of the digestive tract, a progressive diminution of the total enzyme activity was

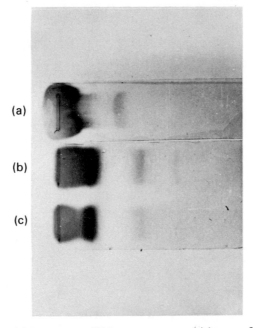

Figure 5. (a) Leyomyoma; (b) leyomyosarcoma; (c) leyomyofibroma.

established from the oesophagus (8.8 μmoles/mg N/min) to the colon, where it was only 0.8 μmole/mg N/min. The number of fast-moving bands was different. The intensities of both the fast-moving and slow-moving fractions were reduced (Fig. 3).

4. *Malignant melanoma.* In the pigmented form of this tumour, as compared to the amelanotic tumour, the enzyme activity was reduced by almost 50%. The fast-moving anodic fractions with pigmented melanomas were reduced in number, and had different locations than those with amelanotic melanoma (Fig. 4).

5. *Leyomyoma, leyomyosarcoma.* In uterine tumours of muscular origin there were intensive bands near the starting line and several less intense fast-moving anodic ones (Fig. 5).

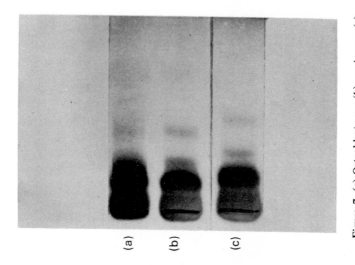

Figure 7. (a) Osteoklastoma; (b) seminoma; (c) lymphogranuloma.

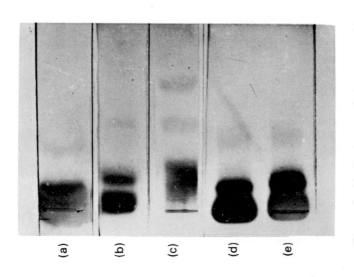

Figure 6. (a) Carcinoma of the uterus; (b) carcinoma of the breast; (c) carcinoma of the thyroid gland; (d) carcinoma of the larynx; (e) carcinoma of the bladder.

6. *Carcinomas located in other organs.* This group included cancer of the uterus, breast, thyroid, bladder and larynx. The highest aldolase activity was found in both uterine and laryngeal cancer (8.7 and 7.7 μmoles/mg N/min, respectively) and the lowest in cancer of the thyroid (0.6 μmole/mg N/min). In the latter tumour, in spite of the low total activity, the fast-moving anodic fractions were clearly conspicuous. In tumours of the bladder and larynx, the fast-moving anodic fractions were small in number and very faint; however, in both tumours a cathodic fraction was revealed at a distance of about 2 cm from the start (Fig. 6).

7. *Other malignant tumours.* In this group there are the isoenzyme fractions with osteoklastoma, seminoma and lymphogranuloma. In osteoklastoma there were five aldolase fractions, one intensive haemoglobin fraction and one fraction of bilirubin, situated near the anode. The number of fractions in the other tumours is also five, but they have different locations (Fig. 7).

DISCUSSION

Agarose gel electrophoresis and the modified method for development of aldolase isoenzyme activity have made possible the separation of a considerable number of isoenzyme fractions. Their further immunochemical and chromatographic identification is the object of future studies.

We think that the improvement in the development technique is due to the participation in the substrate both of glyceraldehyde-3-phosphate dehydrogenase and diaphorase. The G-3-P dehydrogenase elevates the activity of the aldolase [22], whereas the added diaphorase appears as a very good proton conductor of the reduced form of NAD upon other compounds (Nitro BT),

In agreement with other authors [18, 25], the total aldolase activity in all investigated malignant tumours was found to be reduced. The changes in aldolase activity in the tumours of the digestive tract were of particular interest. The aldolase activity progressively diminishes from the oesophagus towards the colon.

The isoenzyme patterns of the different tumours varied both in respect to number and electrophoretic position. It is noteworthy that in most tumours the fractions with a higher activity were chiefly situated near the starting line, i.e. by their electrophoretic localization they corresponded to the fractions of aldolase A. These results are similar to the data obtained by Leese *et al.* [12] and Schapira *et al.* [20, 21].

ACKOWLEDGEMENTS

Thanks are due to Dr. G. Tenchev and his staff for their kind assistance in preparing the illustrative materials. We are indebted to Dr. S. Danev and C. D.

Chacarov for their friendly help and suggestions. We are indebted also to Dr. P. Blagoeva for her kind assistance.

REFERENCES

1. Anstall, H., Lapp, C. and Trujilio, J., *Science, N.Y.* **154** (1966) 657.
2. Aronson, T. and Gronwall, A., *Scand. J. clin. Lab. Invest.* **9** (1957) 338.
3. Baron, D., Foxwell, C. and Buck, G. M., Aldolase Isoenzymes in Normal and Malignant Tissue. British Empire Cancer Campaign for Research, 1966, p. 367.
4. Christen, Ph., Rensing, U., Schmid, A. and Leuthardt, F., *Helv. chim. Acta* **49** (1966) 1872-75.
5. Dikov, A., *Z. klin. Chem. klin. Biochem.* **5** (1968) 386.
6. Dikov, A. and Tschankov, I., *Z. klin. Chem. klin. Biochem.* (1969) in press.
7. Dikov, A., Tschankov, I. and Samardschiew, A., *Z. klin. Chem. klin. Biochem.* **5** (1968) 391.
8. Dikov, A. and Romanov, M., *Z. klin. Chem. klin. Biochem.* (1969) in press.
9. Foxwell, C., Cran, E. and Baron, D., *Biochem. J.* **100** (1966) 44.
10. Ishihara, M. and Biffen, J., Aldolase and Lactic Acid Dehydrogenase Isoenzymes (Human Normal and Malignant Uterine, Ovarian, Cervical and Gastric Tissues). British Empire Cancer Campaign for Research, 1966, p. 20.
11. Kulganek, W. and Klaschka, W., *Vop. med. Chem.*, No. 4 (1961) 434.
12. Leese, C., Gilbert, D., Giffon, P., Yasin, R., Biffen, J. and Ishihara, M., Aldolase and Lactic Acid Dehydrogenase Isoenzymes (Human Normal and Malignant Gastro-intestinal Tissues). British Empire Cancer Campaign for Research, 1966, p. 19.
13. Penhoet, E., Rajkumar, T. and Rutter, W. J., *Proc. natn. Acad. Sci. U.S.A.* **56** (1966) 1275-82.
14. Penhoet, E., Kochman, M., Valentine, R. and Rutter, W. J., *Biochemistry* **6** (9) (1967) 2940-49.
15. Pietruszko, R. and Baron, D., *Biochim. biophys. Acta* **132** (1967) 203-6.
16. Rensing, U., Schmid, A. and Leuthardt, F., *Hoppe-Seyler's Z. physiol. Chem.* **348** (1967) 921-28.
17. Rensing, U., Schmid, A., Christen, Ph. and Leuthardt, F., *Hoppe-Seyler's Z. physiol. Chem.* **348** (1967) 1001-4.
18. Ruberti, A., Castellani, E. and Tobaldini, G., *Arcispedale S. Anna Ferrara* Suppl. 19 (6) (1966) 1079-93.
19. Russanov, E. and Balevska, P., *Bull. Inst. Physiol.* **VII** (1964) 199.
20. Schapira, F., Dreyfus, J. C. and Schapira, G., *Enzymol. Biol. Clin.* **7** (1966) 98-108.
21. Schapira, F., *Eur. J. Cancer* **1-2** (1966) 131.
22. Tai-Wan Kwon and Olcott, H., *Biochem. biophys. Res. Commun.* **19** (1965) 3.
23. Warburg, O. and Chrisian, W., *Biochem. Z.* **314** (1943) 400-408.
24. Winsten, S., Jackson, J. and Wolf, P., *Clin. Chem.* **12** (1966) 497-504.
25. Ziegenbein, R., *Klin. Wschr.* **43** (24) (1965) 1337-40.

FEBS Symposium, Volume 18, 1970, pp. 329-334

The Variation and Sub-Band Formation of Lactate Dehydrogenase Isoenzymes in Formed Elements of Human Blood

V. HULE

Department of Clinical Biochemistry, J. E. Purkyne
University, Brno, Czechoslovakia

Recent studies clearly indicate that human lactate dehydrogenase (LDH) of leukocytes and thrombocytes can be separated into five active fractions [1-7]. Pfleiderer and Wachsmuth [8] also detected in normal human erythrocytes five different zones of lactate dehydrogenase activity, using an electrophoretic method on cellulose acetate strips, as did Dioguardi *et al.* [9]. From the data of Anderson *et al.* [3], and Vesell and Bearn [12], it appears that only four isozymes can be detected.

The use of different electrophoretic methods may, however, lead to inconsistent results. The causes of these inconsistencies could be of importance in correctly reporting the quantitation of the LDH isozymes and also, perhaps, in the study of possible additional forms, occurring as sub-bands. Our investigations by means of agar gel electrophoresis provide some examples of different conditions which entirely distort the isozyme patterns obtained.

MATERIALS AND METHODS

Hemolysates. Ten ml of blood with EDTA 10%, 1:50, obtained from 20 healthy fasting donors and from 185 patients on long-term anticoagulant therapy and with different diseases, comprising 800 specimens, were used. The blood was centrifuged for 15 min in a refrigerated centrifuge at 1600 g and the plasma and buffy coat were carefully removed by aspiration. At this stage some of the erythrocytes were sacrificed in order to make the removal of the buffy coat as complete as possible. The erythrocytes were washed six times with 0.15M saline (1:3 by volume). Hemolysis was by means of distilled water in the ratio of 1:3 and was completed with saponin. Most of the stroma was removed by centrifugation. When hemolysis was performed at the ratio 1:1, the stroma remained in the hemolysates.

Leukolysates. Peripheral white cells were obtained from 10 healthy fasting donors, from 32 patients with chronic lymphatic leukemia, and from 15 patients with chronic myeloid leukemia. The white cells were separated in siliconized

Enzyme. LDH, lactate dehydrogenase, L-lactate: NAD oxidoreductase (EC 1.1.1.27).

glasses by mixing EDTA blood with 6% dextran in normal saline, and centrifuging at 500 r.p.m. for 5 min. After obtaining the white-cell-rich supernatant, the cells were packed by centrifugation at 2000 g. The red cells were hemolyzed through contact with distilled water for 30 s. The white cells were then centrifuged and washed twice with saline. A 50% suspension of leukocytes was made in distilled water, saponin was added, and the suspension homogenized by grinding for 5 min with a pestle and, after a lapse of 1 h, centrifuging for 30 min at 15,000 g in a refrigerated high-speed centrifuge. The supernatant was then used as a crude enzyme extract.

Thrombolysates. The thrombocytes were isolated from 1600 specimens. After collection of 5 ml EDTA blood, the blood was centrifuged at 1500 r.p.m. (350 g) for 5 min, the supernatant aspirated with a plastic dropper and placed in a silicone-coated tube, and recentrifuged at 15,000 r.p.m. for 20 min in a refrigerated high-speed centrifuge. The plasma was aspirated and the button of platelets resuspended and washed once in saline with EDTA (1 mM). Examination of the platelet suspension by phase contrast microscopy indicated that no aggregation had occurred. Platelet preparations obtained were sometimes contaminated with erythrocytes. A value of not more than 2 erythrocytes/1000 thrombocytes was chosen as the minimum standard of purity. The erythrocytes were removed by differential lysis in hypotonic saline. No red cells were observed in the platelet preparations with the Bürker-type counting chamber following the differential lysis procedure. The final pellet was resuspended in 0.5 ml of distilled water, saponin was added, and the suspension homogenized by grinding for 5 min with a pestle.

Electrophoretic separation. This procedure was carried out applying 800 V to six slides for 55 min, using barbital buffer, ionic strength 0.05, pH 8.6, at 12°C. The material was applied in quantities of 0.03 ml. Isozymes separated were detected by staining the agar gel with nitroblue tetrazolium salt (kindly supplied by Dr. B. Večerek) and phenazine methosulphate according to a previously published method [10]. The isozymes are called 1-5, starting from the anode, according to the conventional nomenclature established by Wieland and Pfleiderer [11]. The total LDH activity was determined, using Warburg's optical test, the reagents were supplied as Boehringer sets (the reaction being monitored at 340 nm in the Spektromom 231). For quantitation of isozyme activity the slides were scanned in an MG Berlin instrument for paper electrophoresis.

RESULTS

In *hemolysates* the isozyme patterns obtained usually showed four isozymes, the fourth in a negligible amount. When the time of separation was prolonged to 75 min, the sub-band LDH-2' appeared. After 120 min, the sub-band LDH-3' could also be noticed. While the sub-band LDH-2' ran to the cathode, the sub-band LDH-3' ran to the anode (see Fig. 1).

Five lactate dehydrogenase isozymes in *leucocytes* in chronic lymphatic and myeloid leukemia, as well as in normal granulocytes, were always present. We did not observe any sub-band formation under our conditions. The quantitative studies confirmed the previous results [4]. They showed that the activity of isozyme 5 in lymphocytes is low (7.3 ± 4.1%); in neutrophilic granulocytes high (24.1 ± 6.2%). In eosinophilic granulocytes the isozyme LDH-5 showed a low activity.

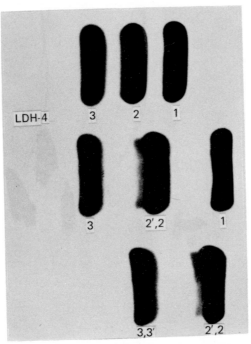

Figure 1. LDH isozyme patterns in hemolysates. Top, from right to left: isozyme LDH-1, LDH-2, LDH-3; LDH-4 activities were rather low. Middle: The time of separation was prolonged to 75 min. The sub-band LDH-2' appeared on the cathodic side. Lower: LDH isozyme 2, sub-band LDH-2', on the anodic side of isozyme LDH-3, sub-band LDH-3' appearing and being separated after 120 min from isozyme LDH-3.

In *thrombocytes* the fifth isozyme was demonstrated as well. Fig. 2 exhibits a comparison of different agars for isozymograms of human thrombocytes. With normal agar, only three isozymes were obtained, while with Difco "Bacto" and with "Oxoid No. 3" agar, there were four isozymes. Only with Difco "Noble special" agar and with "Agarose" was the fifth isozyme, LDH-5, placed in evidence. In 95 patients the examination was repeated 8-14 times within a period of two years. In ten patients (10.5%) a constantly higher activity of LDH-5 was observed and between 4 to 10% variation of the total activity seen, while in other cases the activity was very low. A unique observation was made in

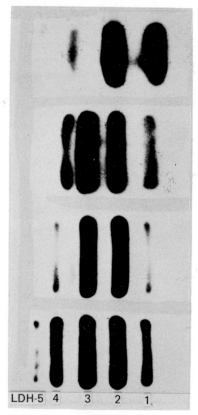

Figure 2. The isozymes of LDH in human thrombocytes on different agars. Top, from right to left: isozyme LDH-1, LDH-2, LDH-3 on normal purified agar. Lower: isozymograms of LDH on Difco "Bacto agar", on "Oxoid No. 3" and on "Agarose" (Nutritional Biochemical Co.). The isozyme pattern is quite distorted.

one subject. In his sample of thrombocytes, a double fifth isozyme was found (see Fig. 3). The same finding was made the next day in the same sample in triplicate, but the examination one year before, and one made two and three months after this observation, showed only one isozyme. In another subject a double first isozyme, LDH-1, appeared. These were the first cases out of 185 patients on long-term anticoagulant therapy and from 3381 specimens of thrombocytes.

DISCUSSION

Our methodical procedure showed high sensitivity for the activity of slow moving isozyme LDH-5. In human hemolysates without reticulocytosis, we observed only four isozymes (1-4) as did Vesell and Bearn [12] on starch gel and

Andersen *et al.* [3] on agar gel. In our opinion only the presence of young erythrocytes or leucocytes leads to the appearance of isozyme LDH-5. Sub-bands 2' and 3' appeared, as a rule, if the time of separation on agar gel was prolonged.

Our findings in lymphocytes of chronic leukemia are in agreement with our previous paper [4] and with the findings of Andersen *et al.* [3], which were not previously known to us. The activity of isozyme LDH-5 in lymphocytes is less pronounced than in granulocytes, i.e. neutrophilic granulocytes (in eosinophils we observed a low isozyme LDH-5 activity).

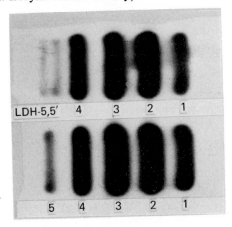

Figure 3. The splitting of isozyme 5 of human thrombocytes. Upper: A case with double isozyme 5. Splitting from unknown cause. Lower: A case with high activity of LDH-5 in human thrombocytes.

Carefully prepared thrombocytes often exhibited the isozyme LDH-5. The main conditions are fresh, carefully prepared suspensions without drastic procedures. The type of agar is very important. We noticed the isozyme LDH-5 only on Difco "special agar Noble" and on agarose, but never on the remaining three types of agar. Bezkorovainy and Rafelson [5] also observed the fifth isozyme; Vesell [6] only in sucrose solutions, not in saline; Schneider *et al.* [13] only by chromatography on DEAE cellulose, but not by polyacrylamide disc electrophoresis.

In one case we observed a double isozyme LDH-5, in another case a double isozyme LDH-1. The appearance differs from the sub-bands which we observed in hemolysates. These anomalies have not yet been elucidated. The sub-bands showed lower activities than the main band, but in the cases mentioned the activity of each double isozyme was on the same level. We have not discovered any corresponding finding in the relevant literature. In the available literature the factors influencing the LDH isozymograms are well demonstrated by the papers of Ressler *et al.* [14-16], and Rosalki and Montgomery [17].

ACKNOWLEDGEMENTS

Thanks are due to Mrs. Miluše Teclová and Mrs. Libuše Vaňková for excellent technical assistance, and to Mr. F. L. Szarka for the photographs.

REFERENCES

1. Dioguardi, N. and Agostoni, A., *Enzymol. Biol. Clin.* 2 (1962) 116.
2. Dioguardi, N., Agostoni, A., Fiorelli, G. and Lomanto, B., *J. Lab. clin. Med.* 61 (1963) 713.
3. Andersen, V., Gerhardt, W. and Clausen, J., *in* "Protides of the Biological Fluids," Vol. 11 (edited by H. Peeters), Elsevier, Amsterdam, 1964, p. 514.
4. Hule, V., *Clinica chim. Acta* 17 (1967) 349.
5. Bezkorovainy, A. and Rafelson, M. E., *J. Lab. clin. Med.* 64 (1964) 212.
6. Vesell, E. S., *Science, N.Y.* 150 (1965) 1735.
7. Hule, V., *Clinica chim. Acta* 13 (1966) 431.
8. Pfleiderer, G. and Wachsmuth, E. D., *Biochem. Z.* 334 (1961) 185.
9. Dioguardi, N., Agostoni, A., Fiorelli, G. and Mannucci, P. M., *Enzymol. Biol. Clin.* 4 (1964) 31.
10. Hule, V., *Čas. Lék. česk.* 105 (1966) 74.
11. Wieland, T. and Pfleiderer, G., *Ann. N.Y. Acad. Sci.* 94 (1961) 898.
12. Vesell, E. S. and Bearn, A. G., *J. gen. Physiol.* 45 (1962) 552.
13. Schneider, W., Schumacher, K. and Gross, R., Abstracts 5th FEBS Meeting (Prague) 1968, p. 169.
14. Ressler, N., Schulz, J. L. and Joseph, R. R., *Nature, Lond.* 198 (1963) 888.
15. Ressler, N., Schulz, J. L. and Joseph, R. R., *J. Lab. clin. Med.* 62 (1963) 571.
16. Ressler, N., *Nature, Lond.* 215 (1967) 284.
17. Rosalki, S. B. and Montgomery, A., *Clinica chim. Acta* 17 (1967) 440.

FEBS Symposium, Volume 18, 1970, pp. 335-339

Genetic Aspects of Human α-Amylase Heterogeneity

R. LAXOVÁ and J. KAMARÝT

*Department of Genetics and Cytology and Department of
Biochemistry, Institute of Paediatric Research,
Brno, Czechoslovakia*

α-Amylase* electrophoretic fractionation in human biological material revealed the presence of two distinct amylase fractions of salivary and pancreatic origin. Both of these may appear in agar-gel electrophoresis as one or two bands. The number of various serum and urine isoamylases is different in different individuals and is genetically determined [1]. The genetically-determined enzyme diversity in "normal" individuals is an expression of enzyme polymorphism. The study of human polymorphisms is one of the methods which aids us in gaining at least some knowledge of the genetic character of man. It is also conceivable that man's genetic biochemical constitution is determined by the character, composition and location of his enzymes, of which polymorphisms are of some significance.

MATERIAL AND METHODS

Human serum and urinary amylase heterogeneity has been studied using our own agar-gel electrophoretic technique, already reported elsewhere [2].

RESULTS AND DISCUSSION

The isoamylases are—in relation to the serum protein electrophorogram—situated between the β- and γ-globulins. The doubtful zones of non-enzymatic origin reported by some authors [3, 4] in the electrophoretic position of the albumins or α- or γ-globulins, are artifacts and may be distinctly differentiated by the application of native starch during activity detection [5].

The fractional activity of serum salivary and pancreatic amylases changes with diseases of the salivary glands and the pancreas [6].

Blood serum salivary amylase attains an average of 54% (s.d. ± 6.4) and pancreatic isoamylase 46% (s.d. ± 6.4) of the total activity in healthy individuals.

---

* α-1,4-glucan-4-glucanohydrolase.

The renal clearance of salivary isoamylase, which has been investigated in ten healthy middle-aged adults during a 24-h period, represents half the pancreatic isoamylase clearance. The total amylase clearance is approximately fifty times lower than that of endogenous creatinin (forming merely 2% of this). The

AMYLASE HETEROGENEITY VARIANTS IN MAN

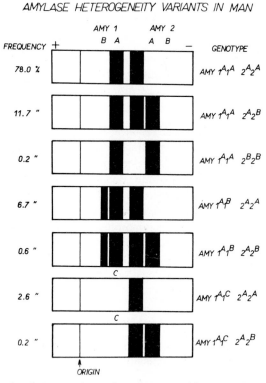

Figure 1. Amylase heterogeneity variants in man, with nomenclature and population incidence of the individual phenotypes.

maximum renal isoamylase excretion in healthy individuals occurs during the afternoon hours, irrespective of diet and daily regime [7].

The foregoing findings enable the proportion of the salivary and pancreatic incretoric activities to be approximately calculated. The salivary glands account for approximately 0.1%, and the pancreas 0.2%, of the total enzyme production.

Serum and urinary amylase heterogeneity has been studied in two separate population groups.

Firstly in a group of 410 unrelated subjects representing a random sample of the South Moravian population. This has enabled incidence, gene and allele frequencies to be estimated. A certain genotype, with a given number of bands of amylolytic activity, is codominantly inherited.

The nomenclature and population incidence of the individual phenotypes is evident from Fig. 1. The salivary fraction has been called Amy 1 and the pancreatic one Amy 2. Each of these fractions may be phenotypically characterized by one (A) (genotypically this individual is a homozygote (AA)), or two (AB) (genotypically this individual is a heterozygote (AB)) bands of activity. Individuals, each with two salivary and pancreatic bands (Amy 1^A1^B, Amy 2^A2^B), are henceforth double heterozygotes. In a small percentage of the subjects no salivary amylase activity can be detected in the serum or urine. We have called this variant Amy 1^C, although we are aware of the fact that this is probably not as simple as it seems.

Our group of unrelated subjects has made possible gene and allele frequency estimations. Probably two genes are involved, each on its own locus: gene Amy^1 for salivary isoamylase is characterized by the alleles: Amy 1^a with a frequency of 0.958 in the population; Amy 1^b for the less frequent salivary band with a frequency of 0.035; and probably Amy 1^c (the missing salivary isoamylase activity) with a frequency of 0.007 (Fig. 2).

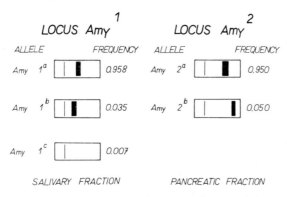

Figure 2. Allele frequency estimation of salivary and pancreatic isoamylases in a group of 410 unrelated subjects of the South Moravian population.

The second gene, Amy^2, for pancreatic isoamylase is characterized by two alleles: Amy 2^a, for the more frequently occurring pancreatic band, frequency 0.950; and Amy 2^b, for the less frequently occurring pancreatic band, frequency 0.05.

These loci are probably not linked but the analysis is the subject of a further study.

The allele frequencies have been calculated according to Hardy-Weinberg's law and the observed numbers are in good agreement with the expected ones (Table 1).

Table 1. Observed and expected values for pancreatic amylase variants.

Genotype	Observed subjects	Expected subjects	Observed frequency		Expected frequency
Amy 2^A2^A	366	370.025	0.8927	p^2	0.9025
Amy 2^A2^B	42	38.95	0.1024	$2pq$	0.0950
Amy 2^B2^B	2	1.025	0.0049	q^2	0.0025
	410	410.000	1.0000		1.0000

The second investigated group included 300 families, 200 of which were families with twins. These studies have enabled the heredity mechanism to be analysed.

In our material, comprising 200 twin families, 42 parents possessed one of the less frequent variants. Approximately 50% of the offspring inherited the variant from parents, which is in accordance with a dominant mode of inheritance.

The 42 parents had a total of 92 offspring, of which 50 inherited the variant and 42 did not. This is an almost ideal result according to Pearson's law:

$$42 \text{ families}$$
$$92 \text{ offsprings}$$
$$50 \atop +\ \ \ \ 42 \atop -$$

$$\chi^2 = \frac{\Sigma d^2}{m} = \frac{(4)^2 + (4)^2}{46} = \frac{32}{46} = 0.7$$

$$p_1 = (0.5 - 0.3)$$

0.7 for one degree of freedom corresponds to P = 0.5 − 0.3. The difference from the ratio 50:50 (1:1) which is the expected ratio for dominant heredity, is therefore insignificant.

However, the fact that both alleles (one from each parent) are evident and manifested in the phenotype, changes the simple dominant mode of inheritance to codominancy. This type of inheritance is similar to that in blood group AB.

Where twin zygoticity is concerned, monozygotic twin concordance is equal to 100%, thus making H equal to 1, for heterogeneity variants.

$$H = \frac{K_{MZ} - K_{DZ}}{100 - K_{DZ}} = \frac{100 - 38.1}{100 - 38.1} = 1$$

Both the salivary and pancreatic locus will probably be useful for linkage studies.

REFERENCES

1. Kamarýt, J. and Laxová, R., *Humangenetik* 3 (1966) 41.
2. Kamarýt, J. and Laxová, R., *Humangenetik* 1 (1965) 579.
3. McGeachin, R. L. and Lewis, J. P., *J. biol. Chem.* 234 (1959) 795.
4. Dreiling, D. A., Janowitz, H. D. and Josephberg, L. D., *Ann. intern. Med.* 58 (1963) 235.
5. Kamarýt, J., *Z. klin. Chem. klin. Biochem.* 6 (1968) 96.
6. Kamarýt, J., Macku, M. and Nováková, J., *Vnitř. Lék.* 14 (1968) 349.
7. Kamarýt, J., *Z. klin. Chem. klin. Biochem.* 7 (1969) 51.

FEBS Symposium, Volume 18, 1970, pp. 341-345

Studies on Alkaline Phosphatase in Liver and Intestine after Bile Duct Ligation

H. BOERNIG, A. HORN, W. MUELLER, U. HORN
and G. GUELDNER

*Institut fur Physiologische Chemie der Friedrich-Schiller-Universitat
Jena, East Germany*

The activity of alkaline phosphatase (EC 3.1.3.1) in rat liver is increased many-fold after bile duct ligation. In a previous paper we were able to show that an approximately ten-fold increase occurred with the isoenzyme localized in the plasma membrane [1]. We also found the enzyme activity in the small intestine to be activated. Although the mechanism of this increase in enzyme activity is not known, there are some examples, such as in cell cultures [2] and leucocytes [3], where alkaline phosphatase is inducible by adrenal glucocorticoid activity. In rat liver [4] and small intestine of mice [5] a close relationship between enzyme activity and adrenal glands has also been described.

As glucocorticoid hormones are metabolized mainly by the liver [6] and partly excreted by the bile, we decided to investigate the possibility that the increase in enzyme activity after bile duct ligation might be due to hormone accumulation acting either by stimulation or induction of the enzyme. Table 1

Table 1. The effect of bile duct ligation, laparotomy or adrenalectomy on the specific activity of alkaline phosphatases in liver and small intestine homogenates.

Conditions		Liver mU/mg protein*	Small intestine U/mg protein
Normal		6.40	2.53
Adrenalectomy	1 day	6.30	2.50
	2 days	5.90	2.07
Ligature	10 h	60.0	3.54
Laparotomy	10 h	20.0	3.89
Adrenalectomy + ligature	1 day 10 h	21.6	1.79

* U = μmole P_i liberated per min.

shows the influence of ligation and adrenalectomy on the specific activity of the alkaline phosphatase in liver and small intestine.

Within the first two days after adrenalectomy, neither the liver nor the intestinal enzyme show any changes in activity. The enzyme activity increases and reaches a maximum in liver and small intestine ten and six hours, respectively, after bile duct ligation. The ten-fold increase observed in the liver drops after adrenalectomy to the level of the control. The intestinal enzyme is not affected by adrenalectomy; however, the levels in the controls are as high as in ligated rats. The increase in activity in both organs is therefore dependent on an intact function of the adrenal gland.

Of the glucocorticoids formed in the adrenal glands, only cortisol, corticosterone, progesterone, and testosterone are excreted via the bile, and may possibly be accumulated in cases of stasis. We therefore investigated the influence of these hormones on the liver and intestinal enzymes.

As shown in Table 2 the ratio of specific activity of hormone-treated animals to that of control animals indicates that the activity of the intestinal phosphatase is increased by all hormones (ratio < 1) and that of liver phosphatase by cortisol only.

Table 2. The effect of hormones on the specific activity of alkaline phosphatases in liver and small intestine.

Hormone (mg/100 g b.w.)	Ratio of specific activities $\frac{\text{treated}}{\text{control}}$	
	Liver	Intestine
2.4 Testosterone	1.06	1.87
1.5 Progesterone	1.30	2.30
1.5 Corticosterone	1.15	2.66
1.5 Cortisol	1.69	2.37

Hormones administered subcutaneously to normal rats 10 h before sacrifice. Controls injected with the corresponding solvent.

To avoid interference from endogenous hormones and secondary reactions, the experiments were repeated with adrenalectomized animals (see Table 3). In fact, the figures of Table 3 show that only cortisol increases the ratio treated/control in both the small intestine and liver.

For the intestinal enzyme the ratio is approximately the same as after ligation. The change in activity of the intestinal enzyme may be due to stress reaction and is assumed to be mediated through cortisol released non-specifically. In the liver of adrenalectomized animals the ten-fold increase of alkaline phosphatase activity observed after ligature cannot be attained even after repeated injections of 2 mg cortisol per 100 g body weight. This indicates that the effective hormone is not one of the group of hormones investigated or that

Table 3. The effect of hormones on the specific activity of alkaline phosphatases in liver and small intestine.

Hormone (mg/100 g b.w.)	Ratio of specific activities $\frac{\text{treated}}{\text{control}}$	
	Liver	Intestine
2.4 Testosterone	0.50	1.20
1.5 Progesterone	0.86	1.05
2.0 Cortisol	2.34	—
5.2 Cortisol	2.17	1.70

Hormones administered subcutaneously to adrenalectomized rats 10 h before sacrifice. Controls injected with the corresponding solvent.

another additional factor is involved which increases the activity after bile duct ligation. In order to determine which of the two possibilities is operative, cortisol injections were given to adrenalectomized animals after bile duct ligation. The changes in activity of alkaline phosphatase were then measured at different time intervals. Figure 1 shows that under these conditions the changes

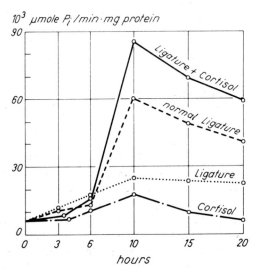

Figure 1. Progress curves for alkaline phosphatase activity in liver homogenate of adrenalectomized rats.

in activity are in general the same as in normal animals after bile duct ligation, with the exception that the curve peak at 10 h reaches even higher levels.

Thus, in liver the change in enzyme activity requires the hormonal factor cortisol plus an additional factor which is formed only under the specific conditions of bile duct ligation. The question now arises as to whether the foregoing changes in enzyme activity should be regarded as cortisol-mediated enzyme induction or activation.

We therefore investigated the effect of protein synthesis inhibitors on enzyme activity in liver and intestine following bile duct ligation.

As seen in Fig. 2, the increase of enzyme activity in the liver after bile duct ligation is reduced considerably by application of actinomycin D, i.e. the increase is dependent on protein synthesis.

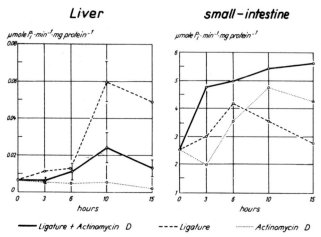

Figure 2. Progress curves for alkaline phosphatase activity in homogenates of liver and small-intestine mucosa homogenates after bile duct ligation and/or s.c. injections of actinomycin D.

The inhibitor has no marked effect on basic enzyme activity. Intestinal activity appears to follow a quite different pattern, and the increase cannot be suppressed by actinomycin D. On the contrary, actinomycin D itself effects an increase of activity. There are reports to the effect that some alkaline phosphatases of different origin show similar behaviours with other inhibitors of protein synthesis, such as ethionine and puromycin. To summarize, the changes in activity of the intestinal enzyme following bile duct ligation may be regarded as a non-specific activation which is dependent on cortisol and independent of protein synthesis. On the other hand, in the liver the process is a protein synthesis-dependent induction of the enzyme. In addition to cortisol, another as yet unknown factor, which is formed under the specific conditions of bile duct ligation, is also necessary.

The induction of alkaline phosphatase by hormones is of considerable interest as this enzyme is the only inducible one that has been localized in the cell membranes of higher animals. As there is a close correlation between alkaline phosphatase and $Na^+ - K^+$ dependent ATPase, the hormone effect on membrane permeability may be due to a hormone →enzyme interaction.

REFERENCES

1. Boernig, H., Stepan, J., Horn, A., Giertler, R., Thiele, G. and Večerek, B., *Hoppe-Seyler's Z. physiol. Chem.* **348** (1967) 1311.
2. Cox, R. P. and McLeod, C. M., *Nature, Lond.* **190** (1961) 85; *J. gen. Physiol.* **45** (1962) 439.
3. McLoy, E. E. and Ebadi, M., *Biochem. biophys. Res. Commun.* **26** (1967) 265.
4. Vail, V. and Kochiakian, C. D., *Am. J. Physiol.* **150** (1947) 580; Kochiakian, C. D. and Vail, V., *J. biol. Chem.* **169** (1947) 1.
5. Moog, F., *J. exp. Zool.* **124** (1953) 329; Moog, F. and Richardson, D., *J. exp. Zool.* **130** (1955) 29.
6. Hubener, D. J. and Staib, W. H., "Biochemie der Nebennierenrinden-Hormone", Georg Thieme Verlag, Stuttgart, 1965, S. 76.

Author Index